JUSTUS LIEBIG

CHEMISTRY
IN ITS APPLICATIONS TO AGRICULTURE AND PHYSIOLOGY

Elibron Classics
www.elibron.com

Elibron Classics series.

ISBN 0-543-88550-X (paperback)
ISBN 0-543-88549-6 (hardcover)

This Elibron Classics Replica Edition is an unabridged facsimile of the edition published in 1843 by Taylor and Walton, London.

CHEMISTRY

IN ITS APPLICATIONS

TO

AGRICULTURE AND PHYSIOLOGY.

BY

JUSTUS LIEBIG, M.D. PH. D., F.R.S., M.R.I.A.,

PROFESSOR OF CHEMISTRY IN THE UNIVERSITY OF GIESSEN; KNIGHT OF THE HESSIAN ORDER, AND OF THE IMPERIAL ORDER OF SAINT ANN; MEMBER OF THE ROYAL ACADEMY OF SCIENCES OF STOCKHOLM; CORRESPONDING MEMBER OF THE ROYAL ACADEMIES OF SCIENCES OF BERLIN AND MUNICH; OF THE IMPERIAL ACADEMY OF ST. PETERSBURGH; OF THE ROYAL INSTITUTION OF AMSTERDAM, ETC. ETC.

EDITED FROM THE MANUSCRIPT OF THE AUTHOR.

BY LYON PLAYFAIR, PH. D., F.G.S.,

HONORARY MEMBER OF AND CONSULTING CHEMIST TO THE ROYAL AGRICULTURAL SOCIETY OF ENGLAND.

Third Edition,

REVISED AND ENLARGED.

LONDON:

PRINTED FOR TAYLOR AND WALTON,

UPPER GOWER-STREET.

MDCCCXLIII.

LONDON:
BRADBURY AND EVANS, PRINTERS, WHITEFRIARS.

AUTHOR'S PREFACE.

Many views and principles which I had endeavoured to develop in reference to nutrition, and especially to the cultivation of vegetables, were strongly opposed, immediately on the appearance of the first edition of this work. I could not, however, resolve to make any material change in the immediately succeeding edition, because I did not consider the scientific investigation of the important questions at issue as finally determined, and because I thought that I ought to trust the decision of them to experience alone.

Many of the objections raised were founded upon a want of mutual understanding; others related to positions and assertions having no connection with the peculiar object of the book. I have set these aside by the omission of all passages thus called in question.

In the three years which have elapsed between this edition and the first, I have not neglected any opportunity of subjecting to a rigorous and careful examination the principles which I had developed of the nutritive properties of plants, and their application to agriculture. I have endeavoured to make myself acquainted with the condition of practical farming, and with what it requires, by a journey through the agricultural districts of England

and Scotland; and during this interval a long series of experiments were carried on in the laboratory of this place, with the sole object of giving a firmer basis to my exposition of the causes of the advantageous results attending the practice of rotation of crops, and also of effectually banishing all doubts concerning their accuracy.

In my "Chemistry in its applications to Physiology and Pathology," I have subjected the process of nutrition of the animal organism to a stricter investigation; and I am now, for the first time, since the completion of these labours, in a situation to give a simple and determinate expression to my view of the origin of animal excrements, and of the cause of their beneficial effects on the growth of all vegetables.

Now that the conditions which render the soil productive and capable of affording support to plants, are ascertained, it cannot well be denied, that from Chemistry alone further progress in Agriculture is to be expected.

Every unprejudiced person will, I trust, be finally convinced by this third edition, that I have earnestly endeavoured to perfect my views, and have striven, with the best intentions, to ascertain truth and obviate error.

JUSTUS LIEBIG.

GIESSEN,
August, 1843.

TO

THE BRITISH ASSOCIATION

FOR THE

ADVANCEMENT OF SCIENCE.

ONE of the most remarkable features of modern times is the combination of large numbers of individuals representing the whole intelligence of nations, for the express purpose of advancing science by their united efforts, of learning its progress, and of communicating new discoveries. The formation of such associations is, in itself, an evidence that they were needed.

It is not every one who is called by his situation in life to assist in extending the bounds of science; but all mankind have a claim to the blessings and benefits which accrue from its earnest cultivation. The foundation of scientific institutions is an acknowledgment of these benefits, and this acknowledgment proceeding from whole nations may be considered as the triumph of mind over empiricism.

Innumerable are the aids afforded to the means of life, to manufactures, and to commerce, by the truths which assiduous and active inquirers have discovered and rendered capable of practical application. But it is not the mere practical utility of these truths which is of importance. Their influence upon mental culture is most beneficial; and the new views acquired by the knowledge of them enable the mind to recognise, in the phenomena of nature, proofs of an Infinite Wisdom, for the unfathomable profundity of which language has no expression.

At one of the meetings of the Chemical Section of the "British Association for the Advancement of Science," the honourable task of preparing a Report upon the state of Organic Chemistry was imposed upon me. In the present work I present the Association with a part of this report.

I have endeavoured to develop, in a manner correspondent to the present state of science, the fundamental principles of Chemistry in general, and the laws of Organic Chemistry in particular, in their applications to Agriculture and Physiology; to the causes of fermentation, decay, and putrefaction; to the vinous and acetous fermentations, and to nitrification. The conversion of woody fibre into wood and mineral-coal, the nature of poisons, contagions, and miasms, and the causes of their action on the living organism, have been elucidated in their chemical relations.

I shall be happy if I succeed in attracting the attention of men of science to subjects which so well merit to engage their talents and energies. Perfect Agriculture is the true foundation of all trade and industry—it is the foundation of the riches of states. But a rational system of Agriculture cannot be formed without the application of scientific principles; for such a system must be based on an exact acquaintance with the means of nutrition of vegetables, and with the influence of soils and actions of manure upon them. This knowledge we must seek from chemistry, which teaches the mode of investigating the composition and of studying the characters of the different substances from which plants derive their nourishment.

The chemical forces play a part in all the process of the living animal organism; and a number of transformations and changes in the living body are exclusively dependent on their influence. The diseases incident to the period of growth of man, contagion, and contagious matters, have their analogues in many chemical processes. The investigation of the chemical connexion subsisting between those actions proceeding in the living body, and the transformations presented by chemical compounds, has also been a subject of my inquiries. A perfect exhaustion of this subject, so highly important to medicine, cannot be expected without the co-operation of physiologists. Hence I have merely brought forward the purely chemical part of the inquiry, and hope to attract attention to the subject.

Since the time of the immortal author of the "Agricultural Chemistry," no chemist has occupied himself in studying the applications of chemical principles to the growth of vegetables, and to organic processes. I have endeavoured to follow the path marked out by Sir Humphry Davy, who based his conclusions only on that which was capable of inquiry and proof. This is the path of true philosophical inquiry, which promises to lead us to truth—the proper object of our research.

In presenting this report to the British Association I feel myself bound to convey my sincere thanks to Dr. Lyon Playfair, of St. Andrew's, for the active assistance which has been afforded me in its preparation by that intelligent young chemist during his residence in Giessen. I cannot suppress the wish that he may succeed in being as useful, by his profound and well-grounded knowledge of chemistry, as his talents promise.

JUSTUS LIEBIG.

GIESSEN,
September 1, 1840.

CONTENTS.

PAGE

Object of the Work 1

PART THE FIRST.

ON THE CHEMICAL PROCESSES IN THE NUTRITION OF VEGETABLES.

CHAPTER

I.—The Constituent Elements of Plants 3

II.—The Origin and Assimilation of Carbon . . . 5

III.—On the Origin and Action of Humus . . . 29

IV.—On the Assimilation of Hydrogen 36

V.—On the Origin and Assimilation of Nitrogen . . 41

VI.—On the Source of Sulphur 60

VII.—Of the Inorganic Constituents of Plants . . . 66

VIII.—On the Formation of Arable Land 84

IX.—The Art of Culture 96

X.—On Fallow 128

XI.—On the Rotation of Crops 139

XII.—On Manure 173

XIII.—Retrospective View of the Preceding Theories . . 195

Supplementary Chapters.—The Sources of Ammonia . . 215

Is Nitric Acid food for plants? 225

Does the Nitrogen of the Air take part in Vegetation? 234

Giant Sea-weed 236

Appendix to Part I. 238

PART THE SECOND.

ON THE CHEMICAL PROCESSES OF FERMENTATION, DECAY, AND PUTREFACTION.

CHAPTER		PAGE
I.	—Chemical Transformations	259
II.	—On the Causes which effect Fermentation, Decay, and Putrefaction	262
III.	—Fermentation and Putrefaction	270
IV.	—On the Transformation of Bodies which do not contain Nitrogen as a Constituent, and of those in which it is present	275
	On the Transformation of Bodies containing Nitrogen	277
V.	—Fermentation of Sugar	282
	Yeast or Ferment	284
VI.	—Eremacausis, or Decay	291
VII.	—Eremacausis, or Decay of Bodies destitute of Nitrogen: Formation of Acetic Acid	299
VIII.	—Eremacausis of Substances containing Nitrogen.—Nitrification	304
IX.	—On Vinous Fermentation:—Wine and Beer	308
X.	—On Fermentation ascribed to the Growth of Fungi and of Infusoria	326
XI.	—Decay of Woody Fibre	336
XII.	—Vegetable Mould	342
XIII.	—On the Mouldering of Bodies:—Paper, Brown Coal, and Mineral Coal	344
XIV.	—On Poisons, Contagions, and Miasms	353
APPENDIX TO PART II.		391
NDEX		393

ORGANIC CHEMISTRY

IN ITS APPLICATION TO

VEGETABLE PHYSIOLOGY AND AGRICULTURE.

THE object of Organic Chemistry is to discover the chemical conditions essential to the life and perfect development of animals and vegetables, and generally to investigate all those processes of organic nature which are due to the operation of chemical laws. Now, the continued existence of all living beings is dependent on the reception by them of certain substances, which are applied to the nutrition of their frame. An inquiry, therefore, into the conditions on which the life and growth of living beings depend, involves the study of those nutritive substances, as well as the investigation of the sources whence they are derived, and of the changes undergone by them in the process of assimilation.

A beautiful connexion subsists between the organic and inorganic kingdoms of nature. Inorganic matter affords food to plants; and they, on the other hand, yield the means of subsistence to animals. The conditions necessary for animal and vegetable nutrition are essentially different. An animal requires for its development, and for the sustenance of its vital functions, a certain class of substances which can be generated only by organic beings possessed of life. Although many animals are entirely carnivorous, yet their primary nutriment must be derived from plants; for the animals upon which they subsist receive their nourishment from vegetable matter. Plants, on the other hand, find new nutritive material only in inorganic substances. Hence, one great end of vegetable life is to generate matter adapted for the nutrition of animals, out of inorganic substances, which are not fitted for this purpose. Now, the purport of this work is, to elucidate the chemical processes engaged in the nutrition of vegetables.

The first part of it will be devoted to the examination of the matters which supply the nutriment of plants, and of the changes which these matters undergo in the living organism. The chemical compounds which afford to plants their principal constituents, viz. carbon, nitrogen, hydrogen, and sulphur, will here come under consideration, as well as the relations in which the vital functions of vegetables stand to those of the animal economy and to other phenomena of nature.

The second part of the work will treat of the peculiar processes usually described as fermentation, putrefaction, and decay. By the action of these processes, the complete destruction of plants and animals after death is effected. Hence the changes undergone by organic substances in their conversion into inorganic compounds, as well as the causes by which these changes are determined, will become matter of inquiry.

PART I.

THE CHEMICAL PROCESSES IN THE NUTRITION OF VEGETABLES.

CHAPTER I.

THE CONSTITUENT ELEMENTS OF PLANTS.

Carbon and hydrogen invariably occur in all parts of plants. They form constituents of all their organs, and are essential to their existence.

The substances which constitute the principal mass of every vegetable are compounds of carbon with oxygen and hydrogen, in the proper relative proportions for forming water. Woody fibre, starch, sugar, and gum, for example, are such compounds of carbon with the elements of water. In another class of substances containing carbon as an element, oxygen and hydrogen are again present; but the proportion of oxygen is greater than would be required for producing water by union with the hydrogen. The numerous organic acids met with in plants, belong, with few exceptions, to this class.

A third class of vegetable compounds contains carbon and hydrogen, but no oxygen, or less of that element than would be required to convert all the hydrogen into water. These may be regarded as compounds of carbon with the elements of water, and an excess of hydrogen. Such are the volatile and fixed oils, wax, and the resins. Many of them have acid characters.

The juices of all vegetables contain organic acids, generally combined with the inorganic bases, or metallic oxides; for metallic oxides exist in every plant, and may be detected in its ashes after incineration.

NITROGEN is an element of vegetable albumen and gluten; it is a constituent of the acid, and of what are termed the "indifferent substances" of plants, as well as of those peculiar vegetable compounds called "organic bases," which possess all the properties of metallic oxides. The seeds also of all plants contain nitrogenous compounds.

Estimated by its proportional weight, nitrogen forms only a very small part of plants; but it is never entirely absent from any part of them. Even when it does not absolutely enter into the composition of a particular part or organ, it is always to be found in the fluids which pervade it.

The nitrogenous compounds thus invariably present in plants contain a certain quantity of sulphur. When the juices, seeds, or organs of particular kinds of plants are subjected to distillation along with water, peculiar oily substances pass over. These are volatile, and are characterised by their large proportion, both of sulphur and of nitrogen. The volatile oils of the horse-radish and of mustard are examples of this class of bodies.

From the remarks now made, it is obvious that there are two great classes into which all vegetable products may be arranged. The first of these contains nitrogen; in the last this element is absent. The compounds destitute of nitrogen may be divided into those in which oxygen forms a constituent, and those into which it does not enter. The nitrogenous compounds may, in like manner, be divided into three smaller classes. The first of these is distinguished by containing both sulphur and oxygen; the second contains sulphur, but is devoid of oxygen; while the third is composed of bodies from which sulphur is entirely absent.

It follows from the facts thus far detailed, that the development of a plant requires the presence, first, of substances containing carbon, nitrogen, and sulphur, and capable of yielding these elements to the growing organism; secondly, of water and its elements; and lastly, of a soil to furnish the inorganic matters which are likewise essential to vegetable life.

CHAPTER II.

THE ORIGIN AND ASSIMILATION OF CARBON.

COMPOSITION OF HUMUS.

SOME virgin soils, such as those of America, contain vegetable matter in large proportion; and as these have been found eminently adapted for the cultivation of most plants, the organic matter contained in them has naturally been recognised as the cause of their fertility.* To this matter, the term "vegetable mould" or *humus* has been applied. Indeed, this peculiar substance appears to play such an important part in the phenomena of vegetation, that vegetable physiologists have been induced to ascribe the fertility of every soil to its presence. It is believed by many to be the principal nutriment of plants, and is supposed to be extracted by them from the soil in which they grow.

The humus, to which allusion has been made, is described by chemists as a brown substance easily soluble in alkalies, but only slightly so in water, and produced during the decomposition of vegetable matters by the action of acids or alkalies. It has, however, received various names, according to the different external characters and chemical properties which it presents. Thus, *ulmin, humic acid, coal of humus,* and *humin,* are names applied to modifications of *humus.* They are obtained by treating peat, woody fibre, soot, or brown coal with alkalies; by decomposing sugar, starch, or

* When the weight of the soluble parts of this vegetable matter is compared with that of the plants growing upon it, it is seen that only a very small part of their substance could have been procured through its agency. This is the case even in the most fertile soils.—(SAUSSURE, *Recherches sur la Vegetation.*)

sugar of milk by means of acids; or by exposing alkaline solutions of tannic and gallic acids to the action of the air.

The modifications of *humus* which are soluble in alkalies, are called *humic acid;* while those which are insoluble have received the designations of *humin* and *coal of humus.*

The names given to these substances might cause it to be supposed that their composition is identical. But a more erroneous notion could not be entertained; since even sugar, acetic acid, and resin do not differ more widely in the proportions of their constituent elements, than do the various modifications of *humus.*

HUMIC ACID formed by the action of hydrate of potash upon sawdust contains, according to the accurate analysis of Peligot, 72 per cent. of carbon, while the humic acid obtained from turf and brown coal contains, according to Sprengel, only 58 per cent.; that produced by the action of dilute sulphuric acid upon sugar, 57 per cent. according to Malaguti; and that, lastly, which is obtained from sugar or from starch, by means of muriatic acid, according to the analysis of Stein, 64 per cent. Malaguti states, moreover, that HUMIC ACID contains an equal number of equivalents of oxygen and hydrogen, that is to say, that these elements exist in it in the proportions for forming water; while, according to Sprengel, the oxygen is in excess; and Peligot estimates the quantity of hydrogen at 14 equivalents, and the oxygen at only 6 equivalents, making the deficiency of oxygen as great as 8 equivalents. Mulder and Herrmann have shown that decayed willow-wood, peat, or vegetable mould, after being treated with water and alcohol, leave a solid brown substance, which yields to alkalies a peculiar humic acid. This humic acid consists of carbon and the elements of water. But besides these usual constituents, it contains a certain quantity of ammonia, in a state of chemical combination.

It is quite evident, therefore, that chemists have been in the habit of designating by the names of humic acid or humin, all the brown or black-coloured products of the decomposition of organic bodies, according as they were soluble or insoluble in alkalies; although in their composition and mode of origin the substances thus confounded might be in no way allied.

Not the slightest ground exists for the belief that one or other of these artificial products of the decomposition of vegetable matters exists in nature, in the form, and endowed with the properties, of the vegetable constituents of mould: there is not the shadow of a proof that one of them exerts any influence on the growth of plants, either in the way of nourishment or otherwise.

Vegetable physiologists have, without any apparent reason, imputed the known properties of the HUMUS and HUMIC ACIDS of chemists to that constituent of mould which has received the same name, and in this way have been led to their theoretical notions respecting the functions of the latter substance in vegetation.

The opinion that the substance called HUMUS is extracted from the soil by the roots of plants, and that the carbon entering into its composition serves to nourish their tissues, without previously assuming another form, is considered by many as so firmly established that any evidence in its favour has been deemed unnecessary: the obvious difference in the growth of plants according to the known abundance or scarcity of HUMUS in the soil, seemed to afford incontestable proof of its correctness. *

Yet, this position, when submitted to a strict examination, is found to be untenable, and it becomes evident from most conclusive proofs, that *humus* IN THE FORM IN WHICH IT EXISTS IN THE SOIL, does not yield the smallest nourishment to plants.

The adherence to the above incorrect opinion has hitherto rendered it impossible to ascertain the true theory of the nutritive process in vegetables, and has thus deprived us of our best guide to a rational practice in agriculture. Any

* This remark applies more to German than to English botanists and physiologists. In England, the idea that humus, as such, affords nourishment to plants is by no means general; but on the Continent, the views of Berzelius on this subject have been almost universally adopted.—ED.

great improvement in that most important of all arts is inconceivable, without a deeper and more perfect acquaintance with the substances which nourish plants, and with the sources whence they are derived; and no other cause can be discovered to account for the fluctuating and uncertain state of our knowledge on this subject up to the present time, than that modern physiology has not kept pace with the rapid progress of chemistry.

In the following inquiry we shall suppose the HUMUS of vegetable physiologists to be really endowed with the properties recognised by chemists in the brownish-black deposits obtained by precipitating an alkaline decoction of mould or peat by means of acids, and which they name HUMIC ACID.

HUMIC ACID, when first precipitated, is a flocculent substance, is soluble in 2500 times its weight of water, and combines with alkalies, forming with lime and magnesia compounds of the same degree of solubility (Sprengel).

Vegetable physiologists agree in the supposition that by the aid of water HUMUS is rendered capable of being absorbed by the roots of plants. But according to the observation of chemists, humic acid is soluble only when newly precipitated, and becomes completely insoluble when dried in the air, or when exposed in the moist state to the freezing temperature. (Sprengel).

Both the cold of winter and the heat of summer, therefore, are destructive of the solubility of humic acid, and at the same time of its capability of being assimilated by plants. So that, if it is absorbed by plants, it must be in some altered form.

The correctness of these observations is easily demonstrated by treating a portion of good mould with cold water. The fluid remains colourless, and is found to have dissolved less than $\frac{1}{100000}$ part of its weight of organic matters, and to contain merely the salts which are present in rain-water.

Decayed oak-wood, likewise, of which humic acid is the principal constituent, was found by Berzelius to yield to cold water only slight traces of soluble materials; and I have myself verified this observation on the decayed wood of beech and fir.

These facts, which show that humic acid, in its insoluble condition, cannot serve for the nourishment of plants, have not escaped the notice of physiologists; and hence they have assumed that the lime or the different alkalies found in the ashes of vegetables, render soluble the humic acid, and fit it for the process of assimilation.

Alkalies and alkaline earths do exist in the different kinds of soil, in sufficient quantity to form such soluble compounds with humic acid.

Now, let us suppose that humic acid is absorbed by plants in the form of that salt which contains the largest proportion of humic acid, namely, in the form of humate of lime; and then, from the known quantity of the alkaline bases contained in the ashes of plants, let us calculate the amount of humic acid which might be assimilated in this manner. Let us admit, likewise, that potash, soda, and the oxides of iron and manganese have the same capacity of saturation as lime with respect to humic acid, and then we may take as the basis of our calculation the analysis of M. Berthier, who found that 1000 lbs. of dry fir-wood yielded 8.3 lbs. of ashes, and that in every 100 lbs. of these ashes, deducting the chloride of potassium and sulphate of potash, 46.1 lbs. consisted of the basic metallic oxides, potash, soda, lime, magnesia, iron, and manganese.

One Hessian acre * of woodland yields annually, according to Dr. Heyer, on an average, 2650 lbs. of dry fir-wood, which contains 10.07 lbs. of metallic oxides.

Now, according to the estimates of Malaguti and Sprengel, 1 lb. of lime combines chemically with 10.9 lbs. of humic acid; 10.07 lbs. of the metallic oxides would accordingly introduce into the trees nearly 111 lbs. of humic acid, which, admitting humic acid to contain 58 per cent. of carbon,

* One Hessian acre is equal to 40,000 square feet, Hessian, or 26,910 square feet, English measure.

would correspond to 165 lbs. of dry wood. But we have seen that 2650 lbs. of fir-wood are really produced.

Again, if the quantity of humic acid which might be introduced into wheat in the form of humates, is calculated from the known proportion of metallic oxides existing in wheat straw, (the sulphates and chlorides also contained in the ashes of the straw not being included,) it will be found that the wheat growing on 1 Hessian acre would receive in that way 57½ lbs. of humic acid, corresponding to 85 lbs. of woody fibre. But the extent of land just mentioned produces, independently of the roots and grain, 1780 lbs. of straw, the composition of which is the same as that of woody fibre.

It has been taken for granted in these calculations, that the basic metallic oxides which have served to introduce humic acid into the plants do not return to the soil, since it is certain that they remain fixed in the parts newly formed during the process of growth.

Let us now calculate the quantity of humic acid which plants can receive under the most favourable circumstances, viz. through the agency of rain-water.

The quantity of rain which falls at Erfurt, one of the most fertile districts of Germany, during the months of April, May, June, and July, is stated by Schubler to be 17½ lbs. over every Hessian square foot of surface (=0.672 square foot English): 1 Hessian acre, or 26,910 square feet, consequently receive, in round numbers, 700,000 lbs. of rain-water.

If, now, we suppose that the whole quantity of this rain is taken up by the roots of a summer plant, which ripens four months after it is planted, so that not a pound of water evaporates except from the leaves of the plant; and if we further assume that the water thus absorbed is saturated with humate of lime (the most generally diffused of the humates, and that which contains the largest proportion of humic acid); then the plants thus nourished would not receive more than 350 lbs. of humic acid, since one part of humate of lime requires 2000 parts of water for solution.

But the extent of land which we have mentioned produces 2580 lbs. of corn (in grain and straw, the roots not included), or 20,000 lbs. of beet-root (without the leaves and small fibres of the radicle). It is quite evident that the 350 lbs. of humic acid, supposed to be absorbed, cannot account for the quantity of carbon contained in the fibres of the radicle and leaves alone, even if the supposition were correct, that the whole of the rain-water was absorbed by the plants. But since it is known that only a small portion of the rain-water which falls upon the surface of the earth is absorbed by plants and evaporates through their leaves, the quantity of carbon which can be conveyed into them in any conceivable manner, by means of humic acid, must be extremely trifling, in comparison with that actually produced in vegetation.

Other considerations of a higher nature confute the common view respecting the nutritive office of humic acid, in a manner so clear and conclusive that it is difficult to conceive how it could have been so generally adopted.

Fertile land produces carbon in the form of wood, hay, grain, and other kinds of growth, the masses of which differ in a remarkable degree.

2650 lbs. of firs, pines, beeches, &c., grow annually as wood upon one Hessian acre of forest-land with an average soil. The same superficies yields 2500 lbs. of hay.

A similar surface of corn-land gives from 18,000 to 20,000 lbs. of beet-root; or 800 lbs. of rye, and 1780 lbs. of straw,—in all 2580 lbs.

One hundred parts of dry fir-wood contain 38 parts of carbon; therefore, 2650 lbs. contain 1007 lbs. of carbon.

One hundred parts of hay, * dried in air, contain 40.73 parts carbon. Accordingly, 2500 lbs. of hay contain 1018 lbs. of carbon.

---

* 100 parts of hay, dried at 100° C. (212° F.) and burned with oxide of copper in a stream of oxygen gas, yielded 51·93 water, 166·8 carbonic acid, and 6·82 of ashes. This gives 45·87 carbon, 5·76 hydrogen, 41·55 oxygen, and 6·82 ashes. Hay, dried in the air, loses 11·2 p. c. water at 100° C. (212° F.)—Dr. Will.

Beet-roots contain from 89 to 89.5 parts water, and from 10.5 to 11 parts solid matter, which contains 40 per cent. of carbon.*

20,000 lbs. of beet-root contain, therefore, 880 lbs. of carbon, the quantity of this element in the leaves and small roots not being included in the calculation.

One hundred parts of straw,† dried in air contain 38 per cent. of carbon; therefore, 1780 lbs. of straw contain 676 lbs. of carbon. One hundred parts of corn contain 43 parts of carbon; 800 lbs. must therefore contain 344 lbs.—in all 1020 lbs. of carbon.

26,910 square feet	of wood-land produce of carbon	. .	1007 lbs.
,, ,,	meadow-land ,,	. . .	1018 lbs.
,, ,,	arable-land, beet-roots without leaves	.	880 lbs.
,, ,,	corn	. . . .	1020 lbs.

It must be concluded from these incontestable facts, that equal surfaces of cultivated land of an average fertility are capable of producing equal quantities of carbon; yet, how unlike have been the different conditions of the growth of the plants from which this has been deduced!

Let us now inquire whence the grass in a meadow, or the wood in a forest, receives its carbon, since there, carbon has not been given to it as nourishment? and how it happens, that the soil, thus exhausted, instead of becoming poorer, becomes every year richer in this element?

A certain quantity of carbon is taken every year from the forest or meadow, in the form of wood or hay, and, in spite of this, the quantity of carbon in the soil augments; it becomes richer in humus.

It is said that in fields and orchards all the carbon which may have been taken away as leaves, as straw, as seeds, or as fruit, is replaced by means of manure; and yet this soil

* I. 0·8075 of dry beet gave 0·416 water and 1·155 carbonic acid. II. 0·400 gave 0·201 water, and 0·595 carbonic acid.—Dr. Will.

† Straw analysed in the same manner, and dried at 100° C., gave 46·37 p. c. of carbon, 5·68 p. c. of hydrogen, 43·93 p. c. of oxygen, and 4·02 p. c. of ashes. Straw dried in the air at 100° C. lost 18 p. c. of water.—Dr. Will.

produces no more carbon than that of the forest or meadow, where it is never replaced. It cannot be conceived that the laws for the nutrition of plants are changed by culture,—that the sources of carbon for fruit or grain, and for grass or trees, are different.

It is not denied that manure exercises an influence upon the development of plants; but it may be affirmed with positive certainty, that to its carbon is not due the favourable influence which it exercises, because we find that the quantity of carbon produced by manured lands is not greater than that yielded by lands which are not manured. The discussion as to the manner in which manure acts has nothing to do with the present question,—which is the origin of the carbon. The carbon must be derived from other sources; and as the soil does not yield it, it can only be extracted from the atmosphere.

In attempting to explain the origin of carbon in plants, it has never been considered that the question is intimately connected with that of the origin of humus. It is universally admitted that humus arises from the decay of plants. No primitive humus, therefore, can have existed—for plants must have preceded the humus.

Now, whence did the first vegetables derive their carbon? and in what form is the carbon contained in the atmosphere?

These two questions involve the consideration of two most remarkable natural phenomena, which, by their reciprocal and uninterrupted influence, maintain the life of individual animals and vegetables, and the continued existence of both kingdoms of organic nature.

One of these questions is connected with the invariable condition of the air with respect to oxygen. One hundred volumes of air have been found, at every period and in every climate, to contain 21 volumes of oxygen, with such small deviations that they must be ascribed to errors of observation.

Although the absolute quantity of oxygen contained in the atmosphere appears very great when represented by numbers, yet it is not inexhaustible. One man consumes by respiration 25 cubic feet of oxygen in 24 hours; 10 cwt. of charcoal consume 32,066 cubic feet of oxygen during its combustion, so that a single iron furnace consumes annually hundreds of millions of cubic feet; and a small town like Giessen (with about 7000 inhabitants) extracts yearly from the air, by the wood employed as fuel, more than 551 millions of cubic feet of this gas.

When we consider facts such as these, our former statement, that the quantity of oxygen in the atmosphere does not diminish in the course of ages*—that the air at the present day, for example, does not contain less oxygen than that found in jars buried for 1800 years in Pompeii —appears quite incomprehensible, unless some cause exists to replace the oxygen abstracted. How does it happen, then, that the proportion of oxygen in the atmosphere is thus invariable?

The answer to this question depends upon another, namely, what becomes of the carbonic acid produced during the respiration of animals, and by the process of combustion? A cubic foot of oxygen gas, by uniting with carbon so as to form carbonic acid, does not change its volume. The billions of cubic feet of oxygen extracted from the atmosphere, are

---

* If the atmosphere possessed, in its whole extent, the same density as it does on the surface of the sea, it would have a height of 24,555 Parisian feet; but it contains the vapour of water, so that we may assume its height to be one geographical mile = 22,843 Parisian feet. Now, the radius of the earth is equal to 860 geographical miles; hence the

Volume of the atmosphere = 9,307,500 cubic miles.
Volume of oxygen . . = 1,954,578 „
Volume of carbonic acid = 3,862·7 „

A man daily consumes 45,000 cubic inches (Parisian). A man yearly consumes 9505·2 cubic feet. 1000 million men yearly consume 9,505,200,000,000 cubic feet (Parisian).

Without exaggeration we may suppose that double this quantity is consumed in the support of respiration of the lower animals, and in the processes of decay and combustion. From this it follows, that the annual consumption of oxygen amounts to 2.392355 cubic miles, or in round numbers to 2·4 cubic miles. Thus, every trace of oxygen would be removed from the atmosphere in 800,000 years. But it would be rendered quite unfit for the support either of respiration or combustion in a much shorter time. When the quantity of oxygen in the air is diminished 8 per cent., the latter exerts a fatal action upon animal life, and extinguishes the combustion of a burning body.

immediately supplied by the same number of billions of cubic feet of carbonic acid.

The most exact and most recent experiments of De Saussure, made in every season for a space of three years, have shown, that the air contains on an average 0·000415 of its own volume of carbonic acid gas; so that, allowing for the inaccuracies of the experiments, which must diminish the quantity obtained, the proportion of carbonic acid in the atmosphere may be regarded as nearly equal to $\frac{1}{1000}$ part of its weight. The quantity varies according to the seasons; but the yearly average remains continually the same.

We have reason to believe that this proportion was much greater in past ages; and nevertheless, the immense masses of carbonic acid which annually flow into the atmosphere from so many causes, ought perceptibly to increase its quantity from year to year. But we find that all earlier observers describe its volume as from one-half to ten times greater than that which it has at the present time; so that we can hence at most conclude that it has diminished.

It is quite evident that the invariable quantities of carbonic acid and oxygen in the atmosphere, must stand in some fixed relation to one another; a cause must exist which prevents the increase of carbonic acid by removing that which is constantly forming; and there must be some means of replacing the oxygen removed from the air by the processes of combustion and putrefaction, as well as by the respiration of animals.

Both these causes are united in the process of vegetable life.

The facts which we have stated in the preceding pages prove that the carbon of plants must be derived exclusively from the atmosphere. Now, carbon exists in the atmosphere only in the form of carbonic acid, and therefore in a state of combination with oxygen.

It has been already mentioned, that carbon and the elements of water form the principal constituents of vegetables; the quantity of the substances which do not possess this composition being in a very small proportion. Now, the relative quantity of oxygen in the whole mass is less than in carbonic acid; for the latter contains two equivalents of oxygen, whilst one only is required to unite with hydrogen in the proportion to form water. The vegetable products containing oxygen in larger proportion than this, are, comparatively, few in number; indeed, in many the hydrogen is in great excess. It is obvious, that when the hydrogen of water is assimilated by a plant, the oxygen in combination with it must be liberated, and will afford a quantity of this element sufficient for the wants of the plant. If this be the case, the oxygen contained in the carbonic acid is quite unnecessary in the process of vegetable nutrition, and it will consequently escape into the atmosphere in a gaseous form. It is therefore certain, that plants must possess the power of decomposing carbonic acid, since they appropriate its carbon for their own use. The formation of their principal component substances must necessarily be attended with the separation of the carbon of the carbonic acid from the oxygen, which must be returned to the atmosphere, whilst the carbon enters into combination with water or its elements. The atmosphere must thus receive a volume of oxygen for every volume of carbonic acid which has been decomposed.

This remarkable property of plants has been demonstrated in the most certain manner, and it is in the power of every person to convince himself of its existence. The leaves and other green parts of a plant absorb carbonic acid, and emit an equal volume of oxygen. They possess this property quite independently of the plant; for if, after being separated from the stem, they are placed in water containing carbonic acid, and exposed in that condition to the sun's light, the carbonic acid is, after a time, found to have disappeared entirely from the water. If the experiment is conducted under a glass receiver filled with water, the oxygen emitted from the plant may be collected and examined. When no more oxygen gas is evolved, it is a sign that all the dissolved carbonic acid is decomposed; but the operation recommences if a new portion of it is added.

Plants do not emit gas when placed in water either free from carbonic acid, or containing an alkali that protects it from assimilation.

These observations were first made by Priestley and Sennebier. The excellent experiments of De Saussure have further shown, that plants increase in weight during the decomposition of carbonic acid and separation of oxygen. This increase in weight is greater than can be accounted for by the quantity of carbon assimilated; a fact which confirms the view, that the elements of water are assimilated at the same time.

The life of plants is closely connected with that of animals, in a most simple manner, and for a wise and sublime purpose.

The presence of a rich and luxuriant vegetation may be conceived without the concurrence of animal life, but the existence of animals is undoubtedly dependent upon the life and development of plants.

Plants not only afford the means of nutrition for the growth and continuance of animal organization, but they likewise furnish that which is essential for the support of the important vital process of respiration; for, besides separating all noxious matters from the atmosphere, they are an inexhaustible source of pure oxygen, and they thus supply to the air the loss constantly sustained by it. Animals, on the other hand, expire carbon, while plants inspire it; and thus the composition of the atmosphere, the medium in which both exist, is maintained constantly unchanged.

It may be asked—Is the quantity of carbonic acid in the atmosphere, scarcely amounting to 1-10th per cent., sufficient for the wants of the whole vegetation on the surface of the earth,—is it possible that the carbon of plants has its origin from the air alone? This question is very easily answered. It is known that a column of air of 1427 lbs. weight rests upon every square Hessian foot (=0.567 square foot English) of the surface of the earth; the diameter of the earth and its superficies are likewise known, so that the weight of the atmosphere can be calculated with the greatest exactness. The thousandth-part of this is carbonic acid, which contains upwards of 27 per cent. carbon. By this calculation it can be shown, that the atmosphere contains 3085 billion lbs. of carbon—a quantity which amounts to more than the weight of all the plants, and of all the strata of mineral and brown coal existing on the earth. This carbon is, therefore, more than adequate to supply all the purposes for which it is required. The quantity of carbon contained in sea-water is proportionally still greater.

If, for the sake of argument, we suppose the superficies of the leaves and other green parts of plants, by which the absorption of carbonic acid is effected, to be double that of the soil upon which they grow—a supposition much under the truth in the case of woods, meadows, and corn-fields. Let us further suppose, that from a stratum of air two feet thick, resting on an acre (Hessian) of land, that is, from 80,000 cubic feet (Hessian) of air, there is absorbed in every second of time, for eight hours daily, carbonic acid equal to 0.00067 of the volume of the air, or $\frac{1}{1000}$th of its weight; then those leaves would receive above 1000 lbs. of carbon in 200 days.*

But it is inconceivable, that the functions of the organs of a plant can cease for any one moment during its life, as long

* The quantity of carbonic acid which can be extracted from the air in a given time, is shown by the following calculation. During the white-washing of a small chamber, the superficies of the walls and roof of which we will suppose to be 105 square metres, and which receives six coats of lime in four days, carbonic acid is extracted from the air, and the lime is consequently converted, on the surface, into a carbonate. It has been accurately determined that one square decimetre receives in this way a coating of carbonate of lime weighing 0·732 grammes. Upon the 105 square metres already mentioned there must accordingly be formed 7686 grammes of carbonate of lime, which contain 4325·6 grammes of carbonic acid. The weight of one cubic decimetre of carbonic acid being calculated at two grammes, (more accurately 1.97978,) the above-mentioned surface must absorb in four days 2·193 cubic metres of carbonic acid. 2500 square metres (one Hessian acre) would absorb, under a similar treatment, 51½ cubic metres=1818 cubic feet of carbonic acid in four days. In 200 days it would absorb 2575 cubic metres=904,401 cubic feet, which contains 11,353 lbs. of carbonic acid, of which 3304 lbs. are carbon, a quantity three times as great as that which is assimilated by the leaves and roots growing upon the same space.

as those organs are not exposed to the action of a process which may counteract the performance of their proper functions. The roots and other parts of it, possessing the same property, constantly absorb water and carbonic acid. This power is independent of solar light. During the night, carbonic acid is accumulated in all parts of their structure; and the decomposition of the carbonic acid, the assimilation of the carbon, and the exhalation of oxygen, commence from the instant that the rays of the sun strike them. As soon as a young plant breaks through the surface of the ground, it begins to acquire colour from the top downwards; and the true formation of woody tissue commences at the same time.

When the vegetable kingdom in the temperate and cold zones ceases to decompose the carbonic acid generated by the processes of respiration and combustion, the proper, constant, and inexhaustible sources of oxygen gas are the tropics and warm climates, where a sky seldom clouded permits the glowing rays of the sun to shine upon an immeasurably luxuriant vegetation. In our winter, when artificial warmth must replace deficient heat of the sun, carbonic acid is produced in superabundance, and is expended in the nourishment of tropical plants. The great stream of air, which is occasioned by the heating of the equatorial regions and by the revolution of the earth, carries with it in its passage to the equator the carbonic acid generated during our winters; and, in its return to the polar regions, brings with it the oxygen produced by the tropical vegetation.

The experiments of De Saussure have proved, that the upper strata of the air contain more carbonic acid than the lower, which are in contact with plants; and that the quantity is greater by night than by day, when it undergoes decomposition.

Plants thus improve the air, by the removal of carbonic acid, and by the renewal of oxygen, which is immediately applied to the use of man and animals. The horizontal currents of the atmosphere bring with them as much as they carry away, and the interchange of air between the upper and lower strata, caused by their difference of temperature, is extremely trifling when compared with the horizontal movements of the winds. Thus vegetable culture heightens the healthy state of a country, so that a previously healthy country would be rendered quite uninhabitable by the cessation of all cultivation.

The various layers of wood and mineral coal, as well as peat, form the remains of a primeval vegetation. The carbon contained in them must have been originally in the atmosphere as carbonic acid, in which form it was assimilated by the plants which constitute these formations. It follows from this, that the atmosphere must be richer in oxygen at the present time than in former periods of the earth's history. The increase must be exactly equal in volume to the carbonic acid abstracted in the nourishment of a former vegetation, and must, therefore, correspond to the quantity of carbon and hydrogen contained in the carboniferous deposit. Thus, by the deposition of ten cubic feet Hessian (5.51 cubic feet English) of Newcastle splint coal (of the formula $C_{24}H_{13}O$, and specific gravity 1228), the atmosphere must have been deprived of above eighteen thousand cubic feet Hessian (9918 cubic feet English) of carbonic acid, and must have been enriched with the same quantity of oxygen. A further quantity of oxygen amounting to 4480 cubic feet Hessian (2468 English) must have been furnished to the air by the decomposition of water, for 10 cubic feet Hessian of coal contains hydrogen corresponding to this amount. In former ages, therefore, the atmosphere must have contained less oxygen, but a much larger proportion of carbonic acid, than it does at the present time; a circumstance which accounts for the richness and luxuriance of the earlier vegetation. When this became entombed, higher forms of animal life were capable of existing.

But a certain period must have arrived in which the quantity of carbonic acid contained in the air experienced neither increase nor diminution in any appreciable quantity. For if it received an additional quantity to its usual proportion, an increased vegetation would be the natural consequence, and the excess would thus be speedily removed. And, on the other hand, if the gas was less than the normal quantity, the progress of vegetation would be retarded, and the proportion would soon attain its proper standard. When man appeared on the earth, the air was rendered constant in its composition.

The most important function in the life of plants, or, in other words, in their assimilation of carbon, is the separation, we might almost say the generation, of oxygen. No matter can be considered as nutritious, or as necessary to the growth of plants, which possesses a composition either similar to or identical with theirs; because the assimilation of such a substance could be effected without the exercise of this function. The reverse is the case in the nutrition of animals. Hence such substances as sugar, starch, and gum, themselves the products of plants, cannot be adapted for assimilation. And this is rendered certain by the experiments of vegetable physiologists, who have shown that aqueous solutions of these bodies are imbibed by the roots of plants, and carried to all parts of their structure, but are not assimilated; they cannot, therefore, be employed in their nutrition.

In the second part of the work we shall adduce satisfactory proofs that decayed woody fibre (*humus*) contains carbon and the elements of water, without an excess of oxygen; its composition (in 100 parts) differing from that of woody fibre only in its being richer in carbon.

Misled by this simplicity in its constitution, physiologists found no difficulty in discovering the mode of the formation of woody fibre; for they say,* humus has only to enter into combination with water, in order to effect the formation of woody fibre, and other substances similarly composed, such as sugar, starch, and gum. But they forget that their own experiments have sufficiently demonstrated the inaptitude of these substances for assimilation. Yet we could scarcely conceive a form more fitted for assimilation than that of the substances just mentioned. They contain all the elements of woody fibre, and with respect to their composition in 100 parts, they correspond closely with humus; but they do not nourish plants.

* Meyen, *Pflanzenphysiologie,* II. S. 141.

All the erroneous opinions concerning the *modus operandi* of humus have their origin in the false notions entertained respecting the most important vital functions of plants; analogy, that fertile source of error, having, unfortunately, led to the very unapt comparison of the vital functions of plants with those of animals.

Substances, such as sugar, starch, &c., containing carbon and the elements of water, are products of the life of plants which live only whilst they generate them. The same may be said of humus, for it can be formed in plants like the former substances. Smithson, Jameson, and Thomson, found that the black excretions of unhealthy elms, oaks, and horse-chesnuts, consisted of humic acid in combination with alkalies. Berzelius detected similar products in the bark of most trees. Now, can it be supposed that the diseased organs of a plant possess the power of generating the matter to which its sustenance and vigour are ascribed?

How does it happen, it may be asked, that the absorption of carbon from the atmosphere by plants is doubted by many botanists and vegetable physiologists, and that by the greater number the purification of the air by means of them is wholly denied?

These doubts have arisen from an erroneous consideration of the behaviour of plants during the night. The experiments of Ingenhouss were in a great degree the cause of the uncertainty of opinion regarding the influence of plants in purifying the air. His observation that green plants emit carbonic acid in the dark, led De Saussure and Grischow to new investigations, by which they ascertained that under such conditions plants do really absorb oxygen and emit carbonic acid; but that the whole volume of air undergoes diminution at the same time. From the latter fact it follows, that the quantity of oxygen gas absorbed is greater than the volume of carbonic acid separated; for, if both were equal, no diminution could occur. These facts cannot be doubted, but the views based on them have been so false, that nothing, except the total disregard and the utmost ignorance of the chemical relations of plants to the atmosphere, can account for their adoption.

It is known that nitrogen, hydrogen, and a number of other gases, exercise a peculiar, and, in general, an injurious influence upon living plants. Is it, then, probable, that oxygen, one of the most energetic agents in nature, should remain without influence on plants when one of their peculiar processes of assimilation has ceased?

It is true that the decomposition of carbonic acid is arrested by absence of light. But then, namely, at night, a true chemical process commences, in consequence of the action of the oxygen in the air, upon the organic substances composing the leaves, blossoms, and fruit. This process is not at all connected with the life of the vegetable organism, because it goes on in a dead plant exactly as in a living one.

The substances composing the leaves of different plants being known, it is a matter of the greatest ease and certainty to calculate which of them, during life, should absorb most oxygen by chemical action when the influence of light is withdrawn.

The leaves and green parts of all plants containing volatile oils or volatile constituents in general, should absorb more than other parts free from such substances; for these change into resin by the absorption of oxygen. Leaves, also, containing either the constituents of nut-galls, or compounds in which nitrogen is present, ought to absorb more oxygen than those destitute of such matters. The correctness of these inferences has been distinctly proved by the observations of De Saussure; for, whilst the tasteless leaves of the *Agave americana* absorb only 0·3 of their volume of oxygen in the dark, during twenty-four hours, the leaves of the *Pinus Abies*, containing volatile and resinous oils, absorb ten times; those of the *Quercus Robur* containing tannic acid 14 times; and the balmy leaves of the *Populus alba* 21 times that quantity. This chemical action is shown very plainly also in the leaves of the *Cotyledon calycinum*, the *Cacalia ficoides*, and others; for they are sour like sorrel in the morning, tasteless at noon, and bitter in the evening. The formation of acids is effected during the night by a true process of oxidation: they are deprived of their acid properties during the day and evening, and are changed by separation of a part of their oxygen into compounds containing oxygen and hydrogen, either in the same proportions as in water, or even with an excess of hydrogen; for such is the composition of all tasteless and bitter substances.

Indeed the quantity of oxygen absorbed could be estimated pretty nearly by the different periods which the green leaves of plants require to undergo alteration in colour, by the influence of the atmosphere. Those continuing longest green will abstract less oxygen from the air in an equal space of time, than those the constituent parts of which suffer a more rapid change. It is found, for example, that the leaves of the *Ilex aquifolium*, distinguished by the durability of their colour, absorb only 0.86 of their volume of oxygen gas in the same time that the leaves of the poplar absorb 8, and those of the beech 9½ times their volume; both the beech and poplar being remarkable for the rapidity and ease with which the colour of their leaves changes.

When the green leaves of the poplar, the beech, the oak, or the holly, are dried under the air-pump, with exclusion of light, then moistened with water, and placed under a glass globe filled with oxygen, they are found to absorb that gas in proportion as they change in colour. The chemical nature of this process is thus completely established. The diminution of the gas which occurs can only be owing to the union of a large proportion of oxygen with those substances already in the state of oxides, or to the oxidation of such vegetable compounds as contain hydrogen in excess. The fallen brown or yellow leaves of the oak contain no longer tannin, and those of the poplar are destitute of balsamic constituents.

The property possessed by green leaves of absorbing oxygen belongs also to fresh wood, whether taken from a twig or from the interior of the trunk of a tree. When fine chips of such wood are placed in a moist condition under a jar filled with oxygen, the gas is seen to diminish in volume. But wood, dried by exposure to the atmosphere and then moistened, converts the oxygen into carbonic acid, without change of volume; fresh wood, therefore, absorbs most oxygen.*

MM. Petersen and Schödler have shown, by the careful elementary analysis of 24 different kinds of wood, that they contain carbon and the elements of water, with the addition of a certain quantity of hydrogen. Oak wood, recently taken from the tree, and dried at 100° C. (212° F.), contain 49.432 carbon, 6.069 hydrogen, and 44.499 oxygen.

The proportion of hydrogen necessary to combine with 44.499 oxygen in order to form water, is $\frac{1}{8}$ of this quantity, namely, 5.56; it is evident, therefore, that oak wood contains $\frac{1}{12}$ more hydrogen than corresponds to this proportion. In *Pinus larix, P. abies*, and *P. picea*, the excess of hydrogen amounts to $\frac{1}{7}$, and in *Tilia europœa* to $\frac{1}{5}$. The quantity of hydrogen stands in some relation to the specific weight of the wood; the lighter kinds of wood contain more of it than the heavier. In ebony wood (*Diospyros ebenum*) the oxygen and hydrogen are in exactly the same proportion as in water.

The difference between the composition of the varieties of wood, and that of simple woody fibre, depends, unquestionably, upon the presence of constituents, in part soluble, and in part insoluble, such as resin and other matters, containing a large proportion of hydrogen: the hydrogen of such substances being in the analysis of the various woods added to that of the true woody fibre.

It has previously been mentioned that mouldering oak wood contains carbon and the elements of water, without any excess of hydrogen. If, in its present state, its further decay

* When villages situated on the banks of rivers become inundated with floods, this property of wood gives rise to much disease. The wood of the floors and rafters of the building become saturated with water, which evaporates very slowly. The oxygen of the air is absorbed rapidly by the moist wood, and carbonic acid is generated. The latter gas exercises a directly pernicious influence when present in air to the amount of 7 or 8 per cent.

does not alter the volume of the air, it is certain that in the beginning of the process the result must have been different, for the amount of hydrogen present in the fresh wood has been diminished, and this could only have been effected by an absorption of oxygen.

Most vegetable physiologists have connected the emission of carbonic acid during the night with the absorption of oxygen from the atmosphere, and have considered these actions as a true process of respiration in plants, similar to that of animals, and, like it, having for its result the separation of carbon from some of their constituents. This opinion has a very weak and unstable foundation.

The carbonic acid, which has been absorbed by the leaves and by the roots, together with water, ceases to be decomposed on the departure of daylight; it is dissolved in the juices which pervade all parts of the plant, and escapes every moment through the leaves in quantity corresponding to that of the water which evaporates.

A soil in which plants vegetate vigorously, contains a certain quantity of moisture indispensably necessary to their existence. Carbonic acid, likewise, is always present in such a soil, whether it has been abstracted from the air, or has been generated by the decay of vegetable matter. Rain and well water, and also that from other sources, invariably contains carbonic acid.—Plants during their life constantly possess the power of absorbing by their roots moisture, and, along with it, air or carbonic acid. Is it, therefore, surprising that the carbonic acid should be returned unchanged to the atmosphere along with water, in the absence of light; for this is known to be the cause of the fixation of its carbon?

Neither this emission of carbonic acid nor the absorption of oxygen has any connexion with the process of assimilation; nor have they the slightest relation to one another; the one is a purely mechanical, the other a purely chemical process. A cotton wick, inclosed in a lamp containing a liquid saturated with carbonic acid, acts exactly in the same manner as a living plant in the night. Water and carbonic acid are sucked up by capillary attraction, and both evaporate from the exterior part of the wick.

Plants living in a moist soil containing humus exhale much more carbonic acid during the night than those growing in dry situations; they also yield more in rainy than in dry weather; these facts point out to us the cause of the numerous contradictory observations made with respect to the change impressed upon the air by living plants, both in darkness and in common daylight; but these contradictions are unworthy of consideration, as they do not assist in the solution of the main question.

There are other facts which prove in a decisive manner that plants yield more oxygen to the atmosphere than they extract from it. These proofs may easily be obtained, without having recourse to any peculiar arrangements, from observations made on plants living under water.

Pools and ditches, the bottoms of which are covered with growing plants, often freeze upon their surface in winter, so that the water is completely excluded from the atmosphere by a clear stratum of ice; under such circumstances small bubbles of gas are observed to escape continually during the day, from the points of the leaves and twigs. These bubbles are seen most distinctly when the rays of the sun fall upon the ice; they are very small at first, but collect under the ice and form larger bubbles. They consist of pure oxygen gas. Neither during the night, nor during the day when the sun does not shine, are they observed to diminish in quantity. The source of this oxygen is the carbonic acid absorbed by the plants from the water, to which it is again supplied by the decay of vegetable substances contained in the soil. If these plants absorb oxygen during the night, it can be in no greater quantity than that which the surrounding water holds in solution; for the gas, which has been exhaled, is not again absorbed.

Sir H. Davy made an elegant experiment in illustration of the facts just stated. He placed a turf, four inches square, in a porcelain dish which swam on the surface of water impregnated with carbonic acid gas. A glass vessel of the capacity of 230 cubic inches was made to cover the grass, to which water was occasionally supplied by a funnel furnished with a stopcock. The water upon which the porcelain dish swam was daily supplied with new water saturated with carbonic acid, so that a small quantity of that gas must always have been present in the receiver. The volume of air in the receiver was found to increase by exposure to daylight, so much so, that after the lapse of eight days, an increase of thirty cubic inches was observed. The air inside the receiver on being analysed was found to contain 4 per cent. more oxygen than the air of the exterior atmosphere. (Davy's Agricultural Chemistry, Lecture V.) In confirmation of the same facts we may also refer to the excellent experiments of Dr. Daubeny.*

In the preceding part of the work, we have furnished proofs that the carbon of plants is derived from the atmosphere. We have yet to consider the action of humus and of certain mineral matters upon the development of vegetation, and also the source whence plants receive their nitrogen.

---

* On the Action of Light upon Plants, and of Plants upon the Atmosphere, Phil. Trans., Part I., 1836.

CHAPTER III.

ON THE ORIGIN AND ACTION OF HUMUS.

It will be shown in the second part of this work, that all plants and vegetable structures undergo two processes of decomposition after death. One of these is named *fermentation, decay*, or *eremacausis* ;* the other *putrefaction*.

It will likewise be shown, that decay is a slow process of combustion,—a process, therefore, in which the combustible parts of a plant unite with the oxygen of the atmosphere.

The decay of woody fibre (the principal constituent of all plants) is accompanied by a phenomenon of a peculiar kind. This substance, in contact with air or oxygen gas, converts the latter into an equal volume of carbonic acid, and its decay ceases upon the disappearance of the oxygen. If the carbonic acid be removed, and oxygen replaced, its decay recommences, that is, it again converts oxygen into carbonic acid. Woody fibre consists of carbon and the elements of water; and if we judge only from the products formed during its decomposition, and from those formed by pure charcoal, burned at a high temperature, we might conclude that the causes were the same in both: the decay of woody fibre proceeds, therefore, as if no hydrogen or oxygen entered into its composition.

A very long time is required for the completion of this process of combustion, and the presence of water is necessary for its maintenance: alkalies promote it, but acids retard it; all antiseptic substances, such as sulphurous acid, the mercurial salts, empyreumatic oils, &c., cause its complete cessation.

* The word *eremacausis* was proposed by the author some time since, in order to explain the true nature of decay; it is compounded from ἠρέμα, *by degrees*, and καῦσις, *burning*.

Woody fibre in a state of decay is the substance called HUMUS.*

The property of woody fibre to convert surrounding oxygen gas into carbonic acid diminishes in proportion as its decay advances, and at last a certain quantity of a brown coaly-looking substance remains, in which this property is entirely wanting. This substance is called MOULD; it is the product of the complete decay of woody fibre. Mould constitutes the principal part of all the strata of brown coal and peat. By contact with alkalies, such as lime or ammonia, a further decay of mould is occasioned.

Humus acts in the same manner in a soil permeable to air as in the air itself; it is a continued source of carbonic acid, which it emits very slowly. An atmosphere of carbonic acid, formed at the expense of the oxygen of the air, surrounds every particle of decaying humus. The cultivation of land, by tilling and loosening the soil, causes a free and unobstructed access of air. An atmosphere of carbonic acid is therefore contained in every fertile soil, and is the first and most important food for the young plants growing upon it.

In spring, when those organs of plants are absent which nature has appointed for the assumption of nourishment from the atmosphere, the component substances of the seeds are exclusively employed in the formation of the roots. Each new radicle fibril acquired by a plant may be regarded as constituting at the same time a mouth, a lung, and a stomach. The roots perform the functions of the leaves from the first moment of their formation: they extract from the soil their proper nutriment, namely, the carbonic acid generated by the humus.

By loosening the soil surrounding young plants, we favour the access of air, and the formation of carbonic acid; and, on the other hand, the quantity of their food is diminished by every difficulty which opposes the renewal of air. A plant itself effects this change of air at a certain period of its growth.

* The humic acid of chemists is a product of the decomposition of humus by alkalies; it does not exist in the humus of vegetable physiologists.

The carbonic acid, which protects the undecayed humus from further change, is absorbed and taken away by the fine fibres of the roots, and by the roots themselves; this is replaced by atmospheric air, which, by its oxygen, renews the process of decay, and forms a fresh portion of carbonic acid. A plant at this time receives its food both by the roots and by the organs above ground, and advances rapidly to maturity.

When a plant is quite matured, and when the organs by which it obtains food from the atmosphere are formed, the carbonic acid of the soil is no further required.

Deficiency of moisture in the soil, or its complete dryness, does not now check the growth of a plant, provided it receives from the dew and from the atmosphere as much as is requisite for the process of assimilation. During the heat of summer it derives its carbon exclusively from the atmosphere.

We do not know what height and strength nature has allotted to plants; we are acquainted only with the size which they usually attain. Oaks are shown, both in London and Amsterdam, as remarkable curiosities, which have been reared by Chinese gardeners, and are only one foot and a half in height, although their trunks, barks, leaves, branches, and whole habitus, evince a venerable age. The small parsnep grown at Teltow, * when placed in a soil which yields as much nourishment as it can take up, increases to several pounds in weight.

THE SIZE WHICH A PLANT ACQUIRES IN A GIVEN TIME IS PROPORTIONAL TO THE SURFACE OF THE ORGANS DESTINED TO CONVEY FOOD TO IT. When the surfaces of two plants are equal, their increase depends upon the length of time that their absorbing powers remain in activity. The absorbing surfaces of fir trees are active during the greater part of the year, so that (*cæteris paribus*), they increase more than those trees which part with their foliage in autumn. Each leaf furnishes to a plant another mouth and stomach.

The power possessed by roots of taking up nourishment

---

* Teltow is a village near Berlin, where small parsneps are cultivated in a sandy soil; they are much esteemed, and weigh rarely above one ounce.

does not cease as long as nutriment is present. When the food of a plant is in greater quantity than its organs require for their own perfect development, the superfluous nutriment is not returned to the soil, but is employed in the formation of new organs. At the side of a cell already formed, another cell arises; at the side of a twig and leaf, a new twig and a new leaf are developed. These new parts could not have been formed had there not been an excess of nourishment. The sugar and mucilage produced in the seeds, form the nutriment of the young plants, and disappear during the development of the buds, green sprouts, and leaves.

The power of absorbing nutriment from the atmosphere, with which the leaves of plants are endowed, being proportionate to the extent of their surface, every increase in the size and number of these parts is necessarily attended with an increase of nutritive power, and a consequent further development of new leaves and branches. Leaves, twigs, and branches, when completely matured, as they do not become larger, do not need food for their support. For their existence as organs, they require only the means necessary for the performance of the special functions to which they are destined by nature; they do not exist on their own account.

We know that the functions of the leaves and other green parts of plants are to absorb nutritive matters from the atmosphere, and, with the aid of light and moisture, to appropriate their elements. These processes are continually in operation: they commence with the first formation of the leaves, and do not cease with their perfect development. But the new products arising from this continued assimilation are no longer employed by the perfect leaves in their own increase: they serve for the formation of woody fibre, and all the solid matters of similar composition. The leaves now produce sugar, amylin or starch, and acids, which were previously formed by the roots when they were necessary for the development of the stem, buds, leaves, and branches of the plant.

The organs of assimilation, at this period of their life,

receive more nourishment from the atmosphere than they employ in their own sustenance; and when the formation of the woody substance has advanced to a certain extent, the expenditure of the nutriment, the supply of which still remains the same, takes a new direction, and blossoms are produced. The functions of the leaves of most plants cease upon the ripening of their fruit, because the products of their action are no longer needed. They now yield to the chemical influence of the oxygen of the air, generally suffer a change in colour, and fall off.

A peculiar transformation of the matter contained in all plants takes place in the period between blossoming and the ripening of the fruit; new compounds are produced, which furnish constituents to the blossoms, fruit, and seeds.

Transformations of existing compounds are constantly taking place during the whole life of a plant, in consequence of which, and as the results of these transformations, there are produced gaseous matters which are excreted by the leaves and blossoms, solid excrements deposited in the bark, and fluid soluble substances which are eliminated by the roots. Such secretions are most abundant immediately before the formation and during the continuance of the blossoms; they diminish after the development of the fruit. Substances containing a large proportion of carbon are excreted by the roots and absorbed by the soil. Through the expulsion of these matters unfitted for nutrition, the soil receives again with usury the carbon which it had at first yielded to the young plants as food, in the form of carbonic acid.

The soluble matter thus acquired by the soil is still capable of decay and putrefaction, and by undergoing these processes furnishes renewed sources of nutrition to another generation of plants; it becomes HUMUS. The fallen leaves of trees, and the old roots of grass in the meadow, are likewise converted into humus by the same influence.

The carbon contained in the roots of annual plants, such as the corn plants and culinary vegetables, is without doubt derived principally from the atmosphere. But after the removal of the crop, their roots remain in the soil, and, undergoing putrefaction and decay, furnish humus, or that substance which is able to yield carbonic acid to a new vegetation. A soil receives more carbon in this form, than its decaying humus had lost in carbonic acid.

Plants do not exhaust the carbon of a soil in the normal condition of their growth; on the contrary, they add to its quantity. But if it be true that plants give back more carbon to a soil than they take from it, it is evident that their growth must depend upon the reception of nourishment from the atmosphere in the form of carbonic acid. It is well known that springs occurring in gardens of the richest vegetable mould, furnish clear and perfectly colourless water destitute both of humus and of salts of humic acid. It is likewise known that humates cannot be detected in the springs of meadows, in the waters of our rivers, or even in acidulous mineral waters, although they contain a considerable quantity of alkaline salts. Now a simple consideration of these facts proves to us either that the richest vegetable mould is free from humic acid; or that this acid cannot be absorbed by plants through the agency of water. Hence it follows that the common view of the action of humus is erroneous. The water resting upon a meadow is found to be rich in carbonic acid and alkaline bases. Well-water also generally contains much of the former ingredient. The influence, then, of humus or decaying vegetable matter upon vegetation, is explained by these facts in the most clear and satisfactory manner. Humus, therefore, does not nourish plants by being assimilated in its soluble state, but by furnishing a gradual and continued source of carbonic acid. This gas forms an important means of nourishment to the roots of plants, and is constantly formed anew as long as the soil admits the free access of air and moisture, these being the necessary conditions for effecting the decay of vegetable matter.

The verdant plants of warm climates are very often such as obtain from the soil only a point of attachment, and are not dependent on it for their growth. How extremely small are the roots of the various species of *Cactus*,* *Sedum*, and *Sempervivum*, in proportion to their mass, and to the surface of their leaves! Large forests are often found growing in soils absolutely destitute of carbonaceous matter; and the extensive prairies of the Western Continent show that the carbon necessary for the sustenance of a plant may be entirely extracted from the atmosphere. Again, in the most arid and barren sand, where it is impossible for nourishment to be obtained through the roots, we see the milky-juiced plants attain complete perfection. The moisture necessary for the nutrition of these plants is derived from the atmosphere, and when assimilated is secured from evaporation by the nature of the juice itself. Caoutchouc and wax, which are formed in these plants, surround the water, as in oily emulsions, with an impenetrable envelope by which the fluid is retained, in the same manner as milk is prevented from evaporating by the skin which forms upon it. The plants become turgid with their juices.

* The Cactus was probably introduced into Sicily by the Spaniards. It forms as important an article of diet with the inhabitants of that island as the potato does with ourselves. This abundant, cooling, and juicy fruit forms the principal food of the lower classes for three months, and is considered very palatable, although strangers usually find it insipid. The hills of Palermo covered with the Cactus correspond to our corn-fields. It is a very important plant for such districts, because its roots easily enter into the cracks and crevices of the volcanic rocks. These, although destitute of humus, soon acquire it by the decay of the leaves, and thus fertile soils are gradually formed for other plants. (*Auslande*, S. 274, 3rd Oct., 1842.)

CHAPTER IV.

ON THE ASSIMILATION OF HYDROGEN.

THE atmosphere contains the principal food of plants in the form of carbonic acid, in the state, therefore, of an oxide. The solid part of plants (woody fibre) contains carbon and the constituents of water, or the elements of carbonic acid, together with a certain quantity of hydrogen. It has formerly been mentioned that water consists of the two gases, oxygen and hydrogen. We can conceive the wood to arise from a combination of the carbon of the carbonic acid with the elements of water, under the influence of solar light. In this case, 72·35 parts of oxygen, by weight, must be separated as a gas for every 27·65 parts of carbon assimilated by a plant; for this is the composition of carbonic acid in 100 parts. Or, what is much more probable, plants, under the same circumstances, may decompose water, in which case the hydrogen would be assimilated along with carbonic acid, whilst its oxygen would be separated. If the latter change takes place, 9·77 parts of hydrogen must unite with 100 parts of carbonic acid, in order to form woody fibre, and the 72·35 parts by weight of oxygen, which was in combination with the hydrogen of the water, and which exactly corresponds in quantity with the oxygen contained in the carbonic acid, must be separated in a gaseous form.*

Each acre of land, producing 10 cwts. of carbon, gives

* As far as regards the final results, it is a matter of perfect indifference to which of these views we accord the preference. Hence we will use both occasionally. The decomposition of carbonic acid must be effected in the formation of those compounds in which oxygen is either entirely absent or insufficient to form water with the hydrogen.

annually to the atmosphere 2865 lbs., or 32,007 cubic feet of free oxygen gas.*

An acre of meadow, wood, or cultivated land, in general replaces, therefore, in the atmosphere as much oxygen as is exhausted by 10 cwts. of carbon, either in its ordinary combustion in the air, or in the respiratory process of animals.

It has been mentioned in a former part of the work that pure woody fibre contains carbon and the component parts of water, but that ordinary wood contains more hydrogen than corresponds to this proportion. This excess is owing to the presence of the green principle of the leaf, wax, oil, resin, and other bodies rich in hydrogen. Water must be decomposed, in order to furnish the excess of this element, and consequently one equivalent of oxygen must be given back to the atmosphere for every equivalent of hydrogen appropriated by a plant to the production of those substances. The quantity of oxygen thus set at liberty cannot be insignificant, for the atmosphere must receive above 100 cubic feet of oxygen for every pound of hydrogen assimilated.

It has already been stated, that a plant, in the formation of woody fibre, must always yield to the atmosphere the same proportional quantity of oxygen; and that the volume of this gas set free would be the same whether it were due to the decomposition of carbonic acid or of water. It was considered most probable that the former was the case.

From their generating caoutchouc, wax, fats, and volatile oils containing hydrogen in large quantity, and no oxygen, we may be certain that plants possess the property of decomposing water, because from no other body could the unazotised substances obtain their hydrogen. It has also been proved by the observations of Humboldt on the fungi, that water may be decomposed without the assimilation of hydrogen. Water is a remarkable combination of two elements, which have the power to separate themselves from one another, in

* The specific weight of oxygen is expressed by the number 1·1026; hence, 1 cubic metre of oxygen weighs 3·157 lbs., and 2865 lbs. of oxygen correspond to 908 cubic metres, or 32,007 cubic feet.

innumerable processes, in a manner imperceptible to our senses; while carbonic acid, on the contrary, is only decomposable by violent chemical action.

Most vegetable structures contain hydrogen in the form of water, which can be separated as such, and replaced by other bodies; but the hydrogen essential to their constitution cannot possibly exist in the state of water.

All the hydrogen necessary for the formation of an organic compound is supplied to a plant by the decomposition of water. The process of assimilation, in its most simple form, consists in the extraction of hydrogen from water, and of carbon from carbonic acid, in consequence of which, either all the oxygen of the water and of the carbonic acid is separated, as in the formation of caoutchouc, the volatile oils containing no oxygen, and other similar substances, or only a part of it is exhaled.

The known composition of the organic compounds most generally present in vegetables, enables us to state in definite proportions the quantity of oxygen separated during their formation.

36 eq. carbonic acid and 22 eq. hydrogen derived from 22 eq. water . . . . . } = *Woody Fibre.*
with the separation of 72 eq. oxygen.

36 eq. carbonic acid and 36 eq. hydrogen derived from 36 eq. water . . . . . } = *Sugar.*
with the separation of 72 eq. oxygen.

36 eq. carbonic acid and 30 eq. hydrogen derived from 30 eq. water . . . . . } = *Starch.*
with the separation of 72 eq. oxygen.

36 eq. carbonic acid and 16 eq. hydrogen derived from 16 eq. water . . . . . } = *Tannic Acid.*
with the separation of 64 eq. oxygen.

36 eq. carbonic acid and 18 eq. hydrogen derived from 18 eq. water . . . . . } = *Tartaric Acid.*
with the separation of 45 eq. oxygen.

36 eq. carbonic acid and 18 eq. hydrogen derived from 18 eq. water . . . . . } = *Malic Acid.*
with the separation of 54 eq. oxygen.

30 eq. carbonic acid and 24 eq. hydrogen derived from 24 eq. water . . . . . } = *Oil of Turpentine.*
with the separation of 84 eq. oxygen.

It will readily be perceived that the formation of the acids is accompanied with the smallest separation of oxygen; that the amount of oxygen set free increases with the production of the so-named neutral substances, and reaches its maximum in the formation of the oils. Fruits remain acid in cold summers; while the most numerous trees under the tropics are those which produce oils, caoutchouc, and other substances containing very little oxygen. The action of sunshine and influence of heat upon the ripening of fruit is thus, in a certain measure, represented by the numbers above cited.

The green resinous principle of the leaf diminishes in quantity, while oxygen is absorbed, when fruits are ripened in the dark; red and yellow colouring matters are formed; tartaric, citric, and tannic acids disappear, and are replaced by sugar, amylin, or gum. 6 eq. Tartaric Acid, by absorbing 6 eq. oxygen from the air, form Grape Sugar, with the separation of 12 eq. carbonic acid. 1 eq. Tannic Acid, by absorbing 8 eq. oxygen from the air, and 4 eq. water, form 1 eq. of Amylin, or starch, with separation of 6 eq. carbonic acid.

We can explain, in a similar manner, the formation of all the unazotised component substances of plants, whether they are produced from carbonic acid and water, with the separation of oxygen, or by the conversion of one substance into the other, by the assimilation of oxygen and separation of carbonic acid. We do not know in what form the production of these constituents takes place; in this respect, the representation of their formation which we have given must not be received in an absolute sense, it being intended only to render the nature of the process more capable of apprehension: but it must not be forgotten, that if the conversion of tartaric acid into sugar, in grapes, be considered as a fact, it must take place under all circumstances in the proportions above mentioned.

The vital process in plants is, with reference to the point we have been considering, the converse of the chemical processes engaged in the formation of salts. Carbonic acid, zinc, and water, when brought into contact, act upon one another, and HYDROGEN IS SEPARATED, while a white pulverulent compound is formed, which contains carbonic acid, zinc, and the oxygen of the water. A living plant represents the zinc in this process: but the process of assimilation gives rise to compounds, which contain the elements of carbonic acid and the hydrogen of water, whilst OXYGEN IS SEPARATED.

Decay has been described above as the great operation of nature, by which that oxygen which was assimilated by plants during life, is again returned to the atmosphere. During the progress of growth, plants appropriate carbon in the form of carbonic acid, and hydrogen from the decomposition of water, the oxygen of which is set free, together with a part or all of that contained in the carbonic acid. In the process of putrefaction, a quantity of water, exactly corresponding to that of the hydrogen, is again formed by extraction of oxygen from the air; while all the oxygen of the organic matter is returned to the atmosphere in the form of carbonic acid. Vegetable matters can emit carbonic acid, during their decay, only in proportion to the quantity of oxygen which they contain; acids, therefore, yield more carbonic acid than neutral compounds; while fatty acids, resin, and wax, do not putrefy; they remain in the soil without any apparent change.

CHAPTER V.

ON THE ORIGIN AND ASSIMILATION OF NITROGEN.

We cannot suppose that a plant could attain maturity, even in the richest vegetable mould, without the presence of matter containing nitrogen; since we know that nitrogen exists in every part of the vegetable structure. The first and most important question to be solved, therefore, is: How and in what form does nature furnish nitrogen to vegetable albumen, and to gluten, or to fruits and seeds? *

This question is susceptible of a very simple solution.

Plants, as we know, grow perfectly well in a mixture of charcoal and earth, previously calcined, if supplied at the same time with rain-water. Rain-water can contain nitrogen only in three forms, as dissolved atmospheric air, as nitric acid, or as ammonia. Now, the nitrogen of the air cannot be made to enter into combination with any element except oxygen, even by the employment of the most powerful chemical means. We have not the slightest reason for believing that the nitrogen of the atmosphere takes part in the processes of assimilation of plants and animals; on the contrary, we know that many plants emit the nitrogen which is absorbed by their roots, either in the gaseous form, or in solution in water. But there are on the other hand numerous

* "It is certain," says Saussure, "from the experiments which have been made on this point, that plants receive their nitrogen only from such animal vegetable extracts, or from such ammoniacal vapours as they may find in the soil, or extract from the air. When plants are made to vegetate by the aid of water in a confined atmosphere, we may presume that the new parts can only obtain nitrogen at the expense of other parts to which it had formerly been supplied. (*De Saussure*, page 190.)

facts, showing, that the formation in plants of substances containing nitrogen, such as gluten, takes place in proportion to the quantity of this element conveyed to their roots in the state of ammonia, derived from the putrefaction of animal matter.

Ammonia, too, is capable of undergoing such a multitude of transformations, when in contact with other bodies, that in this respect it is not inferior to water, which possesses the same property in an eminent degree. It possesses properties which we do not find in any other compound of nitrogen: when pure, it is extremely soluble in water; it forms soluble compounds with all the acids; and when in contact with certain other substances, it completely resigns its character as an alkali, and is capable of assuming the most various and opposite forms. Formate of ammonia changes, under the influence of a high temperature, into hydrocyanic acid and water, without the separation of any of its elements. Ammonia forms urea, with cyanic acid, and a series of crystalline compounds, with the volatile oils of mustard and bitter almonds. It changes into splendid blue or red colouring matters, when in contact with phloridzin, the bitter constituent of the bark of the root of the apple-tree, with orcin, the sweet principle of the *Lichen dealbatus*, or with erythrin, the tasteless matter of the *Rocella tinctoria*. All blue colouring matters capable of being reddened by acids, and all red colouring substances rendered blue by alkalies, contain nitrogen, but not in the form of a base.

These facts are not sufficient to establish the opinion that it is ammonia which affords all vegetables, without exception, the nitrogen of their constituent substances. Considerations of another kind, however, give to this opinion a degree of certainty which completely excludes all other views of the matter.

Let us picture to ourselves the condition of a well-cultured farm, so large as to be independent of assistance from other quarters. On this extent of land there is a certain quantity of nitrogen contained both in the corn and fruit which it produces, and in the men and animals which feed upon them, and also in their excrements. We shall suppose this quantity to be known. The land is cultivated without the importation of any foreign substance containing nitrogen. Now, the products of this farm must be exchanged every year for money, and other necessaries of life—for bodies, therefore, destitute of nitrogen. A certain proportion of nitrogen is exported in the shape of corn and cattle: and this exportation takes place every year, without the smallest compensation: yet after a given number of years, the quantity of nitrogen will be found to have increased. Whence, we may ask, comes this increase of nitrogen? The nitrogen in the excrements cannot reproduce itself, and the earth cannot yield it. Plants, and consequently animals, must, therefore, derive their nitrogen from the atmosphere. (BOUSSINGAULT.)

It will in a subsequent part of this work be shown that the last products of the decay and putrefaction of animal bodies present themselves in two different forms. They are in the form of ammonia (a combination of hydrogen and nitrogen), in the temperate and cold climates, and in that of nitric acid (a compound containing oxygen), in the tropics and hot climates. The formation of the latter is preceded by the production of the first. Ammonia is the last product of the putrefaction of animal bodies; nitric acid is the product of the transformation of ammonia. A generation of a thousand million men is renewed every thirty years; thousands of millions of animals cease to live, and are reproduced, in a much shorter period. Where is the nitrogen contained in them during life? There is no question which can be answered with more positive certainty. All animal bodies during their decay yield to the atmosphere their nitrogen in the form of ammonia. Even in the bodies buried sixty feet under ground in the churchyard of the Eglise de Innocens, at Paris, all the nitrogen contained in the adipocire was in the state of ammonia. Ammonia is the simplest of all the compounds of nitrogen; and hydrogen is the element for which nitrogen possesses the most predominant affinity.

The nitrogen of putrefied animals is contained in the atmosphere as ammonia, in the state of a gas which is capable of entering into combination with carbonic acid, and of forming a volatile salt. Ammonia in its gaseous form, as well as all its volatile compounds, is of extreme solubility in water. Ammonia, therefore, cannot remain long in the atmosphere, as every shower of rain must effect its condensation, and convey it to the surface of the earth. Hence also, rain-water must at all times contain ammonia, though not always in equal quantity. It must contain more in summer than in spring or in winter, because the intervals of time between the showers are in summer greater; and when several wet days occur, the rain of the first must contain more of it than that of the second. The rain of a thunder-storm, after a long-protracted drought, ought for this reason to contain the greatest quantity conveyed to the earth at one time.

But all the analyses of atmospheric air hitherto made have failed to demonstrate the presence of ammonia, although, according to our view, it can never be absent. Is it possible that it could have escaped our most delicate and most exact apparatus? The quantity of nitrogen contained in a cubic foot of air, as ammonia, is certainly extremely small, but, notwithstanding this, the sum of the quantities of nitrogen from thousands and millions of dead animals is more than sufficient to supply all those living at one time with this element.

From the tension of aqueous vapour at 15° C. (59° F.) = 6,98 lines (Paris measure), and from its known specific gravity at 0° C. (32° F.), it follows that when the temperature of the air is 59° F. and the height of the barometer 28″, 1 cubic metre, or 35·3 cubic feet of aqueous vapour are contained in 48.1 cubic metres, or 1698 cubic feet of air; 35·3 cubic feet of aqueous vapour weigh about $1\frac{1}{2}$ lb. Consequently, if we suppose that the air saturated with moisture at 59° F. allows all the water which it contains in the gaseous form to fall as rain, then 1 pound of rain-water must be obtained from every 1132 cubic feet of air. The whole quantity of ammonia contained in the same number of cubic feet will also be returned to the earth in this one pound of rain-water. But if the 1132 cubic feet of air contain a single grain of ammonia, then the few cubic inches usually employed in an analysis—must contain only 0·0000048 of a grain. This extremely small proportion is absolutely inappreciable by the most delicate and best eudiometer; it might be classed among the errors of observation, even were its quantity ten thousand times greater. But the detection of ammonia must be much more easy when a pound of rain-water is examined, for this contains all the gas that was diffused through 1132 cubic feet of air.

If a pound of rain-water contain only $\frac{1}{4}$th of a grain of ammonia, then a field of 26,910 square feet must receive annually upwards of 80 lbs. of ammonia, or 65 lbs. of nitrogen; for by the observations of Schubler, formerly alluded to, the annual fall must be about 2,520,000 lbs. This is much more nitrogen than is contained in the form of vegetable albumen and gluten, in 2650 lbs. of wood, 2500 lbs. of hay, or 200 cwt. of beet-root, which are the yearly produce of such a field; but it is less than the straw, roots, and grain of corn, which might grow on the same surface, would contain.

Experiments made in this laboratory (Giessen) with the greatest care and exactness, have placed the presence of ammonia in rain-water beyond all doubt. It has hitherto escaped observation, because it was not searched for. All the rain-water employed in this inquiry was collected 600 paces south-west of Giessen, whilst the wind was blowing in the direction of the town. When several hundred pounds of it were distilled in a copper still, and the first two or three pounds evaporated with the addition of a little muriatic acid, a very distinct crystallisation of sal-ammoniac was obtained: the crystals had always a brown or yellow colour.

Ammonia may likewise be always detected in snow-water. Crystals of sal-ammoniac were obtained by evaporating in a vessel with muriatic acid several pounds of snow, which were gathered from the surface of the ground in March, when the snow had a depth of ten inches. Ammonia was set free from these crystals by the addition of hydrate of lime. The inferior layers of snow resting upon the ground contained a quantity decidedly greater than those upon the surface.

It is worthy of observation that the ammonia contained in rain and snow-water possesses an offensive smell of perspiration and putrefying matter,—a fact which leaves no doubt respecting its origin.

Hünefield has proved that all the springs in Greifswalde, Wiek, Eldena, and Kostenhagen, contain carbonate and nitrate of ammonia. Ammoniacal salts have been discovered in many mineral springs in Kissingen and other places. The ammonia of these salts can only arise from the atmosphere.*

Any one may satisfy himself of the presence of ammonia in rain by simply adding a little sulphuric or muriatic acid to a quantity of rain-water, and by evaporating this nearly to dryness in a clean porcelain basin. The ammonia remains in the residue, in combination with the acid employed; and may be detected either by the addition of a little chloride of platinum, or, more simply, by a little powdered lime, which separates the ammonia, and thus renders sensible its peculiar pungent smell. The sensation perceived upon moistening the hand with rain-water, so different from that produced by pure distilled water, and to which the term *softness* is vulgarly applied, is also due to the carbonate of ammonia contained in the former.

The ammonia removed from the atmosphere by rain and other causes, is as constantly replaced by putrefaction of animal and vegetable matters.† A certain portion of that which falls with the rain evaporates again with the water; but another portion is, we suppose, taken up by the roots of plants, and entering into new combinations in the different organs of

* Pharmaceutical chemists are well aware of the existence of ammonia in well water, for they have often to reject as much as one-fourth of the water distilled, before they procure water which is not rendered turbid by corrosive sublimate. But when phosphoric acid or alum is added to the water previous to distillation, the product of the distillation is not affected either by corrosive sublimate or by sugar of lead. (WIEGMANN and POTSTORF's Prize Essay on the Inorganic Ingredients of Plants.)

† "We cannot doubt," says Saussure, "that ammonia exists in the atmosphere, for we know that sulphate of alumina is gradually converted into ammoniacal alum by exposure to the air." (*Rech. sur la Veget.*)

assimilation, produces, by the action of these and of certain other conditions, albumen, gluten, and vegetable casein, or quinine, morphia, cyanogen, and a number of other compounds containing nitrogen. The chemical characters of ammonia render it capable of entering into such combinations, and of undergoing numerous transformations. We have now only to consider whether it really is taken up in the form of ammonia by the roots of plants, and in that form applied by their organs to the production of the azotised matters contained in them. This question is susceptible of easy solution by well-known facts.

In the year 1834, I was engaged with Dr. Wilbrand, professor of botany in the University of Giessen, in an investigation respecting the quantity of sugar contained in different varieties of maple-trees, growing upon unmanured soils. We obtained crystallised sugars from all, by simply evaporating their juices, without the addition of any foreign substance; and we unexpectedly made the observation, that a great quantity of ammonia was emitted from this juice when mixed with lime, in the process of refining, as practised with cane sugar. The vessels which hung upon the trees in order to collect the juice were watched with greater attention, on account of the suspicion that some evil-disposed persons had introduced urine into them, but still a large quantity of ammonia was again found in the form of neutral salts. The juice had no colour, and had no reaction on that of vegetables. Similar observations were made upon the juice of the birch tree; the specimens subjected to experiment were taken from a wood several miles distant from any house, and yet the clarified juice, evaporated with lime, emitted a strong odour of ammonia.

In the manufactories of beet-root sugar, many thousand cubic feet of juice are daily purified with lime, in order to free it from vegetable albumen and gluten, and it is afterwards evaporated for crystallisation. Every person who has entered such a manufactory must have been astonished at the great quantity of ammonia volatilised along with the steam. This ammonia must be contained in the form of an ammoniacal salt, because the neutral juice possesses the same characters as the solution of such a salt in water; it acquires, namely, an acid reaction during evaporation, in consequence of the neutral salt being converted by loss of ammonia into an acid salt. The free acid thus formed is a source of loss to the manufacturers of sugar from beet-root, by changing a part of the sugar into uncrystallisable grape sugar and syrup.

The products of the distillation of flowers, herbs, and roots, with water, and all extracts of plants made for medicinal purposes, contain ammonia. The unripe, transparent, and gelatinous pulp of the almond and peach, emit much ammonia when treated with alkalies. (Robiquet.) The juice of the fresh tobacco leaf contains ammoniacal salts. The water which exudes from a cut vine, when evaporated with a few drops of muriatic acid, also yields a gummy deliquescent mass, which evolves much ammonia on the addition of lime. Ammonia exists in every part of plants, in the roots (as in beet-root), in the stem* (of the maple-tree), and in all blossoms and fruit in an unripe condition.

The juices of the maple and birch contain both sugar and ammonia, and therefore afford all the conditions necessary for the formation of the azotised components of the branches, blossoms, and leaves, as well as of those which contain no nitrogen. In proportion as the development of those parts advances, the ammonia diminishes in quantity, and when they are fully formed, the tree yields no more juice.

The employment of animal manure in the cultivation of grain, and the vegetables which serve for fodder to cattle, is the most convincing proof that the nitrogen of vegetables is derived from ammonia. The quantity of gluten in wheat, rye, and barley, is very variable; these kinds of grain also, even when ripe, contain this compound of nitrogen in very different proportions. Proust found French wheat to contain 12·5 per cent. of gluten; Vogel found that the Bavarian contained 24

* In an experiment performed at my request in Calcutta, it was found that the fresh juice of the palm tree abounded with ammonia.—*Ed.*

per cent.; Davy obtained 19 per cent. from winter, and 24 from summer wheat; from Sicilian 21, and from Barbary wheat, 19 per cent. The meal of Alsace wheat contains, according to Boussingault, 17·3 per cent. of gluten; that of wheat grown in the "Jardin des Plantes," 26·7; and that of winter wheat, 33·3 per cent. Such great differences must be owing to some cause, and this we find in the different methods of cultivation. An increase of animal manure gives rise not only to an increase in the number of seeds, but also to a most remarkable difference in the proportion of the substances containing nitrogen, such as the gluten.

Animal manure exerts a very complex action on plants, but as far as regards the assimilation of nitrogen, it acts only by the formation of ammonia. One hundred parts of wheat grown on a soil manured with cow-dung (a manure containing the smallest quantity of nitrogen), afforded only 11·95 parts of gluten, and 62·34 parts of amylin, or starch; whilst the same quantity, grown on a soil manured with human urine, yielded the maximum of gluten, namely 35·1 per cent., or nearly three times the quantity. Putrefied urine contains nitrogen in the forms of carbonate, phosphate, and muriate of ammonia, and in no other form than that of ammoniacal salts.

"Putrid urine is employed in Flanders as a manure, with the best results. During the putrefaction of urine, ammoniacal salts are formed in large quantity, it may be said exclusively; for under the influence of heat and moisture, urea, the most prominent ingredient of the urine, is converted into carbonate of ammonia. The barren soil on the coast of Peru is rendered fertile by means of a manure called *Guano*, which is collected from several islands in the South Sea.* It is sufficient to add a small quantity of guano to a soil consisting only of sand and clay, in order to procure the richest crop of maize. The soil itself does not contain the smallest

* The guano, which forms a stratum several feet in thickness upon the surface of these islands, consists of the putrid excrements of innumerable sea fowl that remain on them during the breeding season. (See the Chapter on Manures.)

particle of organic matter, and the manure employed is formed only of *urate*, *phosphate*, *oxalate*, and *carbonate of ammonia*, together with a few earthy salts.*

Ammonia, therefore, must have yielded the nitrogen to these plants. Gluten is obtained from corn; vegetable albumen from certain juices, such as from the expressed juice of the grape; vegetable casein occurs in the seeds of the leguminous plants; but although all these have different names and properties, they are identical in composition with the ordinary gluten.

It is then ammonia which yields nitrogen to the vegetable albumen, the principal azotised constituent of plants. Nitrogen is not presented to wild plants in any other form capable of assimilation. Ammonia, by its transformation, furnishes nitric acid to the tobacco-plant, sunflower, *Chenopodium*, and *Borago officinalis*, when they grow in a soil completely free from nitre. Nitrates are necessary constituents of these plants, which thrive only when ammonia is present in large quantity, and when they are also subject to the influence of the direct rays of the sun; an influence necessary to effect the disengagement within their stem and leaves of the oxygen which shall unite with the ammonia to form nitric acid.

The urine of men and of carnivorous animals contains a large quantity of nitrogen, partly in the form of phosphates, partly as urea. Urea is converted during putrefaction into carbonate of ammonia, that is to say, it takes the form of the very salt in rain-water. Human urine is the most powerful manure for vegetables rich in nitrogen; the urine of cattle, sheep, and horses, contains less nitrogen, but yet far more than the solid excrements of these animals. In addition to urea, the urine of herbivorous animals contains hippuric acid, which is decomposed during putrefaction into benzoic acid and ammonia. The latter causes the formation of gluten, but the benzoic acid often remains unchanged: for example, in the *Anthoxanthum odoratum*.

* Boussingault, Ann. de Ch. et de Phys. lxv. p. 319.

The solid excrements of men and of animals contain comparatively very little nitrogen, but this could not be otherwise. The food taken by animals supports them only in so far as it offers to the various organs elements for assimilation which they may require for their increase or renewal. Corn, grass, and all plants, without exception, contain highly azotised substances. The quantity of food required by animals for their nourishment, diminishes or increases in the same proportion as it contains more or less of the substances containing nitrogen. A horse may be kept alive by feeding it with potatoes, a food containing a very small quantity of nitrogen; but life thus supported is a gradual starvation: the animal increases neither in size nor strength, and sinks under every exertion. The quantity of rice consumed by an Indian astonishes the European; but the fact that rice contains less nitrogen than any other kind of grain at once explains the circumstance.

Now, as it is evident that the nitrogen of the plants and seeds used by animals as food must be employed in the process of assimilation, it is natural to expect that the solid excrements of these animals will be deprived of it in proportion to the perfect digestion of the food, and can only contain it when mixed with secretions from the liver and intestines. Under all circumstances, they must contain less nitrogen than the food. When, therefore, a field is manured with animal excrements, a smaller quantity of matter containing nitrogen is added to it than has been taken from it in the form of grass, herbs, or seeds. Therefore, it follows that the favourable activity of such manure cannot be due to its nitrogen.

The liquid manure of animals must, on the other hand, be of the highest value with respect to nitrogen: because it contains all or nearly all the nitrogen originally present in the food consumed. In order to comprehend more clearly the importance of liquid excrements, it is necessary to consider the manner in which they are formed.

It is well known that the body of an adult man, or of an animal in a state of health, remains constantly the same, and neither diminishes nor increases in weight to any appreciable extent. In youth the case is different; for an increase of the body is occasioned. The same is the case in the artificial process of fattening. The body of the old man, on the other hand, gradually diminishes in size.

The quantity of nitrogen and of other constituents in the body cannot therefore increase, although the animal always receives in his food a considerable quantity of that element. From this it follows, that the quantity of nitrogen expelled from the body must be the same as that taken in the food by an animal in a state of nature, freely exposed to exercise; for if this were not the case, the body must acquire a larger proportion of nitrogen, which we know it does not.

When an individual is deprived of food and in the progress of starvation, his body diminishes in weight, in such a manner that all parts, except the membranes and bones, participate in the loss. By what means has the nitrogen of those tissues been expelled from the system?

The emaciation which occurs proves that during every moment in the life of an animal, part of its structure loses its vitality, and assumes the form of dead matter. This, after suffering certain changes, is finally separated from the system by the organs of secretion, namely, the skin, lungs, and kidneys. The daily loss thus experienced is restored by food.

The azotised constituents of the food are transformed into blood, which then nourishes the animal by restoring its wasted tissues to their original condition.

THE UNIFORM WEIGHT OF AN ANIMAL PROVES THAT A QUANTITY OF NITROGEN MUST HAVE BEEN EXPELLED FROM THE SYSTEM, EXACTLY CORRESPONDING TO THE AMOUNT CONTAINED IN THE FOOD CONSUMED. The compounds consisting of carbon and hydrogen, derived from the waste matter, are separated by the lungs and skin; whilst those containing nitrogen are eliminated as urine. When a body increases in weight, a smaller quantity of nitrogenous compounds must be separated by the urine; a diminution in weight indicates, on the other hand, a greater separation of these compounds. These considerations prove that the nitrogen extracted from the atmosphere by plants as food, is again in a great measure returned in the urine of man and other animals.

It is obvious that, by collecting both the solid and liquid excrements of an animal fed upon the produce of a certain surface of land, we are enabled to supply to it nearly the same quantity of nitrogen as that contained in the original produce. Thus we supply to the land a certain quantity of ammonia, in addition to that which may be extracted from the atmosphere by the plants growing upon it.

In a scientific point of view, it should be the care of the agriculturist so to employ all the substances containing a large proportion of nitrogen, which his farm affords in the form of animal excrements, that they shall serve as nutriment to his own plants. This will not be the case unless those substances are properly distributed upon his land. A heap of manure lying unemployed upon his land would serve him no more than his neighbours. The nitrogen in it would escape as carbonate of ammonia into the atmosphere, and a mere carbonaceous residue of decayed plants would, after some years, be found in its place.

Tacitus informs us that the surface of Germany was in his time completely covered with impenetrable forests. But now these no longer exist, and all their constituents have disappeared. The carbon and nitrogen deposited in the soil in the form of humus and ammonia have now returned to the atmosphere.

All putrefying animal matters emit carbonic acid and ammonia as long as nitrogen exists in them. In every stage of their putrefaction an escape of ammonia from them may be induced by moistening them with a potash ley; the ammonia being apparent to the senses by a peculiar smell, and by the dense white vapour exhibited when a solid body moistened with an acid is brought near it. This ammonia evolved from manure is imbibed by the soil either in solution in water, or in the gaseous form, and plants thus receive a larger supply of nitrogen than is afforded to them by the atmosphere.*

* "I filled a large retort," says Sir H. Davy, "capable of containing three pints of water, with some hot fermenting manure, consisting princi-

But it is much less the quantity of ammonia yielded to a soil by animal excrements, than the form in which it is presented by them, that causes their great influence on its fertility. Wild plants obtain more nitrogen from the atmosphere in the form of ammonia than they require for their growth; for the water evaporated through their leaves and blossoms emits, after some time, a putrid smell, a peculiarity possessed only by bodies containing nitrogen. Cultivated plants receive the same quantity of nitrogen from the atmosphere as trees, shrubs, and other wild plants; and this is quite sufficient for the purposes of agriculture. Agriculture differs essentially from the cultivation of forests, inasmuch as its principal object consists in the production of THE CONSTITUENTS OF THE BLOOD; whilst the object of forest culture is confined principally to the production of CARBON. But the presence of ammonia alone does not suffice for the production of the nitrogenous ingredients. Other conditions likewise are quite essential. All the various means of culture are subservient to these two main purposes. A part only of the carbonate of ammonia conveyed by rain to the soil is received by plants, because a certain quantity of it is volatilised with the vapour of water; only that portion of it can be assimilated which sinks deeply into the soil, or which is conveyed directly to the

---

pally of the litter and dung of cattle; and adapted a small receiver to the retort, and connected the whole with a mercurial pneumatic apparatus, so as to collect the condensible and elastic fluids which might rise from the dung. The receiver soon became lined with dew, and drops began in a few hours to trickle down the sides of it. Elastic fluid likewise was generated; in three days 35 cubical inches had been formed, which, when analysed, were found to contain 21 cubical inches of carbonic acid; the remainder was hydro-carbonate mixed with some azote, probably no more than existed in the common air in the receiver. The fluid matter collected in the receiver at the same time amounted to nearly half an ounce. It had a saline taste, and a disagreeable smell, and contained some acetate and CARBONATE OF AMMONIA.

"Finding such products given off from fermenting litter, I introduced the beak of another retort, filled with similar dung very hot at the time, into the soil amongst the roots of some grass in the border of a garden; in less than a week a very distinct effect was produced on the grass; upon the spot exposed to the influence of the matter disengaged in fermentation, it grew with more luxuriance than the grass in any other part of the garden.' —Works of Sir H. Davy, Edited by Dr. John Davy, vol. viii. page 31.

eaves by dew, or is absorbed from the air along with the carbonic acid.

Liquid animal excrements, such as the urine with which the solid excrements are impregnated, contain only a small part of their ammonia in the state of salts, that is, in a form in which it has completely lost its volatility. The greatest part exists in the form of carbonate of ammonia—a salt of great volatility. When the ammonia is presented in the condition of a fixed salt, not the smallest portion of it is lost to plants; it is all dissolved by water, and imbibed by their roots. The evident influence of gypsum upon the growth of grasses—the striking fertility and luxuriance of a meadow upon which it is strewed—depends, in some degree, upon its fixing in the soil the ammonia of the atmosphere, which would otherwise be volatilised, with the water which evaporates.* The carbonate of ammonia contained in rain-water is decomposed by gypsum, in precisely the same manner as in the manufacture of sal-ammoniac. Soluble sulphate of ammonia and carbonate of lime are formed; and this salt of ammonia, possessing no volatility, is consequently retained in the soil. All the gypsum gradually disappears, but its action upon the carbonate of ammonia continues as long as a trace of it exists.†

* I made the following experiment on a small garden plot. Beans and peas were planted in the soil, after it had been well manured by mixing it with fresh horse-dung. The whole surface of the plot was strewed with gypsum to the depth of a line, and then covered so as to be protected from the rain. In dry weather it was duly watered.

The plants soon appeared above ground and flourished with great luxuriance. Before the commencement of the experiment, I had examined both the soil and the gypsum, and found that both were quite free from the smallest trace of carbonates. But on testing some of the gypsum taken from the surface after the lapse of several weeks, I ascertained that the greatest part of it had been converted into carbonate of lime. All the soil to the depth of half a foot now effervesced strongly on the addition of acid.

† It has long been the practice in some parts of the country to strew the floors of stables with gypsum. This prevents the disagreeable odour arising from the putrefaction of stable manure, by decomposing and retaining the ammoniacal salts.—Ed.

"I lixiviated some earth," says Spatzier, "and in the filtered solution, after evaporation, I obtained an appreciable quantity of sulphate of ammonia." —Erdman's *Journal*, 1831, Bd. II. s. 89.

The beneficial influence of gypsum and of many other salts has been compared to that of aromatics, which increase the activity of the human stomach and intestines, and give a tone to the whole system. But plants do not contain nerves: we know of no substance capable of exciting them to intoxication and madness, or of lulling them to sleep and repose. No substance can possibly cause their leaves to appropriate a greater quantity of carbon from the atmosphere, when the other constituents required for the growth of the seeds, roots, and leaves, are wanting.* The favourable action of small quantities of aromatics upon man, when mixed with his food, is undeniable; but aromatics are given to plants without food to be digested, and still they flourish with greater luxuriance.

It is quite evident, therefore, that the common view concerning the influence of certain salts upon the growth of plants evinces only ignorance of its cause.

The action of gypsum, chloride of calcium, and of other salts of lime, really consists in their giving a fixed condition to the nitrogen, or ammonia, introduced to the soil. This nitrogen is indispensable for the nutrition of plants.

In order to form a conception of the effect of gypsum, it may be sufficient to remark that 100 lbs. of burned gypsum fixes as much ammonia in the soil as 6250 lbs. of horse's urine† would yield to it, even on the supposition that all the nitrogen of the urea and hippuric acid were absorbed by the plants without the smallest loss, in the form of carbonate of ammonia. If we furnish to a field 40 lbs. of gypsum, and if we suppose that the tenth part of this enters into plants in the form of sulphate of ammonia, we would actually supply

* Schübler states that white arsenic in small quantity exerts a beneficial action upon vegetation—a fact proved by Lampadius, who manured whole fields with this substance.

† The urine of the horse contains, according to Fourcroy and Vauquelin, in 1000 parts,

Urea	7 parts.
Hippurate of soda . . .	14 „
Salts and water . . .	979 „
	1000 parts.

nitrogen sufficient for 100 lbs. of hay, 50 lbs. of wheat, or 60 lbs. of clover.

Water is absolutely necessary to effect the decomposition of the gypsum, on account of its difficult solubility, (1 part of gypsum requires 400 parts of water for solution,) and also to assist in the absorption of the sulphate of ammonia by the plants: hence it happens, that the influence of gypsum is not observable on dry fields and meadows; while the gaseous carbonate of ammonia formed by the decay of animal manures on such fields, on the other hand, does not fail in producing a favourable effect.

The decomposition of gypsum by carbonate of ammonia does not take place instantaneously; on the contrary, it proceeds very gradually; and this explains why the action of the gypsum lasts for several years.

The well-known advantage derived by manuring fields with burnt clay, and the fertility of ferruginous soils, may be explained in an equally simple manner. The favourable effects produced by these causes have been ascribed to the great attraction for water exerted by dry clay and ferruginous earth; but common dry arable land possesses this property in as great a degree: and besides, what influence can be ascribed to a hundred pounds of water spread over a field, in a condition in which it cannot be serviceable either by the roots or leaves? The true cause is this:—

Peroxide of iron and alumina are distinguished from all other metallic oxides by their power of forming solid compounds with ammonia. The precipitates obtained by the addition of ammonia to salts of alumina or iron are true salts, in which the ammonia is contained as a base. Minerals containing alumina or oxide of iron also possess, in an eminent degree, the remarkable property of attracting ammonia from the atmosphere and of retaining it. Vauquelin, whilst engaged in the trial of a criminal case, discovered that all rust of iron contains a certain quantity of ammonia. Chevalier afterwards found that ammonia is a constituent of all minerals containing iron; that even hematite, a mineral which is not at all porous, contains one per cent. of it. Bouis showed also, that the peculiar odour observed on moistening minerals containing alumina, is partly owing to their exhaling ammonia. Indeed, many kinds of gypsum and some varieties of alumina, pipe-clay for example, emit so much ammonia, when moistened with caustic potash, even after they have been exposed for two days, that reddened litmus paper held over them becomes blue. Soils, therefore, containing oxides of iron, and burned clay, must absorb ammonia, an action which is favoured by their porous condition; they further prevent, by their chemical properties, the escape of the ammonia once absorbed. Such soils, in fact, act precisely as a mineral acid would do, if extensively spread over their surface.

The ammonia absorbed by the clay or ferruginous oxides is separated by every shower of rain, and conveyed in solution to the soil.

Powdered charcoal possesses a similar action, but surpasses all other substances in the power which it possesses of condensing ammonia within its pores, particularly when it has been previously heated to redness. Charcoal absorbs 90 times its volume of ammoniacal gas, which may be again separated by simply moistening it with water. (De Saussure.) Decayed wood approaches very nearly to charcoal in this power; decayed oak wood absorbs 72 times its volume, of this gas, after having been completely dried under the air-pump. We have here an easy and satisfactory means of explaining still further the properties of humus, or wood in a decaying state. It is not only a slow and constant source of carbonic acid, but it is also a means by which the necessary nitrogen is conveyed to plants.*

Nitrogen is found in lichens growing on basaltic rocks. Our fields produce more of it than we have given them as

* When the extract of humus is evaporated with muriatic acid, a residue is obtained which evolves ammonia by the addition of potash. When this extract is subjected to distillation along with water, and the products of distillation received into dilute muriatic acid, the latter is found to contain muriate of ammonia. Humus contains carbonate of ammonia.—*Wiegmann und Polstorf, Preisschrift*, s. 53.

manure, and it exists in all kinds of soils and minerals which were never in contact with organic substances. The nitrogen in these cases could only have been extracted from the atmosphere.

We find this nitrogen in the atmosphere in rain water and in all kinds of soils, in the form of ammonia, as a product of the decay and putrefaction of preceding generations of animals and vegetables. We find likewise that the proportion of azotised matters in plants is augmented by giving them a larger supply of ammonia conveyed in the form of animal manure.

No conclusion can then have a better foundation than this, that it is the ammonia of the atmosphere which furnishes nitrogen to plants.*

Carbonic acid, water, and ammonia, contain the elements necessary for the support of animals and vegetables. The same substances are the ultimate products of the chemical processes of decay and putrefaction. All the innumerable products of vitality resume, after death, the original form from which they sprung.

Thus the destruction of an existing generation becomes the means for the production of a new one, and death becomes the source of life.

But it may be asked—Are the compounds now named the only substances necessary for the support of vegetable life? This question must be answered in the negative.

---

* We refer the reader to the Appendix for the part which nitric acid takes in vegetation, and also on the origin of ammonia.

CHAPTER VI.

ON THE SOURCE OF SULPHUR.

PHYSIOLOGY teaches us that all the tissues of the body, such as muscular fibre, cellular tissue, the organic substance of bones, hair, skin, &c., are formed from the blood—the fluid which circulates through every part of the organism.

The blood, from which all parts of the animal frame are produced, is itself furnished to animals by plants. For although the carnivora subsist wholly on the flesh and blood of the herbivora, they actually receive from the latter the component parts of the plants upon which they were nourished.

Chemists have ascertained that sulphur is contained in the two principal ingredients of blood, named by them FIBRIN and ALBUMEN.

When fresh blood is agitated with a rod or stick, FIBRIN is separated in the form of white elastic fibres. A similar separation of this ingredient takes place when blood is allowed to stand for a certain time. The whole becomes coagulated into a sort of jelly, which gradually contracts, and separates itself into a yellowish-coloured liquid, containing the serum or water of the blood, and into a net-work of very fine threads of fibrin. The latter enclose within them the colouring matter of the blood, just as a sponge would do in similar circumstances.

The ALBUMEN is contained in the serum, and communicates to that fluid the property of coagulating by heat, in a manner similar to the white of an egg, which contains albumen as its principal ingredient.

Fibrin, when removed from the circulation, is found to

be perfectly insoluble in cold water. Albumen, on the other hand, in its natural condition, as it exists in serum or in the white of egg, is soluble in water, and miscible with it in all proportions.

CASEIN, or cheese, the principal ingredient of milk, must also be enumerated as a material used in the formation of blood. Casein is generated in the animal economy, and is the only azotised nutriment furnished by the mother to the offspring.

Now albumen, fibrin, and casein contain sulphur, a circumstance by which they are distinguished from all other component parts of the animal body. This sulphur does not exist in the form of an oxide, such as sulphuric acid or one of its salts. It is well known that the albumen of eggs emits, during its putrefaction, sulphuretted hydrogen gas; and it is owing to this that rotten eggs possess the property of blackening silver or other metals with which they may be brought in contact. During the putrefaction of fibrin and albumen, the same gas is likewise generated. There are many other ways by which we might prove the presence of sulphur in these bodies.

From what source does the animal body derive these three fundamental components? Unquestionably they are obtained from the plants upon which the animals subsist; but in what form, and in what condition, are they contained in plants?

Recent investigations of chemists have enabled us to answer these questions with positive certainty. Plants contain, either deposited in their roots or seeds, or dissolved in their juices, variable quantities of compounds containing sulphur. In these nitrogen is an invariable constituent. Two of the compounds containing sulphur exist in the seeds of cereal plants, and in those of leguminous vegetables, such as peas, lentils, and beans. A third is always present in the juices of all plants; and it is found in the greatest abundance in the juices of those which we use for the purposes of the table.

A very exact inquiry into the properties and composition of these substances has produced a very remarkable result, namely, that the sulphur compound dissolved in the juice of plants, is, in reality, identical with the ALBUMEN contained in the serum of blood, and in the white of an egg; that the sulphur compound in the seeds of the cereals possesses the same properties and composition as the FIBRIN of blood; and that the nutritious constituent of peas, beans, and lentils, is actually of the same nature and composition as the CASEIN of milk. Hence it follows that plants, and not animals, generate the constituents of blood containing sulphur. When these are absent from the food given to an animal, its blood cannot be formed. From this it also follows, that vegetable food will be proportionally nutritious and fit to sustain the vital processes of the animal body, according to the amount of these ingredients contained within it.

There also exist certain families of plants, such as the Cruciferæ, which contain peculiar sulphur compounds much richer in that element than the vegetable constituents of blood. The seeds of black mustard, the horse-radish, garlick, onions, and scurvy-grass, are particularly marked in this respect. From all of these plants we obtain, by simple distillation with water, certain volatile oils, differing from all other organic compounds not containing sulphur, by their peculiar, pungent, and disagreeable odour.

Those compounds containing sulphur are present in the seeds of all plants, as well as in the plants themselves; and as they are particularly abundant in cultivated plants employed for animal nutrition, it is quite obvious that a substance containing sulphur is absolutely essential to the development of such compounds, in order to supply to them their proper proportion of this element.

It is also obvious, that although all other conditions for the nourishment of plants be present, if the compound containing sulphur be either wholly absent or deficient in quantity, the vegetable constituents containing sulphur will either be not at all formed, or they will be generated only in proportion to the quantity of the above compound. The air cannot contain any substances in which sulphur is present, unless indeed we except minute and scarcely appreciable traces of sulphuretted hydrogen. The soil, therefore, must be the only means of furnishing the sulphur so necessary to the growth of plants; and we are ignorant of any form by which it can be assimilated except that in which it enters the roots.

The numerous analyses made of the water of mineral springs, furnish us with a satisfactory explanation of the form in which sulphur occurs in soils. The water of such springs is entirely derived from the rain which falls upon the surface of the earth; the water percolating through the earth, dissolves all soluble materials which it may meet in its course. The substances thus dissolved communicate to the water properties which are not possessed by pure water. Water procured from springs or wells is found to be very rarely deficient in soluble salts of sulphuric acid. The liquid obtained by lixiviating good soil from garden or arable land also contains very appreciable quantities of these salts.

The facts now detailed leave little doubt as to the source whence plants obtain their sulphur. As far as our knowledge extends, they receive their sulphur from the sulphates dissolved in the water absorbed by their roots from the soil.

Ammoniacal salts, particularly sulphate of ammonia, are rarely detected in spring water; but this is owing to the constant presence of supercarbonate of lime, which effects their decomposition, and allows the escape of ammonia during the evaporation of the liquid for the purposes of analysis.

According to our view, sulphate of ammonia is of all compounds containing sulphur the one most fitted for the assimilation of that element. Sulphate of ammonia contains two elements, both of which are equally necessary for the support of vegetable life; these are sulphur and nitrogen, and they form constituents also of vegetable albumen, fibrin, and casein. But what is still more worthy of observation, sulphate of ammonia, viewing it according to the proportion of its elements, or what is termed its empirical formula, (O_3, H_3, S,N,), may be considered as a compound of water with equal equivalents of sulphur and nitrogen. Thus, by the simple removal of the elements of water from this compound, its sulphur and nitrogen might be enabled to pass over into the composition of the plants.

The ingredients of plants containing sulphur are so composed that one equivalent of sulphur exists for every 25 equivalents of nitrogen. Hence it is obvious that much more ammonia must be offered to plants than that contained in the form of sulphate of ammonia, if all the sulphur of the latter is to become a constituent of the organic ingredients alluded to.

This bears a complete analogy to the assimilation of the carbon and nitrogen furnished to plants in the form of carbonate of ammonia. This salt may contain two equivalents of carbon to one equivalent of nitrogen. Hence it is necessary that the carbon of six equivalents of carbonic acid must at the same time be taken up, and enter into combination with the nitrogen, in order to produce the principal nitrogenous constituents which contain one equivalent of nitrogen to eight equivalents of carbon.

The passage of sulphur derived from a sulphate into the composition of vegetable matter, necessarily indicates that the sulphate has been exposed to the action of the same causes as those by which the decomposition of carbonic acid was effected in the plant; and, therefore, that the sulphuric acid has been decomposed into sulphur and oxygen, the former of which is assimilated, whilst the latter is separated. If we suppose the sulphuric acid to be presented in the form of sulphate of potash or soda, the basis of these salts must be set at liberty after the decomposition of their acid.

Now we actually find these bases in all cultivated, and even in most wild plants. They are found either united to organic acids, or, what is still more remarkable, they are found in union with the vegetable compounds containing sulphur. The vegetable casein of peas, beans, and other leguminous plants, is itself insoluble in water; but it is very soluble in the form in which it occurs in the plant. This solubility is due to the soda and potash with which it is united. In like manner, the albumen contained in the juices of plants is combined with an alkali; and we must even suppose that vegetable fibrin, the insoluble ingredient of cereal plants, must have originally been soluble, and have attained its position in the seeds by the agency of alkalies.

The potash and soda of the alkaline sulphates which furnish to plants their sulphur, remain, therefore, either in combination with the ingredients containing that element, or they enter into some new state of combination, or, finally, they are returned to the soil.

Gypsum (sulphate of lime) is the most generally diffused sulphate. Being soluble, it may either pass directly into the plant, or it may be decomposed by the carbonate of ammonia existing in rain water, and pass into the plant in the form of sulphate of ammonia.

A solution of gypsum containing common salt or chloride of potassium, such as sea water, and the water of most springs, may be viewed as a mixture of an alkaline sulphate with chloride of calcium. From this it must be obvious, that when we furnish to a plant at the same time both gypsum and common salt (chloride of sodium), we actually furnish by such a solution the same materials that we would do if we supplied a mixture of sulphate of soda and chloride of calcium. In order to form the constituents containing sulphur, that element and the alkali must be retained by the plant, while the chlorine and calcium will be expelled by the roots.

We know that this process actually does take place in the case of marine plants. The soda or potash is obtained from common salt or chloride of potassium, which suffers decomposition by the presence of sulphate of lime or sulphate of magnesia. It is necessary to suppose that this process also occurs with the cereal and all other plants destitute of lime. Thus we are enabled to explain the use of common salt as a manure; it enables the plant, for which this manure is useful, to extract its sulphur from the soil in which it existed in the form of sulphate of lime.

CHAPTER VII.

OF THE INORGANIC CONSTITUENTS OF PLANTS.*

CARBONIC acid, water, and ammonia, are necessary for the existence of plants, because they contain the elements from which their organs are formed; but other substances are likewise requisite for the formation of certain organs destined for special functions peculiar to each family of plants. Plants obtain these substances from inorganic nature. In the ashes left after the incineration of plants, the same substances are found, although in a changed condition.

Many of the inorganic constituents vary according to the soil in which the plants grow, but a certain number of them are indispensable to their development. All substances in solution in a soil are absorbed by the roots of plants, exactly as a sponge imbibes a liquid, and all that it contains, without selection. The substances thus conveyed to plants are retained in greater or less quantity, or are entirely separated when not suited for assimilation.

Alkaline and earthy phosphates form invariable constituents of the seeds of all kinds of grasses, of beans, peas, and lentils.

---

* "Many authors," says Saussure, "consider that the mineral ingredients of plants are merely accidentally present, and are not at all necessary to their existence, because the quantity of such substances is exceedingly small. This opinion may be true as far as regards those matters which are not always found in plants of the same kind; but there is certainly no evidence of its truth with those invariably present. Their small quantity does not indicate their inutility. The phosphate of lime existing in the animal body does not amount to the fifth part of its weight, yet it cannot be affirmed that this salt is unnecessary for the formation of its bones. I have detected the same compound in the ashes of all plants submitted to examination, and we have no right to suppose that they could exist without it." (*De Saussure*, p. 241.)

These salts are introduced into bread along with the flour, and into beer along with barley. The bran of flour contains a large quantity of ammoniacal phosphate of magnesia. This salt forms large crystalline concretions, often amounting to several pounds in weight, in the *cæcum* of horses belonging to millers; and when ammonia is mixed with beer, the same salt separates as a white precipitate.

Most plants, perhaps all of them, contain organic acids of very different composition and properties, all of which are in combination with bases, such as potash, soda, lime, or magnesia; plants containing free organic acids are few in number. These bases evidently regulate the formation of the acids, for the diminution of the one is followed by a decrease of the other: thus in the grape, for example, the quantity of acid contained in its juice is less when it is ripe than when unripe; and the bases, under the same circumstances, are found to vary in a similar manner. Such constituents exist in small quantity in those parts of a plant in which the process of assimilation is most active, as in the mass of woody fibre; and their quantity is greater in those organs whose office it is to prepare substances conveyed to them for assimilation by other parts. The leaves contain more inorganic matters than the branches, and the branches more than the stem (SAUSSURE). The potato plant contains more potash before blossoming than after it (MOLLERAT).

The acids found in the different families of plants are of various kinds; it cannot be supposed that their presence and peculiarities are the result of accident. The fumaric and oxalic acids in the lichens, the kinovic acid in the *China nova*, the rocellic acid in the *Rocella tinctoria*, the tartaric acid in grapes, and the numerous other organic acids, must serve some end in vegetable life. But if these acids constantly exist in vegetables, and are necessary to their life, which is incontestable, it is equally certain that some alkaline base is also indispensable, in order to enter into combination with the acids; for these are always found in the state of neutral or acid salts. All plants yield by incineration ashes containing carbonic acid; all, therefore, must contain salts of an organic acid.*

Now, as we know the capacity of saturation of organic acids to be unchanging, it follows that the quantity of the bases united with them cannot vary; and for this reason the latter substances ought to be considered with the strictest attention, both by the agriculturist and physiologist.

We have no reason to believe that a plant in a condition of free and unimpeded growth produces more of its peculiar acids than it requires for its own existence; hence, a plant, on whatever soil it grows, must contain an invariable quantity of alkaline bases. Culture alone will be able to cause a deviation.

In order to understand this subject clearly, it will be necessary to bear in mind that any one of many of the alkaline bases may be substituted for another, the action of all being the same. Our conclusion is, therefore, by no means endangered by the existence in one plant of a particular alkali which may be absent in others of the same species. If this inference be correct, the absent alkali or earth must be supplied by one similar in its mode of action, or in other words, by an equivalent of another base. The number of equivalents of these various bases which may be combined with a certain portion of acid must consequently be the same, and therefore the amount of oxygen contained in them must remain unchanged under all circumstances and on whatever soil they grow.†

* Salts of organic acids yield carbonates on incineration, if they contain either alkaline or earthy bases.

† When sulphuric acid is placed in contact with potash, soda, lime, or magnesia, the properties both of the acid and of the alkali disappear, and if the proportions have been just, the compound thus produced is a neutral sulphate of these bases.

100 parts of sulphuric acid require for neutralisation very different quantities of the above bases; thus, to effect this purpose, it is necessary to employ 118 parts of potash, 78 parts of soda, 71·2 parts of lime, and 51·6 parts of magnesia.

In order to produce a neutral salt of nitric acid with 118 parts of potash (the quantity necessary to saturate 100 parts of sulphuric acid), we must employ 135 parts of nitric acid. Now, when we examine how much soda, lime, or magnesia is required to saturate the same quantity of nitric acid (135 parts)

Of course, this argument refers only to those alkaline bases which in the form of organic salts form constituents of the plants. Now, these salts are preserved in the ashes of plants as carbonates, the quantity of which can be easily ascertained. The bases contained in the bark do not any longer belong to the vital organism of the plant.

It has been distinctly shown, by the analysis of De Saussure and Berthier, that the nature of a soil exercises a decided influence on the quantity of the different metallic oxides con-

---

it is found that complete saturation is effected by 78 of soda, 71·2 of lime, 51·6 of magnesia, or exactly the same quantities as in the case of sulphuric acid. It is quite indifferent what acids we use to neutralise their bases, or how much the numbers obtained may differ from those now stated; still the relative proportion remains invariable. If for the saturation of any particular acid 51·6 parts of magnesia have been used, we may be perfectly certain that the same quantity of this acid will be exactly neutralised by 78 parts of soda.

We have now to state the causes which occasion this unequal power of these metallic oxides to neutralise acids. We have also to explain why, to produce the same effect, it is necessary to employ a smaller quantity of soda, and only one half the quantity of magnesia that we would use of potash, and still that the relative quantities are constant with all acids.

A knowledge of the composition of the bases has afforded us a very simple explanation of these causes. All the bases now mentioned contain oxygen combined with a metal; and their capacity of saturation depends upon the quantity of oxygen contained within them.

Although the absolute quantities of the above bases are so very different, they all contain the same quantities of oxygen.

			Oxygen contained.
100 Sulphuric Acid neutralise	118	Potash	= 20
100 " " "	78	Soda	= 20
100 " " "	71·2	Lime	= 20
100 " " "	51·6	Magnesia	= 20

Now, if we neutralise 100 parts of sulphuric acid with potash and soda, or with potash, soda, and lime, or with potash, soda, lime, and magnesia, the sulphuric acid unites with quantities of two, three, or four bases exactly corresponding to their united quantity of oxygen. This may be represented in the following table:—

100 parts sulphuric acid neutralise	Potassium Sodium	20 parts oxygen.
100 " " " "	Potassium Sodium Calcium	20 " oxygen.
100 " " " "	Potassium Sodium Calcium Magnesium	20 " oxygen.

tained in the plants which grow on it; that magnesia, for example, was contained in the ashes of a pine-tree grown at Mont Breven, whilst it was absent from the ashes of a tree of the same species from Mont La Salle, and that even the proportion of lime and potash was very different.

Hence it has been concluded, (erroneously, I believe,) that the presence of bases exercises no particular influence upon the growth of plants: but even were this view correct, it must be considered as a most remarkable accident that these same analyses furnish proof for the very opposite opinion. For although the composition of the ashes of these pine-trees was so very different, they contained, according to the analyses of De Saussure, an equal number of equivalents of metallic oxides; or, what is the same thing, the quantity of oxygen contained in all the bases was in both cases the same.

100 parts of the ashes of the pine-tree from Mont Breven contained—

Carbonate of Potash .	3·60	Quantity of oxygen in the Potash	0·415
,, Lime . .	46·34	,, ,, ,, Lime	7·327
,, Magnesia	6·77	,, ,, ,, Magnesia	1·265
Sum of the carbonates	56·71	Sum of the oxygen in the bases	9·007

100 parts of the ashes of the pine from Mont La Salle contained*—

Carbonate of Potash .	7·36	Quantity of oxygen in the Potash	0·85
,, Lime . .	51·19	,, ,, ,, Lime	8·10
,, Magnesia	00·00		
Sum of the carbonates .	58·55	Sum of the oxygen in the bases	8·95

The numbers 9·007 and 8·95 approach each other as nearly as could be expected even in analyses made for the very purpose of ascertaining the fact above demonstrated; which the analyst in this case had not in view.

Let us now compare Berthier's analyses of the ashes of two fir-trees, one of which grew in Norway, the other in Allevard (département de l'Isère). One contained 50, the other 25

* According to the experiments of Saussure, 1000 parts of the wood of the pine from Mont Breven gave 11·87 parts of ashes; the same quantity of wood from Mont La Salle yielded 11·28 parts.

per cent. of soluble salts. A greater difference in the proportion of the alkaline bases could scarcely exist between two totally different plants, and yet even here the quantity of oxygen in the bases of both was the same.

100 parts of the ashes of firwood from Allevard contained, according to Berthier, (Ann. de Chim. et de Phys. t. xxxii. p. 248,)

Potash and Soda	16·8	in which	3·57	parts must be oxygen.
Lime . .	29·6	"	8·36	" "
Magnesia . .	3·3	"	1·26	" "
	49·7		13·19	

Only part of the potash and soda in these ashes was in combination with organic acids; the remainder was in the form of sulphates, phosphates, and chlorides. One hundred parts of the ashes contain 0·797 sulphuric acid, 3·12 phosphoric acid, and 0·077 hydrochloric acid, which together neutralise a quantity of base containing 0·53 oxygen. This number, therefore, must be subtracted from 13·19. The remainder, 12·66 indicates the quantity of oxygen in the alkaline bases, combined with organic acids in the firwood of Allevard.

The firwood of Norway contained in 100 parts,—

Potash . .	14·1	of which	2·4	parts would be oxygen.
Soda . .	20·7	"	5·3	" "
Lime . . .	13·6	"	3·82	" "
Magnesia .	4·35	"	1·69	" "
	52·75		13·21	

And if we subtract from 13·21 the quantity of oxygen of the bases in combination with sulphuric and phosphoric acid, viz. 0·79, 12·42 parts remain as the amount of oxygen contained in the bases which were in combination with organic acids.

These remarkable approximations cannot be accidental; and if future investigations confirm them in other kinds of plants, no other explanation than that already given can be adopted.

It is not known in what form manganese, and oxide of iron, are contained in plants; but we are certain that potash, soda, and magnesia, can be extracted by means of water from all parts of their structure in the form of salts of organic acids. The same is the case with lime, when not present as insoluble oxalate of lime. It must here be remembered, that in plants yielding oxalic acid, the acid and potash never exist in the form of the neutral oxalate or quadroxalate, but always as a bin-oxalate, on whatever soil they may grow. The potash in grapes is always found as an acid salt, viz. cream of tartar (bitartrate of potash), and never in the form of a neutral compound. As these acids and bases are never absent from plants, and as even the form in which they present themselves is not subject to change, it may be affirmed that they exercise an important influence on the development of the fruits and seeds, and also on many other functions, of the nature of which we are at present ignorant. The quantity of alkaline bases existing in a plant also depends evidently on this circumstance of their existing only in the form of acid salts,—for the capacity of saturation of an acid is constant.

From these considerations we must perceive, that exact and trustworthy examinations of the ashes of plants of the same kind growing upon different soils would be of the greatest importance to vegetable physiology, and would decide whether the facts above mentioned are the results of an unchanging law for each family of plants, and whether an invariable number can be found to express the quantity of oxygen which each species of plant contains in the bases united with organic acids. In all probability such inquiries will lead to most important results; for it is clear that if the production of a certain unchanging quantity of an organic acid is required by the peculiar nature of the organs of a plant, and is necessary to its existence, then potash or lime must be taken up by it in order to form salts with this acid; that if these do not exist in sufficient quantity in the soil, other alkaline bases, of equal value, must supply their place; and that the progress of a plant must be wholly arrested when none are present.

Seeds of the *Salsola kali*, when sown in common garden soil, produce a plant containing both potash and soda; while the plants grown from the seeds of this contain only salts of potash, with mere traces of muriate of soda.* (CADET.)

The existence of vegetable alkalies in combination with organic acids gives great weight to the opinion that alkaline bases in general are connected with the development of plants.

If potatoes are grown where they are not supplied with earth, the magazine of inorganic bases, (in cellars, for example), a true alkali, called Solanin, of very poisonous nature, is formed in the sprouts extending towards the light, while mere traces of such a substance can be discovered in the roots, herbs, blossoms, or fruits of potatoes grown in fields (OTTO). In all the species of the *Cinchona,* kinic acid is found; but the quantity of quinina, cinchonina, and lime contained in them is most variable. From the fixed bases in the products of incineration, however, we may estimate pretty accurately the quantity of the peculiar organic bases. A maximum of the first corresponds to a minimum of the latter, as must necessarily be the case if they mutually replace one another according to their equivalents. We know that different kinds of opium contain meconic acid in combination with very different quantities of narcotina, morphia, codeia, &c., the quantity of one of these alkaloids diminishing on the increase of the others. Thus the smallest quantity of morphia is accompanied by a maximum of narcotina. Not a trace of meconic acid† can be discovered in

* "We planted," say Wiegmann and Polstorf, "several plants in a flower-pot filled with common earth from the garden, and watered them with a weak solution of chloride of potassium, having previously ascertained that the earth contained mere traces of metallic chlorides. Subjected to this treatment, the plants flourished very luxuriantly, so much so that they completely covered the flower-pot, stretching considerably over its sides. We now transplanted them into the ordinary soil, and did not supply them any longer with chloride of potassium; but, in the following year, they shrunk and died during the period of blossoming. It follows, from the experiments which we have detailed, that both kinds of plants required metallic chlorides for their proper nourishment, but that it is quite indifferent whether the chlorine be united with sodium or potassium." (*Preischrift über die anorganischen Bestandtheile der Pflanzen.*)

† Robiquet did not obtain a trace of meconate of lime from 300 lbs. of opium, whilst in other kinds the quantity was very considerable. (*Ann. de Chim.* liii. p. 425.)

many kinds of opium, but there is not on this account an absence of acid, for the meconic is here replaced by sulphuric acid. Here, also, we have an example of what has been before stated; for in those kinds of opium where both these acids exist, they are always found to bear a certain relative proportion to one another.

Now if it be found, as appears to be the case in the juice of poppies, that an organic acid may be replaced by an inorganic, without impeding the growth of a plant, we must admit the probability of this substitution taking place in a much higher degree in the case of the inorganic bases.

When roots find their more appropriate base in sufficient quantity, they will take up less of another.

These phenomena do not show themselves so frequently in cultivated plants, because they are subjected to special external conditions, for the purpose of the production of particular constituents or of particular organs.

By sprinkling with the juice of the *Phytolacca decandra*, the soil in which a white hyacinth is growing in a state of blossom, its white blossoms assume in one or two hours a red colour, which again disappears after a few days under the influence of sunshine, and they become white and colourless as before.* The juice in this case evidently enters into all parts of the plant, without being at all changed in its chemical nature, or without its presence being apparently either necessary or injurious. But this condition is not permanent, and when the blossoms have again become colourless, none of the colouring matter remains; and if it should occur that any of its elements were adapted for the purposes of nutrition of the plant, then these alone would be retained, whilst the rest would be excreted in an altered form by the roots.

Exactly the same thing must happen when we sprinkle a plant with a solution of chloride of potassium, nitre, or nitrate of strontia; they will enter into the different parts of the plant, just as the coloured juice mentioned above, and will be found in its ashes if it should be burnt at this period. Their

* Biot, in the Comptes rendus des Séances de l'Académie des Sciences, à Paris, premier Sémestre, 1837, p. 12.

presence is merely accidental; but this does not furnish ground for any conclusion against the necessity of the presence of other bases in plants. The experiments of Macaire-Princep have shown, that plants made to vegetate with their roots in a weak solution of acetate of lead, and then in rain-water, yield to the latter all the salt of lead which they had previously absorbed. They return, therefore, to the soil all matters unnecessary to their existence. Again, when a plant, freely exposed to the atmosphere, rain and sunshine, is sprinkled with a solution of nitrate of strontia, the salt is absorbed, but it is again separated by the roots and removed further from them by every shower of rain that falls upon the soil, so that at last not a trace of it is to be found in the plant. (DAUBENY.)

Let us consider the composition of the ashes of two fir-trees as analysed by an acute and most accurate chemist. One of these grew in Norway, on a soil of invariable composition, but to which soluble salts, and particularly common salt, were conveyed in great quantity by rain-water. How did it happen that its ashes contained no appreciable trace of salt, although we are certain that its roots must have absorbed it after every shower?

We can explain the absence of salt in this case by means of the direct and positive observations referred to, which have shown that plants have the power of returning to the soil all substances unnecessary to their existence; and the conclusion to which all the foregoing facts lead us, when their real value and bearing are apprehended, is that the alkaline bases existing in the ashes of plants must be necessary to their growth, since if this were not the case they would not be retained.

The perfect development of a plant, according to this view, is dependent on the presence of alkalies or of alkaline earths; for when these substances are totally wanting its growth will be arrested, and when they are only deficient it must be impeded.

In order to apply these remarks, let us compare two kinds of trees, the wood of which contains unequal quantities of alkaline bases, and we shall find that one of these may grow luxuriantly in several soils upon which the other is scarcely able to vegetate. For example, 10,000 parts of oak-wood yield 250 parts of ashes, the same quantity of fir-wood only 83, of linden-wood 500, of rye 440, and of the herb of the potato plant 1500 parts.*

Firs and pines find a sufficient quantity of alkalies in granitic and barren sandy soils in which oaks will not grow; and wheat thrives in soils favourable for the linden-tree, because the bases necessary to bring it to complete maturity exist there in sufficient quantity. The accuracy of these conclusions, so highly important to agriculture and to the cultivation of forests, can be proved by the most evident facts.

All kinds of grasses, and the *Equisetaceæ*, for example, contain in the outer parts of their leaves and stalk, a large quantity of silicic acid and potash in the form of acid silicate of potash. The proportion of this salt does not vary perceptibly in the soil of corn-fields, if it be again conveyed to them as manure in the form of putrefying straw. But this is not the case in a meadow, and hence we never find a luxuriant crop of grass † on sandy and calcareous soils containing little potash, evidently because one of the constituents indispensable to the growth of the plants is wanting. Soils formed from basalt, grauwacke, and porphyry, are, *cæteris paribus*, the best for meadow-land, on account of the large quantity of potash they contain. The potash abstracted by the plants is restored during the annual irrigation. The potash contained in the soil itself is inexhaustible in comparison with the quantity removed by plants.

A harvest of grain is obtained every thirty or forty years from the soil of the Luneburg heath, by strewing it with the ashes of the heath-plants (*Erica vulgaris*) growing upon it. These plants, during the long period just mentioned, collect

* Berthier, Annales de Chimie et de Physique, t. xxx. p. 248.

† It would be of importance to examine what alkalies are contained in the ashes of the sea-shore plants which grow in the humid hollows of downs, and especially those of the millet-grass. If potash is not found in them, it must certainly be replaced by soda, as in the *Salsola*, or by lime, as in the *Plumbagineæ*.

the potash and soda contained in the soil and conveyed to them by rain-water; and it is by means of these alkalies that oats, barley, and rye, to which they are indispensable, are enabled to grow on this sandy heath.

The woodcutters in the vicinity of Heidelberg have the privilege of cultivating the soil for their own use, after felling the trees used for making tan. Before sowing the land thus obtained, the branches, roots, and leaves, are in every case burned, and the ashes used as a manure, which is found to be quite indispensable for the growth of the grain. The soil itself upon which the oaks grow in this district consists of sandstone; and although the trees find in it a quantity of alkaline earths sufficient for their own sustenance, yet in its ordinary condition it is incapable of producing grain.

The most decisive proof of the use of strong manure was obtained at Bingen (a town on the Rhine), where the produce and development of vines were highly increased by manuring them with such nitrogenous manures as shavings of horn, &c.; but after some years the formation of the wood and leaves decreased to the great loss of the possessor, to such a degree that he has long had cause to regret his departure from the usual methods. By the manure employed by him, the vines had been too much hastened in their growth; in two or three years they had exhausted the potash in the formation of their fruit, leaves, and wood, so that none remained for the future crops, his manure not having contained any potash.

There are vineyards on the Rhine the plants of which are above a hundred years old, and all of these have been cultivated by manuring them with cow-dung, a manure containing a large proportion of alkaline ingredients, although very little nitrogen. All the alkalies, in fact, contained in the food consumed by a cow are again immediately discharged in the liquid excrements.

The leaves and small branches of trees contain the greatest quantity of ashes and of alkalies; and the quantity of them annually removed from a wood, for the purpose of being employed as litter,* contain much more of the alkalies than all the old wood cut down. The bark and foliage of oaks, for example, contain from 6 to 9 per cent. of alkalies, the needles of firs and pines, 8 per cent.

With every 2650 lbs. of firwood yearly removed from an acre of forest, only 7 or 8 lbs. of alkalies are abstracted from the soil, calculating the ashes at 0·83 per cent. The leaves, however, cover the soil, and being very rich in alkalies, in comparison with the wood, retain those alkalies on the surface, which would otherwise so easily penetrate with the rain through the sandy soil. By their decay an abundant provision of alkalies is supplied to the roots of the trees, and a fresh supply is rendered unnecessary.

The ashes of the tobacco plant, of the vine, of peas, and of clover, contain a large quantity of lime. Such plants do not flourish on soils devoid of lime. By the addition of salts of lime to such soils, they become fitted for the growth of these plants; for we have every reason to believe that their development essentially depends upon the presence of lime. The presence of magnesia is equally essential, there being many plants, such as the different varieties of beet and potatoes, from which it is never absent.

The supposition that alkalies, metallic oxides, or inorganic matter in general, may be produced by plants, is entirely refuted by these well-authenticated facts.

It is thought very remarkable, that the plants of the grass tribe, fitted for the food of man, follow him like the domestic animals. But saline plants seek the sea-shore or saline springs, and the Chenopodium the dunghill, from similar causes. Saline plants require common salt, and the plants growing only on dunghills need ammonia and nitrates, and they are attracted to places where these can be found, just as the

* [This refers to a custom some time since very prevalent in Germany, although now discontinued. The leaves and small twigs of trees were gleaned from the forests by poor people, for the purpose of being used as litter for their cattle. The trees, however, were found to suffer so much in consequence, that their removal is now strictly prohibited. The cause of the injury was that stated in the text.—Ed.]

dung-fly is to animal excrements. So likewise none of our corn plants can bear perfect seeds, that is, seeds yielding flour, without a large supply of phosphate of magnesia and ammonia, substances which they require for their maturity. And hence, these plants grow only in a soil where these three constituents are found combined, and no soil is richer in them than those where men and animals dwell together; where the urine and excrements of these are found corn plants appear, because their seeds cannot attain maturity unless supplied with the constituents of those matters.

When we find sea plants near our salt-works, several hundred miles distant from the sea, we know that their seeds have been carried there in a very natural manner, namely, by wind or by birds, which have spread them over the whole surface of the earth, although they grow only in those places in which they find the conditions essential to their life.

Numerous small fish, of not more than two inches in length (*Gasterosteus aculeatus*), are found in the salt-pans of the graduating-house at Nidda (a village in Hesse Darmstadt). No living animal is found in the salt-pans of Neuheim, situated about 18 miles from Nidda; but the water there contains so much carbonic acid and lime, that the walls of the graduating-house are covered with stalactites. Hence the eggs conveyed to this place, by whatever cause, do not find the conditions necessary for their development, although they did so in the former place.

How much more wonderful and inexplicable does it appear, that bodies remaining fixed in the strong heat of a fire, have under certain conditions the property of volatilising and, at ordinary temperatures, of passing into a state, of which we cannot say whether they have really assumed the form of a gas or are dissolved in one! Steam or vapours in general have a very singular influence in causing the volatilisation of such bodies, that is, of causing them to assume the gaseous form. A liquid during evaporation communicates the power of assuming the same state in a greater or less degree to all substances dissolved in it, although they do not of themselves possess that property.

Boracic acid is a perfectly fixed substance; it suffers no change of weight appreciable by the most delicate balance, when exposed to a white heat, and therefore it is not volatile. Yet its solution in water cannot be evaporated by the gentlest heat, without the escape of a sensible quantity of the acid with the steam. Hence it is that a loss is always experienced in the analysis of minerals containing this acid, when liquids in which it is dissolved are evaporated. The quantity of boracic acid which escapes with a cubic foot of steam, at the temperature of boiling water, cannot be detected by our most sensible re-agents; but nevertheless the many hundred tons annually brought from Italy as an article of commerce, are procured by the uninterrupted accumulation of this apparently inappreciable quantity. The hot steam issuing from the interior of the earth, passes through cold water in the lagoons of Castel Nuovo and Cherchiago; in this way the boracic acid is gradually accumulated, till at last it may be obtained in crystals by the evaporation of the water. It is evident, from the temperature of the steam, that it must have come out of depths in which human beings and animals never could have lived, and yet it is very remarkable and highly important that ammonia is never absent from it. In the large works in Liverpool, where natural boracic acid is converted into borax, many hundred pounds of sulphate of ammonia are obtained at the same time.

THIS AMMONIA HAS NOT BEEN PRODUCED BY THE ANIMAL ORGANISM, BUT EXISTED BEFORE THE CREATION OF HUMAN BEINGS, BEING A PART, A PRIMARY CONSTITUENT, OF THE GLOBE ITSELF.

The experiments instituted under Lavoisier's guidance by the *Direction des Foudres et Salpêtres*, have proved that during the evaporation of the saltpetre ley, the salt volatilises with the water, and causes a loss which could not before be explained. It is known also that, in sea-storms, leaves of plants in the direction of the wind are covered with crystals of salt, even at the distance of from 20 to 30 miles from the sea. But it does not require a storm to cause the volatilisation of the salt, for the air hanging over the sea always contains enough of this substance to render turbid a solution of nitrate of silver, and every breeze must carry this away. Now, as thousands of tons of sea-water annually evaporate into the atmosphere, a corresponding quantity of the salts dissolved in it, viz. of common salt, chloride of potassium, magnesia, and the remaining constituents of the sea-water, will be conveyed by wind to the land.

This volatilisation is a source of considerable loss in salt-works, especially where the proportion of salt in the water is small. This has been completely proved at the salt-works of Nauheim, by the very intelligent director of that establishment, M. Wilhelmi. He hung a plate of glass between two evaporating-houses, distant about 1200 paces from each other, and found in the morning, after the drying of the dew, that the glass was covered with crystals of salt on one or the other side, according to the direction of the wind.

By the continual evaporation of the sea, its salts* are spread over the whole surface of the earth; and being subsequently carried down by the rain, furnish to vegetation those salts necessary to its existence. This is the origin of the salts

* According to Marcet, sea-water contains in 1000 parts,

26·660	Chloride of Sodium.
4·660	Sulphate of Soda.
1·232	Chloride of Potassium.
5·152	Chloride of Magnesium.
1·5	Sulphate of Lime.
39·204	

According to M'Clemm, the water of the North Sea contains in 1000 parts,

24·84	Chloride of Sodium.
2·42	Chloride of Magnesium.
2·06	Sulphate of Magnesia.
1·31	Chloride of Potassium.
1·20	Sulphate of Lime.

In addition to these constituents, it also contains inappreciable quantities of carbonate of lime, magnesia, iron, manganese, phosphate of lime, iodides, and bromides, and organic matter, together with ammonia and carbonic acid. (Liebig's *Annalen der Chemie*, Bd. xxxvii. s. 3.)

found in the ashes of plants, in those cases where the soil could not have yielded them.

In a comprehensive view of the phenomena of nature, we have no scale for that which we are accustomed to name, small or great; all our ideas are proportioned to what we see around us; but how insignificant are they in comparison with the whole mass of the globe! that which is scarcely observable in a confined district appears inconceivably large when regarded in its extension through unlimited space. The atmosphere contains only a thousandth part of its weight of carbonic acid; and yet small as this proportion appears, it is quite sufficient to supply the whole of the present generation of living beings with carbon for a thousand years, even if it were not renewed. Sea-water contains $\frac{1}{12400}$ of its weight of carbonate of lime; and this quantity, although scarcely appreciable in a pound, is the source from which myriads of marine mollusca and corals are supplied with materials for their habitations.

Whilst the air contains only from 4 to 6 ten-thousandth parts of its volume of carbonic acid, sea-water contains 100 times more, (10,000 volumes of sea-water contain 620 volumes of carbonic acid—Laurent, Bouillon—Lagrange). Ammonia* is also found in this water; so that the same conditions which sustain living beings on the land are combined in this medium, in which a whole world of other plants and animals exist.

The roots of plants are constantly engaged in collecting from the rain those alkalies which formed part of the sea-water, and also those of the water of the springs penetrating the soil. Without alkalies and alkaline bases most plants could not exist, and without plants the alkalies would disappea gradually from the surface of the earth.

When it is considered that sea-water contains less than one-millionth of its own weight of iodine, and that all combinations of iodine with the metallic bases of alkalies are highly soluble in water, some provision must necessarily be supposed to exist in the organisation of sea-weed and the different

* When the solid saline residue obtained by the evaporation of sea-water is heated in a retort to redness, a sublimate of sal-ammoniac is obtained. MARCET.

kinds of Fuci, by which they are enabled during their life to extract iodine in the form of a soluble salt from sea-water, and to assimilate it in such a manner, that it is not again restored to the surrounding medium. These plants are collectors of iodine, just as land plants are of alkalies; and they yield us this element in quantities such as we could not otherwise obtain from the water without the evaporation of whole seas.

We take it for granted, that the sea plants require metallic iodides for their growth, and that their existence is dependent on the presence of those substances. With equal justice, then, we conclude, that the alkalies and alkaline earths, always found in the ashes of land plants, are likewise necessary for their development.

CHAPTER VIII.

ON THE FORMATION OF ARABLE LAND.

THE hardest rocks and stones gradually lose their coherence when exposed to the influence of certain agencies. Soils consist of the *debris* of rocks which have suffered this change.

The disintegration of minerals and rocks is effected partly by mechanical, and partly by chemical causes. It has been remarked in all the mountainous districts of perpetual snow, that the most refractory rocks crumble into fragments,* which are either rounded by the action of glaciers, or are thoroughly pulverised into dust. The rivers and streams arising out of the glaciers are rendered turbid with this mineral debris, which they deposit on reaching the plains and valleys; thus fertile soils are formed.

"As often as I have seen beds of mud, sand, and shingle, accumulated to the thickness of many thousand feet, I have felt inclined to exclaim, that causes such as the present rivers and the present beaches could never have ground down such masses. But, on the other hand, when listening to the rattling noise of these torrents, and calling to mind that whole races of animals have passed away from the surface of the globe, during the period throughout which, night and day, these stones have gone rattling onwards in their course, I have thought to myself, Can any mountains, any continent, withstand such waste?"†

* "I frequently observed, both in Terra del Fuego and within the Andes, that where the rock was covered during the greater part of the year with snow, it was shivered in a very extraordinary manner into small angular fragments. Scoresby has observed the same fact in Spitzbergen; he says, 'The invariably broken state of the rocks appeared to have been the effects of frost.'"—Darwin's Natural History of the Voyage of the Beagle, page 388.

† Darwin, Nat. Hist. of the Voyage of the Beagle, page 386.

In addition to these mechanical causes of waste, we have to consider the influence exerted by chemical forces in effecting the disintegration of rocks, such as the action of the oxygen and carbonic acid of the air, as well as that of water, upon their constituent parts. Whilst we apply the term WASTE to the effects produced by mechanical agencies, we shall confine the term DISINTEGRATION to the effects produced by chemical forces. The latter causes may be very gradual in their operation, not being limited in regard to time. Hence we cannot refuse to acknowledge the existence of their action, even though the effect produced may not be sensible during the life of an individual.

Many years are necessary before the polished surface of an exposed fragment of granite loses its polish; but in process of time this is effected, and the large fragment falls to pieces under the influence exerted upon its constituents by the chemical forces.

The action of water is so much connected with that of oxygen and of carbonic acid, that it is scarcely possible to consider their effects apart.

Many kinds of rocks, such as basalt and clay-slate, contain as an ingredient protoxide of iron. This oxide has a great tendency to absorb oxygen from the air, becoming the higher oxide known as peroxide of iron. This property is especially apparent in our rich ferruginous soils. The surface of such soils to a certain depth is of a red or brownish-red colour, an indication that it contains peroxide of iron; whilst the black or brownish-black colour of the subsoil indicates the presence of the protoxide of the same metal. It often happens that the subsoil is thrown upon the surface in the course of subsoil ploughing, and the consequence on such soils is, that their fertility is destroyed for a certain number of years. The injury thus received continues until all the surface soil again becomes red, that is, until all the protoxide of iron is converted into the peroxide.

It is known that a crystallised salt of iron loses its coherence on exposure to air, and crumbles into a powder by the absorption of oxygen. In a similar manner the disintegration of most minerals is effected, for their ingredients are susceptible of entering into union with oxygen. In consequence of the formation of new compounds, the coherence of the original body is destroyed. If the minerals contain metallic sulphurets, such as the pyrites in granite, these are gradually converted into sulphates.

Most kinds of rocks, such as felspar, basalt, clay-slate, porphyry, and the numerous members of the limestone formation, consist of compounds of silica, with alumina, lime, potash, soda, iron, and protoxide of manganese.

Before we can properly comprehend the action of water and of carbonic acid upon minerals, it is necessary to recollect the properties of silica and of its compounds with alkaline bases.

Quarz forms a very pure variety of silica, and, in this condition, it is quite insoluble both in cold and in hot water, is without taste, and does not exert any action on vegetable colours. The principal property of silica in this state is, that it unites with alkalies, forming saline compounds, which are termed silicates. Window and plate glass consist of mixtures of silicates of the alkaline bases, potash, soda, and lime. In such compounds the alkali is generally completely neutralised. The property of neutralising metallic oxides and alkalies belongs only to acids, and it is owing to this that silica has received the name of silicic acid.

Silica is a very feeble acid, for we have already mentioned that, in its crystallised form, it is destitute both of taste and of solubility in water; but it dissolves by continued digestion in alkaline leys.

We may easily obtain compounds of silica with potash and soda, by melting it either with a pure alkali, or with an alkaline carbonate. By this treatment white glasses are obtained, differing in properties according to their amount of soluble ingredients. When the glass contains 70 per cent. of silica and 30 per cent. of potash or soda, it becomes soluble in boiling water. Its solution may be spread over a surface of wood or of iron, and then dries into a vitreous substance which has received the name of soluble glass. When there is a smaller proportion of alkali than the above quantity, or, in other words, when there is a larger proportion of silica, the resulting glass diminishes in solubility in a greater or less degree.

All silicates soluble in water are decomposed by acids. If the solution of the silicate contains silica corresponding to more than $\frac{1}{30}$ the weight of the water, the addition of an acid causes the formation of a precipitate of a very gelatinous appearance. This precipitate, being a compound of silica with water, is termed the hydrate of silica. But, if the solution contains less silica than the above proportion, no precipitate is formed on the addition of an acid, the whole remaining perfectly clear. This circumstance proves that silica, in the state in which it is precipitated by an acid, possesses a certain degree of solubility in pure water. Indeed, by washing with water the gelatinous precipitate of silica formerly alluded to, its volume diminishes, and silica may be detected in solution by evaporating the water which has passed through.

From these facts we perceive, that silica possesses two distinct chemical characters. In the form in which it is separated from a silicate, it possesses quite different properties from those which it has when in the state of sand, quarz, or rock crystal. When sufficient water is present during its separation from a base, to effect its solution, the whole remains dissolved; in certain conditions, silica is more soluble in water than gypsum.

On drying, silica loses almost completely its solubility in water. The solution of silica in acids acquires, at a certain degree of concentration after cooling, such a gelatinous consistence that the vessel containing it may be turned upside down without spilling a drop of the transparent jelly. By drying it still further, the water which retained it in the gelatinous condition, escapes along with that which had served to hold it in solution. When the water has been once removed in this way, the silica is no longer soluble in water. But, although it has thus lost its solubility, it does not acquire all the properties of crystallised silica, such as sand and quarz, for it still possesses the power of dissolving in alkalies and alkaline carbonates at the ordinary temperature of the air, and this power it retains even when it has been heated to redness.

There is scarcely any other mineral substance which can be compared to silica for the possession of such remarkable properties as those now described.

Most of the insoluble silicates containing alkaline bases are decomposed by the action of hot water, particularly when that water contains an acid. In the middle of the last century, the ignorance of this fact led chemists to believe that water might be converted into an earth.

When water is distilled in glass vessels, it is found to contain always a certain quantity of earthy substances, which may be detected by evaporation, even if the water has been subjected to many repeated distillations. Lavoisier proved that part of the glass was dissolved in this operation by the boiling water; and further, that the diminution in the weight of the glass vessel corresponded exactly to the quantity of earthy residue left by the evaporation of the water. When the distillation of water is effected in metallic vessels no such residue can be obtained.

The action of water upon the silicates contained in glass may be observed in the opacity which gradually comes over the windows of hot-beds, these being exposed in a great degree to the influence of the air. This action is more marked in the windows of stables, where the carbonic acid formed by the processes of respiration of the animals, and by the decay of animal matter, accelerates the decomposition.

Silica being an acid of a very feeble character, the decomposition of the soluble silicates is effected even by carbonic acid.

A solution of soluble glass may be converted into a gelatinous mass by saturating it with carbonic acid gas. The same decomposition must take place in very dilute solutions, although we cannot detect in them any separation of silica, which remains dissolved in the water.

The decomposition of silicates by the combined action of water and of acids proceeds with a rapidity proportional to the quantity of alkalies contained in them.

We find numerous examples in the inorganic kingdom of a continued and progressing process of decomposition of the silicates contained in the various kinds of rocks; this decomposition is effected by the action of carbonic acid, and of water.

A consideration of the preceding observations shews clearly that porcelain clay or kaolin has been formed by the decomposing action of water on the silicates of potash and soda contained in felspar or felspathic rocks. Felspar,* may be viewed as a combination of silicate of alumina with silicate of potash; the last of which being gradually removed by water, leaves behind the porcelain clay.

It has been shown by Forchammer, that felspar may be decomposed by water of 150° C. (302° F.), and at a pressure corresponding to this temperature. The water becomes strongly alkaline, and is found to contain silica in solution. The hot springs in Iceland possess a high temperature, and come from a great depth, where they must have been subjected to high pressure. Forchammer has shown by analysis that the water of these springs contains the constituents of soda felspars, and of magnesian silicates, minerals of very frequent occurrence in trap districts. There cannot be a doubt that a conversion of crystalline felspar into clay must

---

* COMPOSITION OF FELSPATHIC MINERALS.

	Felspar.	Albit.	Labrador.	Anorth.
Silica	65·9	69·8	55·8	44·5
Alumina	17·8	18·8	26·5	34·5
Potash	16·3	—	—	—
Soda	—	11·4	4·0	—
Magnesia	—	—	—	5·2
Lime	—	—	11·0	15·7
Protoxide of iron	—	—	1·3	0·7

The chemical formula of felspar is Al_2, O_3 3 $Si O_3$ + Ko, $Si O_3$. This formula, when multiplied by three, may be divided into porcelain clay, 3 Al_2, O_3, 4 $Si O_3$, and into soluble silicate of potash, 3 Ko, 8 $Si O_3$.

be proceeding to a great extent at the bottom of these springs.*

Ordinary water containing carbonic acid acts in precisely the same manner as water at a high temperature, and at a high pressure.

Polstorf and Wiegmann boiled some white sand with a mixture of nitric and muriatic acids, and after completely removing the acid by washing the sand with water, they exposed it thus purified to the action of water saturated with carbonic acid gas. After the expiration of thirty days, this water was subjected to analysis, and was found to contain in solution, silica, carbonate of potash, and also lime and magnesia; thus proving that the silicates contained in the sand were unable to withstand the continued action of water containing carbonic acid, although the same silicates had resisted the short action of the aqua regia.

Certain of the alkaline silicates found in nature contain in their crystalline state water in chemical combination. In this class are the zeolites, analcime, mesotype, sodalite, apophyllite, &c.; the felspars, properly so called, are always anhydrous.

These silicates differ very much in their behaviour to acid reagents. When mesotype, or a mineral corresponding to it in composition, is kept in the state of a fine powder in contact with cold muriatic acid, it increases in bulk to a thick jelly. The mineral being exposed to the action of the acid at the ordinary temperature, those constituents which are soluble in the acid are taken up by it, whilst the greatest part of the silica remains undissolved. Labrador spar (calcareous felspar) behaves similarly when treated with acids; but the minerals adularia and albite (potash and soda felspars) are not attacked by acids under similar circumstances.

The difference in properties, with respect to reagents,

* The dry residue of 28 ounces of the water consisted of—

Gypsum	0·453
Sulphate of Soda, Magnesia	0·827
Common Salt	2·264
Soda	1·767
Silica	0·506

enable us to decompose very complex kinds of rocks into their constituent parts. C. Gmelin used a process in the analysis of phonolite, or clinkstone rock, by which we may separate and determine the amount of the minerals capable of disintegration contained in different kinds of rocks or soils submitted to examination. For example, phonolite from Abterode in Hegau was found to contain*—

2 097 of a mineral analogous to mesotype, and soluble in acids.
11·142 of felspar, insoluble in acids.

The constituents of both of these are as follows :—

	The portion soluble in acids.	Insoluble residue.
Silica	38·574	66·291
Alumina	24·320	16·510
Potash	3·079	9·249
Soda	12·656	4·960
Lime	1·802	A trace.
Peroxide of iron	11·346	2·388
Peroxide of manganese	2·194	0·896
Titanic acid	0·620	
Water	4·209	
Organic substances	0·405	

In a similar manner, Frick has analysed clay slate, and Lowe the basalt and lava from Mount Etna.

Basalt contains in 100 parts { 4·615 Magnetic Iron Ore. 39·800 Zeolite †. 55·885 Augit ‡.

By treating clay slate from Bendorf with muriatic acid, it was decomposed into—

26·46 parts soluble in muriatic acid.
73·54 parts insoluble in muriatic acid.

* Poggendorf's Annalen, Bd. x. p. 357.

† Zeolite contains—

Silica	38·83
Alumina	28·77
Lime	10·45
Soda	13·81
Potash	1·42
Water	6·72

‡ Augite is a silicate of lime and magnesia.

The composition of these were as follows :—

	Soluble part of clay slate.	Insoluble part of clay slate.
Silica	22·39	77·06
Alumina	19·35	15·99
Peroxide of iron	27·61	1·53
Magnesia	7·00	0·57
Lime	2·42	0·33
Potash without soda	2·37	3·94
Water, carbonic acid, and loss	18·86	0·39
Oxide of copper		0·19

From these analyses we may deduce some highly important results.

It is known that felspar is unable to resist the solvent action of water saturated with carbonic acid, although it is scarcely affected by being left in contact with cold muriatic acid for twenty-four hours. The analyses given above show that the most widely diffused rocks contain a mixture of silicates, which, being soluble in cold muriatic acid, must be much more easily attacked than felspar by water holding in solution carbonic acid.

All minerals and rocks containing silicates of alkaline bases are incapable of resisting the continued solvent action of carbonic acid dissolved in water. The alkalies, with lime and magnesia, will either dissolve alone, or the former will enter into solution along with silica, while the alumina remains behind, mixed or combined with silica. Disintegrated phonolite from Abterode, formed by the action of air and moisture on the solid mineral (the analysis of which is given above), behaves to acids in a manner quite different from the latter.

The mineral clinkstone contains more than 20 per cent. of ingredients soluble in muriatic acid, whilst the same mineral, when disintegrated, does not contain more than 5 per cent. of soluble constituents.*

* The soluble part of disintegrated clinkstone contains—

Silica	13·396
Alumina	5·660
Potash (Soda)	1·074
Lime	—
Peroxide of Iron	63·396
Peroxide of Manganese	11·132
Titanic Acid	3·396

The insoluble portion of disintegrated clinkstone contains—

Silica	66·462
Alumina	16·810
Potash	9·569
Soda	4·281
Lime	1·523
Peroxide of Iron	2·989
Peroxide of Manganese	0·172

The insoluble portion of disintegrated phonolite is scarcely altered in composition; in the soluble portion, iron and manganese form the principal constituents: these two oxides exist in the soluble portions of the undisintegrated mineral in the proportion of 11·346 : 2·194; and in the disintegrated mineral, 100 parts contain 63·39 of peroxide of iron to 11·3 of peroxide of manganese, or nearly the same proportion as the former.

In the process of disintegration, therefore, the alkalies, lime, and magnesia have been dissolved and carried away by water along with silica and alumina; and the residue contains only $\frac{1}{15}$ the amount of the alkalies originally present. But as long as the mineral contains a trace of an alkali, or of any base soluble in carbonic acid, water containing that gas continues to exercise an action upon it, and effects a progressive disintegration of its constituents.

Forchammer considers that the yellow clay, which occurs so frequently in Denmark, consists of granite, the felspar of which has been altered, whilst its mica remains unchanged, and its quarz forms the sand of the clay.

The magnetic and titanic oxides of iron existing in granite is still found in the clay as peroxide of iron and titanic acid.

The blue clays arise from syenite and greenstone; for in these mica is absent (Forchammer).

The great layers of clay at Halle have had their origin in the disintegration of porphyry.*

The white basis of the clay is easily distinguished by moistening it; while the felspar may be recognised by its yellow colour (Mitscherlich). The silica, dissolved by the potash or soda, is sometimes found deposited in a crystalline form on the crystals of felspar; this is often observed in the trachyte of the Seven Mountains near Bonn (Mitscherlich). Most sandstones contain, mixed with them, silicates with alkaline bases.

* The decomposed felspar, porcelain clay of Morl, near Halle, consists of—

Silica	71·42
Alumina	26·07
Peroxide of iron . . .	1·93
Lime	0·13
Potash	0·45

In the sandstone of the Holy Mountain near Heidelberg, small unchanged fragments of felspar are observed, which are partly changed into clay and form white points in the sandstone.

The analysis of the porcelain clays proves that the felspars from which they were formed have not reached their utmost limit of disintegration, for they still contain potash. The porcelain clays are those which are refractory in the fire, and do not melt when exposed to the strongest heat of our furnaces. The difficult fusibility of the porcelain clays depends upon their proportion of the alkaline bases, potash, soda, lime, magnesia, and protoxide of iron.* When we compare the other kinds of clay with the porcelain clays, we find that the infusible clays, or clays poor in potash, are of rare occurrence. The clays diffused through the most kinds of rocks, those occurring in arable land, and those in the beds of clay interspersed with the layers of brown and mineral coal, contract when exposed to heat, and become vitrified in a strong fire. Loam also melts in a similar manner. When the oxides of iron are not present in the clays, their fusibility is in direct proportion to the amount of their alkaline ingredients. Clays arising from the disentegration of the potash felspars, are free from lime; those formed from Labrador spar (the principal component of basalt and lava), contain lime and soda.

The limestones containing much clay are proportionally the richest in alkaline ingredients. The marls and stones used for cement belong to this class of minerals. They differ from other limestones by possessing the property, after burning, of hardening when in contact with water. During the burning of marl and of many other natural cements, the constituents of the clay and lime act chemically upon each other, giving

---

* COMPOSITION OF PORCELAIN CLAYS.

	St. Yvreux.	Meissen.
Silica	46·8	52·8
Alumina	37·3	31·2
Potash	2·5	2·2

	Schneeberg.
Silica	43·6
Alumina	37·7
Peroxide of iron	1·5
Potash and water	12·5

rise to an anhydrous apophyllite, or an analogous compound of silicate of potash and silicate of lime, which, being brought in contact with water, forces the latter into chemical combination in a manner similar to burnt gypsum, and crystallises along with it.* When a fragment of chalk is moistened with a solution of silicate of potash, the latter forms a new compound on the surface, and this becomes hard and stony. The lime of the chalk takes the place of potash in the silicate of potash, and a certain quantity of potash is set at liberty in the form of a carbonate. (KUHLMANN.)

The preceding remarks prove very clearly that arable land has had its origin in the chemical and mechanical actions exerted upon rocks and minerals rich in alkalies and alkaline earths, by which means their coherence has been gradually destroyed. It is scarcely necessary to furnish any further proofs that all clays, whether they be pure or mixed with other minerals, so as to form soils, suffer progressive and continued changes. These changes consist in the giving of a soluble form to the alkalies and alkaline bases, by the combined action of water and of carbonic acid. This gives rise to the formation of soluble silicates, or if these are decomposed by the carbonic acid, to the hydrate of silica, which, being in its peculiar soluble condition, may be taken up by the roots of plants.

The influence of air, carbonic acid, and moisture, upon the constituents of rocks, is best observed in certain uninhabited districts of South America, where huntsmen and herds are the discoverers of rich mines of silver. By the action of the weather, the constituents of the ores of silver are gradually dissolved and carried away by winds and by rains; the nobler metals resist the destruction and remain on the surface. It is well known that metallic silver veins are found in sharp angular projections from the surface of the rock.†

* Formula of Apoplyllite—Ko, 2 Si O_3 + 8 Ca O, Si O_3 + 16 aq.

† Darwin states that the mine at Chanuncillo, from which many hundred thousand pounds sterling have been obtained in a few years, was discovered by a man who, in throwing a stone after a mule, found it heavier than an ordinary stone; it was a piece of solid silver, and was a fragment of a projecting vein of that metal.

CHAPTER IX.

THE ART OF CULTURE.

THE conditions necessary for the life of all vegetables have been considered in the preceding part of the work. Carbonic acid, ammonia, and water yield elements for all the organs of plants. Certain inorganic substances—salts and metallic oxides—serve peculiar functions in their organism, and many of them must be viewed as essential constituents of particular parts.

The atmosphere and the soil offer the same kind of nourishment to the leaves and roots. The former contains a comparatively inexhaustible supply of carbonic acid and ammonia; the latter, by means of its humus, generates constantly fresh carbonic acid, whilst, during the winter, rain and snow introduce into the soil a quantity of ammonia, sufficient for the development of the leaves and blossoms.

The complete, or it may be said, the absolute insolubility in cold water of vegetable matter in progress of decay, (humus,) appears on closer consideration to be a most wise arrangement of nature. For if humus possessed even a smaller degree of solubility than that ascribed to the substance called humic acid, it must be dissolved by rain-water. Thus, the yearly irrigation of meadows would remove a great part of it from the ground, and a heavy and continued rain would impoverish a soil. But humus is soluble only when combined with oxygen; it can be taken up by water, therefore, only as carbonic acid.

When moisture is absent, humus may be preserved for centuries; but when moistened with water, it converts the surrounding oxygen into carbonic acid. As soon as the action of the air ceases, that is, as soon as it is deprived of oxygen, the humus suffers no further change. Its decay proceeds only when plants grow in a soil containing it; for they absorb by their roots the carbonic acid as it is formed. The soil receives again from living plants the carbonaceous matter it thus loses, so that the proportion of humus in it does not decrease.

The stalactitic caverns in Franconia, and those in the vicinity of Baireuth, and Streitberg, lie beneath a fertile arable soil; the abundant decaying vegetables or humus in this soil, being acted on by moisture and air, constantly evolve carbonic acid, which is dissolved by the rain. The rain-water thus impregnated permeates the porous limestone, which forms the walls and roofs of the caverns, and dissolves in its passage as much carbonate of lime as corresponds to the quantity of carbonic acid contained in it. Water and the excess of carbonic acid evaporate from this solution when it has reached the interior of the caverns, and the limestone is deposited on the walls and roofs in crystalline crusts of various forms. There are few spots on the earth where so many circumstances favourable to the production of humate of lime are combined, if the humus actually existed in the soil in the form of humic acid. Decaying vegetable matter, water, and lime in solution, are brought together, but the stalactites formed contain no humic acid; they are of a glistening white or yellowish colour, in part transparent, like calcareous spar, and may be heated to redness without becoming black.

The subterranean vaults in the old castles near the Rhine, in the "Bergstrass," and in the Wetterau, are constructed of sandstone, granite, or basalt, and present appearances similar to the limestone caverns. The roofs of these vaults or cellars are covered externally to the thickness of several feet with vegetable mould, which has been formed by the decay of plants. The rain falling upon them, sinks through the earth, and dissolves the mortar by means of the carbonic acid derived from the mould; and this solution evaporating in the interior of the vaults, covers them with small thin stalactites, which are quite free from humic acid.

In such a filtering apparatus, built by the hand of Nature, we have placed before us the result of experiments which have been continued for hundreds or thousands of years. Now, if water possessed the power of dissolving a hundred-thousandth part of its own weight of humic acid or of humate of lime, and if humic acid were present, we should find the inner surface of the roofs of these vaults and caverns covered with these substances; but we cannot detect the smallest trace of them. We must feel convinced that humic acid is absent both from the soils of fields and of gardens, when we consider that humic acid gives to water a dark brown colour, whereas well and spring water is quite clear and colourless, and leaves after evaporation only a residue of salts formed by mineral acids, without humic acid. The water of wells and of springs is actually rain-water which, in passing through the soil, must exert all its solvent action on the humates. If humate of potash existed in soils, all the spring and river water collected at a certain depth ought to contain traces of it. But even the mineral waters from the springs of Selter and Fachinger, containing alkaline carbonates, are destitute of a trace of humic acid; although these waters arise in a marshy soil abounding in vegetable matter. There could scarcely be found more clear and convincing proofs of the absence of the humic acid of chemists from common vegetable mould.

The common view adopted respecting the *modus operandi* of humic acid does not afford any explanation of the following phenomenon:—A very small quantity of humic acid dissolved in water gives to it a yellow or brown colour.. Hence it would be supposed that a soil would be more fruitful in proportion as it was capable of giving this colour to water, that is, of yielding it humic acid. But it is very remarkable that cultivated plants do not thrive in such a soil, and that all manure must have lost this property before it can exercise a favourable influence upon their vegetation. Water from barren peat soils and marshy meadows, upon which few plants flourish, contains much of this humic acid; but all agriculturists and gardeners agree that the most suitable and best manure for cultivated plants is that which has completely lost the property of giving a colour to water.

The soluble substance, which gives to water a brown colour, is a product of the putrefaction of all animal and vegetable matters; its formation is an evidence that there is not oxygen sufficient to begin, or at least to complete, the decay. The brown solutions containing this substance are decolorised in the air by absorbing oxygen, and a black coaly matter precipitates—the substance named "coal of humus." Now if a soil were impregnated with this matter, the effect on the roots of plants would be the same as that of entirely depriving the soil of oxygen; plants would be as little able to grow in such ground as they would if hydrated protoxide of iron were mixed with the soil. All plants die in soils and water destitute of oxygen; absence of air acts exactly in the same manner as an excess of carbonic acid. Stagnant water on a marshy soil excludes air, but a renewal of water has the same effect as a renewal of air, because water contains it in solution. When the water is withdrawn from a marsh, free access is given to the air, and the marsh is changed into a fruitful meadow.

In a soil to which air has no access, or at most but very little, the remains of animals and vegetables do not decay, for they can only do so when freely supplied with oxygen; but they undergo putrefaction, for which air is present in sufficient quantity. Putrefaction is known to be a most powerful deoxidising process, the influence of which extends to all surrounding bodies, even to the roots and the plants themselves. All substances from which oxygen can be extracted yield it to putrefying bodies; yellow oxide of iron passes into the state of black oxide, sulphate of iron into sulphuret of iron, &c.

The frequent renewal of air by ploughing, and the preparation of the soil, especially its contact with alkaline metallic oxides, the ashes of brown coal, burnt lime or limestone, change the putrefaction of its organic constituents into a pure process of oxidation; and from the moment at which all the organic matter existing in a soil enters into a state of oxidation or decay, its fertility is increased. The oxygen is no longer employed for the conversion of the brown soluble matter into the insoluble coal of humus, but serves for the formation of carbonic acid. This change takes place very slowly, and in some instances the oxygen is completely excluded by it; and whenever this happens, the soil loses its fertility. Thus, in the vicinity of Salzhausen (a village in Hesse Darmstadt, famed for its mineral springs,) upon the meadows of Grünschwalheim, unfruitful spots are seen here and there covered with a yellow grass. If a hole be bored from twenty to twenty-five feet deep in one of these spots, carbonic acid is emitted from it with such violence that the noise made by the escape of the gas may be distinctly heard at the distance of several feet. Here the carbonic acid rising to the surface displaces completely all the air, and consequently all the oxygen, from the soil; and without oxygen neither seeds nor roots can be developed; a plant will not vegetate in pure nitrogen or carbonic acid gas.

Humus supplies young plants with nourishment in the form of carbonic acid by the roots, until their leaves are matured sufficiently to act as exterior organs of nutrition; its quantity heightens the fertility of a soil by yielding more nourishment in this first period of growth, and consequently by increasing the number of organs of atmospheric nutrition. Humus acts in this respect as a source of carbon to plants; but vegetable mould contains other substances which are equally necessary to plants. Vegetable mould contains invariably carbonate of ammonia, besides the salts and alkalies left behind by the putrefaction of former plants.* Those plants which receive

* Some vegetable mould taken from the interior of a hollow oak, yielded $\frac{41}{1000}$ of residue after incineration; of this residue 100 parts contained 24 parts of soluble salts with alkaline bases, 10·5 parts of earthy phosphates, 10 parts of earthy carbonates, and 32 parts of silica. The aqueous extract gave 66 per cent. of soluble salts. (SAUSSURE.) One thousand parts of the extract obtained by hot water from vegetable mould formed by the decay of the *Rhododendron Ferrugineum* gave 140 parts of ashes, which contained, according to Saussure:

their first food from the substance of their seeds, such as bulbous plants, could completely dispense with humus; its presence is useful only in so far as it increases and accelerates their development, but it is not necessary—indeed, an excess of it at the commencement of their growth is in a certain measure injurious.

The amount of food capable of being extracted by young plants from the atmosphere, in the form of carbonic acid and ammonia, is limited; they cannot assimilate more than the air contains. Now, if the quantity of their stems, leaves, and branches has been increased by the excess of food yielded by the soil at the commencement of their development, they will require in a given time for the completion of their growth, and for the formation of their blossoms and fruits, more nourishment from the air than it can afford, and consequently they will not reach maturity. In many cases, the nourishment afforded by the air under these circumstances suffices only to complete the formation of the leaves, stems, and branches. The same result then ensues as when ornamental plants are transplanted from the pots in which they have grown to larger ones, in which their roots are permitted to increase and multiply. All their nourishment is employed for the increase of their roots and leaves; they grow luxuriantly, but do not blossom. When, on the contrary, we take away part of the branches, and of course their leaves with them, from dwarf trees, since we thus prevent the development of new branches, an excess of nutriment is artificially procured for the trees, and is employed by them in the increase of the blossoms and enlargement of the fruit. It is to effect this purpose that vines are pruned.

A new and peculiar process of vegetation ensues in all

Carbonate of potash	14
Chloride of potassium	23
Sulphate of potash	15
Earthy phosphates	17·25
Earthy carbonates	21·50
Silica	3·25
Metallic oxides and loss	5·00

perennial plants, such as shrubs, fruit and forest trees, after the complete maturity of their fruit. The leaves of annual plants at this period of their growth change in colour; while the leaves of trees and shrubs, on the contrary, remain in activity until the commencement of the winter. The formation of the layers of wood progresses, the wood becomes harder and more solid, but after August no more new wood is formed; all the carbonic acid which the plants now absorb is employed for the production of nutritive matter for the following year: instead of woody fibre, starch is formed, and is diffused through every part of the plant by the autumnal sap (sève d'Août).* According to the observations of M. Heyer, the starch thus deposited in the body of the tree can be recognised in its known form by the aid of a good microscope. The barks of several aspens and pine-trees† contain so much of this substance, that it can be extracted from them as from potatoes by trituration with water. It exists also in the roots and other parts of perennial plants. A very early winter, or sudden change of temperature, prevents the formation of this provision for the following year; the wood, as in the case of the vine-stock, does not ripen, and its growth is in the next year very limited.

From the starch thus accumulated, sugar and gum are produced in the succeeding spring, while from these the unnitrogenised constituents of the leaves and young sprouts are in their turn formed. After potatoes have germinated, the quantity of starch in them is found to be diminished. The juice of the maple-tree loses sugar and ceases to be sweet, when its buds, blossoms, and leaves attain their maturity.

The branch of a willow, which contains a large quantity of granules of starch in every part of its woody substance, puts forth both roots and leaves in pure distilled rain-water; but in proportion as it grows, the starch disappears, it

---

* Hartig, in Erdmann und Schweigger-Seidels Journal, V. 217. 1835.

† It is well known that bread is made from the bark of pines in Sweden during famines.

being evidently exhausted for the formation of the roots and leaves.

Upon the blossoming of the sugar-cane, likewise, part of the sugar disappears; and it has been ascertained, that the sugar does not accumulate in the beet-root until after the leaves are completely formed.

These well-authenticated observations remove every doubt as to the functions performed by sugar, starch, and gum, in the development of plants; and it ceases to be enigmatical, why these three substances exercise no influence on the growth or process of nutrition of a matured plant, when applied to it as food.

The accumulation of starch in plants during the autumn has been compared, although certainly erroneously, to the fattening of hibernating animals before their winter sleep; but in these animals every vital function, except the process of respiration, is suspended, and they only require, like a lamp slowly burning, a substance rich in carbon and hydrogen to support the process of combustion in the lungs. On their awaking from their torpor in the spring, the fat has disappeared, but has not served as nourishment. It has not caused the least increase in any part of their body, neither has it changed the quality of any of their organs. With nutrition, properly so called, the fat in these animals has not the least connexion.

The annual plants form and collect their future nourishment in the same way as the perennial; they store it in their seeds in the form of vegetable albumen, starch and gum, which are used by the germs for the formation of their leaves and first fibres of the radicle. The proper nutrition of the plants, their increase in size, begins after these organs are formed.

Every germ and every bud of a perennial plant is the engrafted embryo of a new individual, while the nutriment accumulated in the stem and roots corresponds to the albumen of the seeds.

Nutritive matters are, correctly speaking, those substances which, when presented from without, are capable of sustaining the life and all the functions of an organism, by furnishing to the different parts the materials for the production of their peculiar constituents.

In animals, the blood is the source of the material of the muscles and nerves; by one of its component parts, the blood supports the process of respiration, by others, the peculiar vital functions; every part of the body is supplied with nourishment by it, but its own production is a special function, without which we could not conceive life to continue. If we destroy the activity of the organs which produce it, or if we inject the blood of one animal into the veins of another, at all events, if we carry this beyond certain limits, death is the consequence.

The smallest particles of sugar, when left to themselves, crystallize, that is, they obey a power strictly chemical. It is evident that starch and woody fibre are more highly organised compounds than sugar, for they possess a form which they could not have obtained by the mere power of cohesion. We may suppose that starch and woody fibre were originally gum and sugar, or that both have been formed from sugar; but certain conditions must be necessary for the conversion of sugar into starch, so that it will not be effected when these conditions fail.

Other substances must be present in a plant, besides the starch, sugar and gum, if these are to take part in the development of the germ, leaves, and first fibres of the radicle. There is no doubt that a grain of wheat contains within itself the component parts of the germ and of the fibres of the radicle. These component parts are starch and gluten; and it is evident that neither of them alone, but that both simultaneously assist in the formation of the root, for they both suffer changes under the action of air, moisture, and a suitable temperature. The starch is converted into sugar, and the gluten also assumes a new form, and both acquire the capability of being dissolved in water, and of thus being conveyed to every part of the plant. Both the starch and the gluten are completely consumed in the formation of the first part of the roots and leaves; an excess of either could not be used in the formation of leaves, or in any other way.

The conversion of starch into sugar during the germination of grain is ascribed to a vegetable principle called *diastase*, which is generated during the act of commencing germination. But this mode of transformation can also be effected by gluten, although it requires a longer time. Seeds, which have germinated, always contain much more diastase than is necessary for the conversion of their starch into sugar, for five parts by weight of starch can be converted into sugar by one part of malted barley. This excess of diastase can by no means be regarded as accidental, for, like the starch, it aids in the formation of the first organs of the young plant, and disappears with the sugar.

Carbonic acid, water, and ammonia, are the food of fully-developed plants; starch, sugar, and gum, serve, when accompanied by an azotised substance, to sustain the embryo, until its first organs of nutrition are unfolded. The nutrition of a fœtus and development of an egg proceed in a totally different manner from that of an animal which is separated from its parent; the exclusion of air does not endanger the life of the fœtus, but would certainly cause the death of the independent animal. In the same manner, pure water is more advantageous to the growth of a young plant than that containing carbonic acid, but after a month the reverse is the case (Saussure).

The formation of sugar in maple-trees does not take place in the roots, but in the woody substance of the stem. The quantity of sugar in the sap augments until it reaches a certain height in the stem of the plant, above which point it remains stationary.

Just as germinating barley produces a substance which, in contact with starch, causes it to lose its insolubility and to become sugar, so in the roots of the maple, at the commencement of vegetation, a substance must be formed, which, being dissolved in water, permeates the wood of the trunk, and converts into sugar the starch, or whatever it may be, which it finds deposited there. It is certain, that when a hole is bored into the trunk of a maple-tree just above its roots, filled with sugar, and then closed again, the sugar is dissolved by the ascending sap. It is further possible that this sugar may be disposed of in the same manner as that formed in the trunk; at all events, it is certain that the introduction of it does not prevent the action of the juice upon the starch; and since the quantity of the sugar present is now greater than can be exhausted by the leaves and buds, it is excreted from the surface of the leaves or bark. Certain diseases of trees, for example that called honey-dew, evidently depend on the want of the due proportion between the quantity of the azotised and that of the unazotised substances which are applied to them as nutriment.

If now we direct our attention to the particular organs of a plant, we find every fibre and every particle of wood surrounded by a juice containing an azotised matter; while the starch, granules, and sugar, are enclosed in cells formed of a substance containing nitrogen. Indeed everywhere, in all the juices of the fruits and blossoms, we find a substance destitute of nitrogen, accompanied by one containing that element.

The wood of the stem cannot be formed, *quasi* wood, in the leaves, but another substance must be produced which is capable of being transformed into wood. This substance must be in a state of solution, and accompanied by a compound containing nitrogen; it is very probable that the wood and the vegetable gluten, the starch granules and the cells containing them, are formed simultaneously, and in this case a certain fixed proportion between them would be a condition necessary for their production.

In the buds and young leaves, we find salts with alkaline bases; we find also the azotised constituents accompanied by salts of phosphoric acid: we must, therefore, suppose that these substances execute some functions necessary to the support of the vital processes of plants. We may suppose that, in the absence of certain constituents of the soil, the compounds of plants containing nitrogen and sulphur could not be formed, and that without the presence of such compounds and of alkaline bases, carbonic acid could not be taken up and decomposed.

According to this view, the assimilation of the substances generated in the leaves will (*cæteris paribus*) depend on the quantity of nitrogen contained in the food. When a sufficient quantity of nitrogen is not present to aid in the assimilation of the substances destitute of it, these substances will be separated as excrements from the bark, roots, leaves, and branches. The exudations of mannite, gum, and sugar, in strong and healthy plants cannot be ascribed to any other cause.*

Analogous phenomena are presented by the process of digestion in the human organism. In order to restore the loss sustained by every part of the body in the processes of respiration and perspiration, the organs of digestion require to be supplied with food, consisting of substances containing nitrogen, and of others destitute of it, in definite proportions, and also with certain mineral substances to effect their transformation into blood. If the substances destitute of nitrogen preponderate, either they will be expended in the formation of fat, or they will pass unchanged through the organism. This is particularly observed in those people who live almost exclusively upon potatoes; their excrements contain a large quantity of unchanged granules of starch. Potatoes, which when mixed with hay alone are scarcely capable of supporting the strength of a horse, form with bread and oats a strong and wholesome fodder.

It will be evident from the preceding considerations, that

* M. Trapp, in Giessen, possesses a *Clerodendron fragrans* growing in the house; it exudes on the surface of its leaves, in September, large colourless drops of sugar-candy, which form regular crystals upon drying;—I am not aware whether the juice of this plant contains sugar. Langlois has lately observed, during the dry summer in 1842, that the leaves of the linden-tree became covered with a thick and sweet liquid, in such quantity that for several hours of the day it ran off the leaves like drops of rain. Many kilogrammes might have been collected from a moderately sized linden-tree. This sweet juice contained principally grape sugar and mannite. (*Annales de Chimie et Physique*, iii. Série, tom. vii. p. 348.)

the products generated by a plant may vary exceedingly, according to the substances given it as food. A superabundance of carbon in the state of carbonic acid conveyed through the roots of plants, without being accompanied by nitrogen, cannot be converted either into gluten, albumen, or wood; but either it will be separated in the form of excrements, such as sugar, starch, oil, wax, resin, mannite, or gum, or these substances will be deposited in greater or less quantity in the wide cells and vessels.

The quantity of gluten, vegetable albumen, and mucilage, will augment when plants are supplied with an excess of food containing nitrogen; and ammoniacal salts will remain in the sap; when, for example, in the culture of the beet, we manure the soil with a highly nitrogenous substance, or when we suppress the functions of the leaves by removing them from the plant.

We know that the ananas is scarcely eatable in its wild state, and that it shoots forth a great quantity of leaves when treated with rich animal manure, without the fruit on that account acquiring a large amount of sugar; that the quantity of starch in potatoes increases when the soil contains much humus, but decreases when the soil is manured with strong animal manure, although then the number of cells increases, the potatoes acquiring in the first case a mealy, in the second a soapy, consistence. Beet-roots taken from a barren sandy soil, contain a maximum of sugar, and no ammoniacal salts; and the Teltowa parsnep loses its mealy state in a manured land, because there all the circumstances necessary for the formation of cells are united.

An abnormal production of certain component parts of plants presupposes a power and capability of assimilation to which the most powerful chemical action cannot be compared. The best idea of it may be formed by considering that it surpasses in power the strongest galvanic battery, with which we are not able to separate the oxygen from carbonic acid. The affinity of chlorine for hydrogen, and its power of decomposing water under the influence of light, and of setting at liberty its oxygen, cannot be considered as at all equalling the power and energy with which a leaf separated from a plant decomposes the carbonic acid absorbed by it.

In living plants and in their seeds, there exists a peculiar power different from all other causes of increase of mass. This power, however, only shows itself in action when exposed to the influence of heat or of light. In spring, when the heat of the sun penetrates the earth, the asparagus may put forth shoots of many feet in length quite independently of the action of light. But the constituents of these shoots were formerly constituents of the roots. A conversion of pre-existing compounds into new products, and their assumption of new forms, can proceed without light, although not without heat. But this is not a true increase of mass, or an increase in the quantity of carbon. The latter process only takes place under the influence of light.

The common opinion that only the direct solar rays can effect the decomposition of carbonic acid in the leaves of plants, and that reflected or diffused light does not possess this property, is wholly an error, for exactly the same constituents are generated in a number of plants, whether the direct rays of the sun fall upon them, or whether they grow in the shade. They require light, and indeed sun-light, but it is not necessary that the direct rays of the sun should reach them. Their functions certainly proceed with greater intensity and rapidity in sunshine than in the diffused light of day; but there is nothing more in this than the similar action which light exercises on ordinary chemical combinations; it merely accelerates in a greater or less degree the action already subsisting.

Thus chlorine and hydrogen combining form muriatic acid. This combination is effected in a few hours in common day-light, but it ensues instantly, with a violent explosion, under exposure to the direct solar rays, whilst not the slightest change in the two gases takes place in perfect darkness. When the oil formed from olefiant gas is exposed in a vessel with chlorine gas to the direct solar rays, chloride of carbon is immediately produced; but the same compound can be obtained with equal facility in the diffused light of day, a longer time only being required. When this experiment is performed in the way first mentioned, two products only are observed (muriatic acid and perchloride of carbon); whilst by the latter method a class of intermediate bodies are produced, in which the quantity of chlorine constantly augments, until at last the whole oil is converted into the same two products as in the first case. Here, also, not the slightest trace of decomposition takes place in the dark. Nitric acid is decomposed in common daylight into oxygen, and peroxide of nitrogen; and chloride of silver becomes black in the diffused light of day, as well as in the direct solar rays;—in short, all actions of a similar kind proceed in the same way in diffused light as well as in the solar light, the only difference consisting in the time in which they are effected. It cannot be otherwise in plants, for the mode of their nutriment is the same in all, with the exception of certain parasites which obtain their carbon, either not at all, or only partially, from the original source; and their component substances afford proof that their food has suffered absolutely the same change, whether they grow in the sunshine or in the shade.*

All the carbonic acid, therefore, which we supply to a plant will undergo a transformation, provided its quantity be not greater than can be decomposed by the leaves. We know that an excess of carbonic acid kills plants, but we know also that nitrogen to a certain degree is not essential for the decomposition of carbonic acid. All the experiments hitherto instituted prove, that fresh leaves placed in water impregnated with

* The impossibility of bringing to blossom and seed mosses and other cryptogamous plants, in ordinary daylight, induced Mr. Nöller, an excellent botanist and chemist in Darmstadt, to form the opinion that the green light from the leaves formed a necessary condition of their life. He planted numerous kinds of these plants in mouldered wood placed in little glass tubes, and covered the whole with a green glass globe. The experiment established his view in a beautiful manner. All these elegant plants developed under these conditions with the greatest luxuriance, and put forth both blossoms and seeds.

carbonic acid, and exposed to the influence of solar light, emit oxygen gas, whilst the carbonic acid disappears. Now in these experiments no nitrogen is supplied at the same time with the carbonic acid; hence no other conclusion can be drawn from them than that a simultaneous introduction of nitrogen is not necessary for the decomposition of carbonic acid,—for the exercise, therefore, of one of the functions of plants. And yet the presence of a substance containing this element appears to be indispensable for the assimilation of the products newly formed by the decomposition of the carbonic acid, and their consequent adaptation for entering into the composition of the different organs.

The carbon abstracted from the carbonic acid acquires in the leaves a new form, in which it is soluble and transferable to all parts of the plant. In this new form the carbon aids in constituting several new products; these are named sugar when they possess a sweet taste, gum or mucilage when tasteless, and excrementitious matters when expelled by the roots or other parts.

Hence it is evident that the quantity and quality of the substances generated by the vital processes of a plant will vary according to the proportion of the different kinds of food with which it is supplied. The development of every part of a plant in a free and uncultivated state depends on the amount and nature of the food afforded to it by the spot on which it grows. A plant is developed on the most sterile and unfruitful soil as well as on the most luxuriant and fertile; the only difference which can be observed being in its height and size, in the number of its twigs, branches, leaves, blossoms, and fruit. Whilst the individual organs of a plant increase on a fertile soil, they diminish on another where those substances which are necessary for their formation are not so bountifully supplied; and the proportion of the constituents containing nitrogen, and those free from it, varies with the amount of nitrogenous matters in the food of plants.

The development of the stem, leaves, blossoms, and fruit of plants is dependent on certain conditions, the knowledge of which enables us to exercise some influence on certain of their internal constituents as well as on their size. It is the duty of the natural philosopher to discover what these conditions are; for the fundamental principles of agriculture must be based on a knowledge of them. There is no profession which can be compared in importance with that of agriculture, for to it belongs the production of food for man and for animals; on it depends the welfare and development of the whole human species, the riches of states, and all commerce. There is no profession in which the application of correct principles is productive of more beneficial effects, or is of greater and more decided influence. Hence it appears quite unaccountable, that we may vainly search for one leading principle in the writings of agriculturists and vegetable physiologists.

The methods employed in the cultivation of land are different in every country, and in every district; and when we inquire the causes of these differences, we receive the answer that they depend upon circumstances. (*Les circonstances font les assolements.*) No answer could show ignorance more plainly.

In addition to the general conditions, such as heat, light, moisture, and the component parts of the atmosphere, all of which are necessary for the growth of all plants, certain substances are found to exercise a peculiar influence on their development. These substances either are already contained in the soil, or are supplied to it in the form of the matters known under the general name of manure. But what does the soil contain, and what are the components of the substances used as manure? Until these points are satisfactorily determined, a rational system of agriculture cannot exist. The power and knowledge of the physiologist, of the agriculturist and chemist, must be united for the complete solution of these questions; and, in order to attain this end, a commencement must be made.

The GENERAL object of agriculture is to produce in the most advantageous manner certain qualities, or a maximum size, in certain parts or organs of particular plants. Now, this object can be attained only by the application of such substances as we know to be indispensable to the development of these parts or organs, or by supplying the conditions necessary to the production of the qualities desired.

The rules of a rational system of agriculture should enable us, therefore, to give to each plant that which it requires for the attainment of the object in view.

The SPECIAL object of agriculture is to obtain an abnormal development and production of certain parts of plants, or of certain vegetable matters, employed as food for man and animals, or for the purposes of industry.

The means employed vary according to the objects which it is desired to attain. Thus, the mode of culture employed for the purpose of procuring fine pliable straw for Tuscan hats, is the very opposite to that which must be adopted in order to produce a maximum of corn from the same plant. Peculiar methods must be used for the production of nitrogen in the seeds, others for giving strength and solidity to the straw, and others again must be followed when we wish to give such strength and solidity to the straw as will enable it to bear the weight of the ears.

We must proceed in the culture of plants in precisely the same manner as we do in the fattening of animals. The flesh of the stag and roe, or of wild animals in general, is quite devoid of fat, like the muscular flesh of the Arab; or it contains only small quantities of it. The production of flesh and fat may be artificially increased, as in the case of domestic animals. The food given to animals increases the activity of certain organs, and is itself capable of being transformed into fat. We add to the quantity of food, or we lessen the processes of respiration and perspiration by preventing motion.

The increase or diminution of the vital activity of vegetables depends only on heat and solar light, which we have not arbitrarily at our disposal: all that we can do is to supply substances adapted for assimilation by the power already present in the organs of the plant. But what then are these substances? They may easily be detected by the examination of a soil always fertile in the existing cosmical and atmospheric conditions; for it is evident that the knowledge of its state and composition must enable us to discover the conditions under which such a soil is rendered fertile. It is the duty of the chemist to explain the composition of a fertile soil, but the discovery of its proper physical state or condition belongs to the agriculturist; our present business lies only with the former.

Arable land is originally formed by the crumbling of rocks, and its properties depend on the nature of their principal component parts. Sand, clay, and lime, are the names given to the principal constituents of the different kinds of soil.

Pure sand and pure limestones, in which there are no other inorganic substances except siliceous earth, carbonate or silicate of lime, form absolutely barren soils. But argillaceous earths form always a part of fertile soils. Now, from whence come the argillaceous earths in arable land, what are their constituents, and what part do they play in favouring vegetation? They are produced by the disintegration of aluminous minerals, among which the common potash and soda felspars, Labrador spar, mica, and the zeolites, are those which most commonly undergo this change. These minerals are found mixed with other substances in granite, gneiss, mica-slate, porphyry, clay-slate, grauwacke, and the volcanic rocks, basalt, clinkstone, and lava. In the grauwacke we have pure quartz, clay-slate, and lime; in the sandstones, quartz and loam. The transition limestone and the dolomites contain an intermixture of clay, felspar, porphyry, and clay-slate; and the mountain limestone is remarkable for its quantity of argillaceous earths. Jura limestone contains 3—20, that of the Wurtemberg Alps 45—50 per cent. of these earths. And in the *muschelkalk* and in the *calcaire grossier* they exist in greater or less quantity.

It is known that the aluminous minerals are the most widely diffused on the surface of the earth, and, as we have already mentioned, they are never absent from fertile soils;

and, if they should happen to be absent in soils capable of cultivation, this only happens when certain of their constituents are supplied by other sources. Argillaceous earth must, therefore, contain something which enables it to exercise an influence on the life of plants, and to assist in their development. The property on which this depends is that of its invariably containing alkalies and alkaline earths, with sulphates and phosphates.

Alumina exercises only an indirect influence on vegetation, by its power of attracting and of retaining water and ammonia; it is itself very rarely found in the ashes of plants,* but silica is often present, having in most cases entered the plants by means of alkalies. In order to form a distinct conception of the quantities of alkalies in aluminous minerals, it must be remembered that felspar contains 17¾ per cent. of potash, albite 11·43 per cent. of soda, and mica 3—5 per cent.; and that zeolite contains 13—16 per cent. of alkalies†. The late analysis of Ch. Gmelin, Löwe, Fricke, Meyer, and Redtenbacher, have also shown, that basalt and clinkstone contain from ¾ to 3 per cent. of potash, and from 5—7 per cent. of soda; that clay-slate contains from 2·75—3·31 per cent. of potash, and loam from 1½—4 per cent. of potash.

If, now, we calculate from these data, and from the specific weights of the different substances, how much potash must be contained in a layer of soil, formed by the disintegration of 26,910 square feet (1 Hessian acre) of one of these rocks to the depth of 20 inches, we find that a soil of

Felspar	contains	.	.	. 1,152,000 lbs.	
Clinkstone	,,	from	200,000 to	400,000	,,
Basalt	,,	,,	47,500 ,,	75,000	,,
Clay-slate	,,	,,	100,000 ,,	200,000	,,
Loam	,,	,,	87,000 ,,	300,000	,,

The alkalies, potash and soda, are present in all clays; at

* Hydrate of alumina, when mixed with extract of humus, decolorises this substance and renders insoluble the colouring matter. (*Wiegmann und Polstorf.*)

† Recent investigations have shown that potash felspars always contain a certain quantity of soda, and that soda felspars always contain potash.

least, they have been found in all the argillaceous earths in which they have been sought. The fact that they contain potash may be proved in the clays of the transition and stratified mountains, as well as in the recent formations surrounding Berlin, by simply digesting them with sulphuric acid, by which process alum is formed. (MITSCHERLICH.) It is well known also to all manufacturers of alum, that the leys contain a certain quantity of this salt ready formed, the potash of which has its origin from the ashes of the stone and brown coal, which contain much argillaceous earth.

A thousandth part of loam mixed with the quartz in new red sandstone, or with the lime in the different limestone formations, affords as much potash to a soil only twenty inches in depth as is sufficient to supply a forest of pines growing upon it for a century. A single cubic foot of felspar is sufficient to supply an oak wood, covering a surface of 26,910 square feet, with the potash required for five years.

Land of the greatest fertility contains argillaceous earths and other disintegrated minerals, with chalk and sand in such a proportion as to give free access to air and moisture. The land in the vicinity of Vesuvius may be considered as the type of a fertile soil, and its fertility is greater or less in different parts, according to its proportion of clay or sand.

This soil being derived from the disintegration of lava, cannot possibly, owing to its origin, contain the smallest trace of vegetable matter; yet every one knows that when lava or volcanic ashes have been exposed for a time to the influence of air and moisture, all kinds of plants grow in them with the utmost luxuriance.

This fertility of lava is owing to the alkalies, alkaline earths, and silica, contained in it, which by exposure to the weather are rendered capable of being absorbed by plants. Thousands of years have been necessary to convert stones and rocks into the soil of arable land, and thousands of years more will be requisite for their perfect reduction, that is, for the complete exhaustion of their alkalies.

We see from the composition of the water in rivers, streamlets, and springs, how little alkali the rain-water is able to extract from a soil, even after a term of years; this water is generally soft, and the common salt, which even the softest invariably contains, proves that the alkaline salts, which are carried to the sea by rivers and streams, are returned again to the land by wind and by rain.

Let us suppose that a soil has been formed by the action of the weather on the component parts of granite, grauwacke, mountain limestone, or porphyry, and that the vegetation upon it has remained the same for thousands of years. Now this soil would become a magazine of alkalies in a condition favourable for their assimilation by the roots of plants.

The interesting experiments of Struve have proved that water impregnated with carbonic acid decomposes rocks containing alkalies, and then dissolves a part of the alkaline carbonates. It is evident that plants also, by producing carbonic acid during their decay, and by means of the acids which exude from their roots in the living state, contribute no less powerfully to destroy the coherence of rocks. Next to the action of air, water, and change of temperature, plants themselves are the most powerful agents in effecting the disintegration of rocks.

Air, water, and change of temperature prepare the different species of rocks for yielding to plants their alkalies. A soil exposed for centuries to all the influences which affect the disintegration of rocks, but from which the alkalies have not been removed, will be able to afford, during many years, the means of nourishment to vegetables requiring alkalies for their growth; but it must gradually become exhausted, unless those alkalies which have been removed are again replaced; a period, therefore, will arrive when it will be necessary to expose it from time to time to a further disintegration, in order to obtain a new supply of soluble alkalies. For, small as is the quantity of alkali essential to plants, it is nevertheless quite indispensable for their perfect development. But when one or more years have elapsed without the removal of any alkalies from the soil, a new harvest may be expected.

The first colonists of Virginia found a soil similar to that mentioned above; harvests of wheat and tobacco were obtained for a century from one and the same field, without the aid of manure; but now whole districts are abandoned and converted into unfruitful pasture-land, which without manure produces neither wheat nor tobacco. From every acre of this land there were removed in the space of one hundred years 12,000 lbs. of alkalies in leaves, grain, and straw; it became unfruitful therefore, because it was deprived of every particle of alkali fit for assimilation, and because that which was rendered soluble again in the space of one year was not sufficient to satisfy the demands of the plants. Almost all the cultivated land in Europe is in this condition; fallow is the term applied to land left at rest for further disintegration. It is the greatest possible mistake to suppose that the temporary diminution of fertility in a soil is owing to the loss of humus: it is the mere consequence of the exhaustion of alkalies, and of other essential ingredients.

Let us consider the condition of the country around Naples, which is famed for its fruitful corn-land; the farms and villages are situated from eighteen to twenty-four miles distant from one another, and between them there are no roads, and consequently no transportation of manure. Now corn has been cultivated on this land for thousands of years, without any part of that which is annually removed from the soil being artificially restored to it. How can any influence be ascribed to humus under such circumstances, when it is not even known whether humus was ever contained in the soil?

The method of culture in that district completely explains he permanent fertility. It appears very bad in the eyes of our agriculturists, but there it is the best plan that could be adopted. A field is ploughed once every three years, and is in the intervals allowed to serve as a sparing pasture for cattle. The soil experiences no change in the two years during which it lies fallow, further than that it is exposed to the influence of the weather, by which a fresh portion of its alkalies are again set free or rendered soluble. The animals fed on these fields yield nothing to them which they did not formerly possess. The weeds upon which the cattle live spring from the soil, and the materials returned to it in the form of excrements must always be less in quantity than those removed as food. The fields, therefore, can have gained nothing from the mere feeding of cattle upon them; on the contrary, the soil must have lost some of its constituents.

Experience has shown in agriculture that wheat should not be cultivated after wheat on the same soil, for it belongs with tobacco to the plants which exhaust a soil. But if the humus of a soil gives it the power of producing corn, how happens it that wheat does not thrive in many parts of Brazil, where the soils are particularly rich in this substance, or in our own climate, in soils formed of mouldered wood; that its stalk under these circumstances attains no strength, and droops prematurely? The cause is this, that the strength of the stalk is due to silicate of potash, and that the corn requires certain phosphates, and these substances a soil of humus cannot afford, since it does not contain them; the plant may, indeed, under such circumstances, become an herb, but will not bear fruit.

Again, how does it happen that wheat does not flourish on a sandy soil, and that a calcareous soil is also unsuitable for its growth, unless it be mixed with a considerable quantity of clay? It is because these soils do not contain alkalies and certain other ingredients in sufficient quantity, the growth of wheat being arrested by this circumstance, even should all other substances be presented in abundance.

It is not mere accident that we find on soils of gneiss, mica-slate, and granite in Bavaria, of clinkstone on the Rhone, of basalt in the Vogelsberge, and of clay-slate on the Rhine and in the Eifel, the finest forests of oaks, which cannot be produced on the sandy or calcareous soils upon which firs and pines thrive. It is explained by the fact that trees, the leaves of which are renewed annually, require for their leaves six to ten times more alkalies than the fir-tree or pine, and hence they do not attain maturity when placed in soils containing very small quantities of alkalies.* When we see oaks growing on a sandy or calcareous soil—or the red-beech, the service-tree, and the wild-cherry for example—thriving luxuriantly on limestone, we may be assured that alkalies are present in the soil, for they are necessary to their existence. Can we, then, regard it as remarkable that oak copse should thrive in America, on those spots on which forests of pines which have grown and collected alkalies for centuries, have been burnt, and to which the alkalies are thus at once restored; or that the *Spartium scoparium, Erysimum latifolium, Blitum capitatum, Senecio viscosus*, plants remarkable for the quantity of alkalies contained in their ashes, should grow with the greatest luxuriance on the localities of conflagrations?†

All plants of the grass kind require silicate of potash. Now this is conveyed to the soil, or rendered soluble in it by the irrigation of meadows. The *equisetaceæ*, the reeds and species of cane containing such large quantities of siliceous earth, or silicate of potash, thrive luxuriantly in marshes, in argillaceous soils rich in potash, and in ditches, streamlets, where the change of water renews constantly the supply of dissolved silica. The amount of silicate of potash removed from a meadow in the form of hay is very considerable. We need only call to mind the melted vitreous mass found on a meadow between Manheim and Heidelberg after a thunderstorm. This mass was at first supposed to be a meteor, but was found on examination (by Gmelin) to consist of silicate

* One thousand parts of the dry leaves of oaks yielded 55 parts of ashes, of which 24 parts consisted of alkalies soluble in water; the same quantity of pine leaves gave only 29 parts of ashes, which contain 4·6 parts of soluble salts. (De Saussure.)

† After the great fire in London, large quantities of the *Erysimum latifolium* were observed growing on the spots where a fire had taken place. On a similar occasion the *Blitum capitatum* was seen at Copenhagen, the *Senecio viscosus* in Nassau, and the *Spartium scoparium* in Languedoc. After the burnings of forests of pines in North America poplars grew on the same soil.

of potash; a flash of lightning had struck a stack of hay, and nothing was found in its place except the melted ashes of the hay.

Alkalies and alkaline earths are not, however, the only substances necessary for the existence of most plants; but other substances besides alkalies are required to sustain the life of plants.

Phosphoric acid has been found in the ashes of all plants hitherto examined, and always in combination with alkalies or alkaline earths. By incinerating the seeds of wheat, rye, maize, peas, beans, and lentils, ashes are obtained quite free from carbonic acid, and consisting entirely of phosphates, with the exception of very small quantities of sulphates and of chlorides.

Plants obtain their phosphoric acid from the soil. It is a constituent of all land capable of cultivation, and even the heath at Lüneburg contains it in appreciable quantity. Phosphoric acid has been detected also in all mineral waters in which its presence has been tested; and in those in which it has not been found it has not been sought for. The most superficial strata of the deposits of sulphuret of lead (*galena*) contain crystallised phosphate of lead (*green lead ore*); clay-slate, which forms extensive strata, is covered in many places with crystals of phosphate of alumina (*Wavellite*); all its fractured surfaces are overlaid with this mineral.

Apatite (phosphate of lime of similar composition to bone earth) is found in every fertile soil. This mineral may be easily recognised, in its crystalline form, in many varieties of rocks. It occurs in this state in the plutonic, volcanic, and metamorphic rocks, although it is usually found only in small quantity. In the plutonic and volcanic rocks it is found in granite (as in the mines of Johann Georgenstadt, Schneeberg, and in the loose gravel near Berlin); in syenite it occurs in small crystals, as at Meissen, and in larger crystals at Friedrichswern, in South Norway. It exists also in hypersthene, as at Elfdalen, in Sweden, and very often in large quantity, as at Meiches, in the Vogelsberge (a district celebrated for its fertility in wheat), and also in the hills of Löbau, in Saxony; Tuhlowitz, in Bohemia, &c. It is found in basalt and other volcanic rocks in various localities; for example, at Wickenstein, at Hamberg, and also at Cabo de Gata, in Spain, and in the volcanic boulders of the Laacher See. Apatite is found also in the metamorphic rocks, and particularly in the talc and chloritic schists; it occurs in large yellow crystals in the micaceous schists of Snarum, in Norway; and in the calcareous deposits of Pargas, in Finland, and in the Lake Baikal; in the deposits of magnetic iron ore in Arendal, and in other places in Sweden and in Norway. It is found also in the oceanic rocks, particularly in the chalk of Cape la Hêve, at Havre, and of the Capes Blancnez and Grisnez, at Calais, and in the layers of limestone at Amberg, &c. (Gustavus Rose.)

The water of the imperial spring at Aachen contains, according to Monheim, 0·142 grains of phosphate of soda in 1 lb.; that of the Quirinus Spring contains the same quantity, and the water of the Roses spring contains 0·133 of the same salt. The water of the fountain at Carlsbad contains 0·0016 grains of phosphate of lime. (Berzelius.) The Ferdinand's spring contains 0·010 phosphate of soda, according to Wolf. The saline springs at Pyrmont, contain 0·022 phosphate of potash, 0·075 phosphate of lime, and 0·1249 grains phosphate of alumina. (Krüger.) When we consider that sea-water contains phosphate of lime in such small quantity that its amount cannot be determined in a pound of water, and yet that from this quantity all the living animals in the sea receive the phosphates contained in their bones and flesh, we must admit that the amount of phosphates in the above mentioned mineral waters is very considerable. It may be shown by calculation that the water of the fountain at Carlsbad must take up many thousand pounds of phosphate of lime in its passage through the layers of rocks.

A few very simple experiments point out the manner in which the earthy phosphates, and particularly phosphate of lime, are taken up by the roots of plants.

Phosphate of lime is insoluble in pure water, but it dissolves readily in water containing common salt, or a salt of ammonia; and in water containing sulphate of ammonia it dissolves as readily as gypsum. Phosphate of lime is also soluble in water containing carbonic acid; in this respect it is ana logousto carbonate of lime.

The soil in which plants grow furnishes their seeds, roots, and leaves, with phosphoric acid, and they in turn yield it to animals, to be used in the formation of their bones, and of those constituents of the brain which contain phosphorus. We may form an idea of the quantity of phosphate of magnesia contained in grain, when we consider that the concretions in the cæcum of horses consist of phosphate of magnesia and ammonia, which must have been obtained from the hay and oats consumed as food. Twenty-nine of these stones were taken after death from the rectum of a horse belonging to a miller, in Eberstadt, the total weight of which amounted to 3 lbs.; and Dr. F. Simon has lately described a similar concretion found in the horse of a carrier, which weighed $1\frac{1}{2}$ lb.

Some plants extract other matters from the soil besides silica, the alkalies, alkaline earths, sulphuric and phosphoric acids, which are essential constituents of the plants ordinarily cultivated. These other matters, we must suppose, supply, in part at least, the place, and perform the functions, of the substances just named. We may thus regard common salt, nitre, chloride of potassium, and other matters, as necessary constituents of several plants.

Clay-slate contains generally small quantities of oxide of copper; and soils formed from micaceous schist contain some metallic fluorides. Now, small quantities of these substances also are absorbed into plants, although we cannot affirm that they are necessary to them.

It appears that in certain cases fluoride of calcium may take the place of the phosphate of lime in the bones and teeth; at least it is impossible otherwise to explain its constant presence in the bones of antediluvian animals, by which they are distinguished from those of a later period. The bones of human skulls found at Pompeii contain as much fluoric acid as those of animals of a former world; for if they be placed in a state of powder in glass vessels, and digested with sulphuric acid, the interior of the vessel will, after twenty-four hours, be found powerfully corroded (LIEBIG); whilst the bones and teeth of animals of the present day contain only traces of it. (BERZELIUS.)

In spring and in the first half of the summer, when the earth is still moist with water, it is quite certain that a greater quantity of alkaline bases and of salts must enter the organism of a plant, than in the height of summer, when there is a deficiency of water, this being the means of carrying the bases to the plant.

In many districts the crops of corn for the whole year depend upon a single shower of rain; for when water is deficient at a certain period of the growth of plants, their future progress is retarded. The introduction of water to a soil is, properly speaking, an introduction of alkalies and of certain salts, which, by means of rain water, become fit to be absorbed by plants. In the middle of summer the air is much more charged with the vapour of water than at other seasons of the year, and, therefore, the hydrogen which is essential to the nourishment of plants, is presented to them in sufficient quantity.

When the soil is deficient in moisture, we observe a phenomenon, which appeared quite inexplicable, before we understood the importance of mineral matters, as means of nourishment to plants. We see the leaves close to the soil (those which had been first developed), lose their vitality, shrink and fall off, after becoming yellow, without the apparent action of any injurious cause. This phenomenon is not perceived, in this form, in moist years, nor is it observed with evergreens, and only rarely with those plants which throw out long deep roots; it is observed only in harvest and in winter with perennial plants.

The cause of this phenomenon is now quite apparent. The matured leaves absorb continually from the air carbonic acid and ammonia, which are converted into the constituents of new leaves, buds, and twigs; but this conversion cannot be effected without the co-operation of alkalies and of other inorganic substances. When the soil is moist these are constantly conveyed to the plants, which retain their green colour in consequence. But in dry weather, the deficiency of water prevents them being absorbed by the plant; and in consequence of this, they are taken from the plant itself. The mineral ingredients in the juice of the fully formed leaves is abstracted from them, and is employed in the formation of the young sprout; and when the seeds become developed, the vitality of the old leaf is completely destroyed. These withered leaves contain mere traces of soluble salts, while the buds and sprouts are remarkably rich in these ingredients.

The reverse of this phenomenon is seen in the case of many kitchen plants, when they are supplied with rich manure containing an excess of mineral ingredients; salts are separated from the surface of their leaves, and cover them with a thin white crust. In consequence of these exudations the plant becomes sickly, the organic activity of the leaves diminishes, the growth of the plant is destroyed, and if this condition lasts, the plant finally dies. These observations are best made on plants with leaves of large dimensions, through which large quantities of water are evaporated.

This disease generally attacks turnips, gourds, and peas, when the soil is drenched with sudden and violent rain, after continued dry weather, at a time when the plants are near, but have not attained maturity; it is also necessary for its occurrence, that dry weather should again happen after the rain.

By the rapid evaporation of the water absorbed by the roots, a larger quantity of salts enters the plants than they are able to use. These salts effloresce on the surface of the leaves, and when they are juicy, act as if the plants had been treated with solutions of salts, in greater quantity than their organism could bear. Of two plants of the same kind the one nearest maturity is most liable to this disease; if the other plant has either been planted at a later period, or if its development has been restrained, the causes, which exercised injurious effects upon the first plant, accelerate the development of the latter. The germ springing out of the earth, the leaf on coming out of the bud, the young stem, and the green sprouts, contain a much larger quantity of salts with alkaline bases, and give ashes on incineration much richer in alkaline ingredients, than parts of the matured plant. The leaves, being the part in which the absorption and decomposition of carbonic acid is effected, are much richer in mineral ingredients than other parts of the plant.

The simple fact that a plant is restrained in growth by the want of rain to convey to it alkalies, proves completely that these alkalies play a most important part in vegetation.

Although it was found by Saussure that wheat before blossoming yielded $\frac{70}{1000}$, in blossom $\frac{54}{1000}$, and after the ripening of the seeds only half this quantity of ashes; it cannot hence be concluded that the ingredients of the soil present in the young and growing plants, were again returned to the soil. Equal quantities of young plants yield twice the amount of ashes that matured plants do; but this evidently arises from the circumstance, that new quantities of organic constituents are added to the carbon, hydrogen, and nitrogen, previously existing in the young plant. The amount of ashes remain the same in both plants, although their relative proportions have become different.

We may feel assured that the alkalies contained in the vine, in the potato, and beet, and found in the juices, united with tartaric, citric, oxalic, and malic acids, are not merely present for the purpose of being used in druggists' shops, or in our household, as acid or as neutral salts. These organic acids must be necessary for the formation of certain constituents in the plants.

We have already come to the conclusion, that the carbon of all plants is derived from carbonic acid; tartaric, oxalic, citric acid, &c., must, therefore, obtain their carbon from the same source. But, can we conceive that the carbon forms a direct and immediate combination with hydrogen for the production of substances so various as sugar, starch, woody fibre, resin, wax, and oil of turpentine? Is it not much more probable that the conversion of the carbon of carbonic acid into the constituent of a plant proceeds in a gradual manner; that by the union of the constituents of water with carbonic acid, a substance is formed, becoming gradually poorer in oxygen; and that the carbon assumes the form of oxalic, tartaric, or other organic acids, before it is converted into sugar, starch, or woody fibre?

According to this view, a ready and simple explanation is furnished of the necessity of alkalic bases to vegetable life; for they are present for the purpose of effecting the conversion of carbonic acid into a living part of a plant. The smallest particles of sugar, or of organic acids, when separated from plants, follow their own peculiar attractions; they form crystals, or they follow the power which induces the cohesion of their atoms, but still their carbon is capable of being converted into a constituent of a living organ; and, although sugar and tartaric acid have been formed by vital agencies, they do not in themselves possess any vital functions.

From the preceding part of this chapter it will be seen that fallow is that period of culture when the land is exposed to progressive disintegration by the action of the weather, for the purpose of liberating a certain quantity of alkalies and silica to be absorbed by future plants.

The careful and frequent working of fallow land will accelerate and increase its disintegration; for the purposes of culture it is quite the same whether the land be covered with weeds, or with a plant which does not extract the potash of the soil.

CHAPTER X.

ON FALLOW.

Agriculture is both an art and a science. Its scientific basis embraces a knowledge of all the conditions of vegetable life, of the origin of the elements of plants, and of the sources whence they derive their nourishment.

From this knowledge fixed rules are formed for the practice of the art, that is, for the necessity or advantage of all the mechanical operations of the farm, by which the land is prepared for the growth of plants, and by which those causes are removed, which might exercise an injurious influence upon them.

Experience acquired in the practice of this art can never stand in contradiction to its scientific principles; because the latter have been deduced from all the observations of experience, and are actually an intellectual expression of it. Neither can Theory ever stand in antagonism to Practice, for it is merely the tracing back of a class of phenomena to their ultimate causes.

A field, upon which we cultivate the same plants successively for a number of years, may become unfertile for these plants in three years; whilst another field may last seven, another twenty, and another one hundred years, without losing its fertility. One field bears wheat but not beans; another bears turnips but not tobacco; and a third yields rich crops of turnips, but does not bear clover.

What is the reason that a field loses gradually its fertility for the same plant? What is the reason that a certain kind of plant flourishes on it, and that another fails?

These questions are proposed by the science of Agriculture.

What means are necessary to enable a field to sustain its fertility for the same plant, and to make it fit for the cultivation of one, two, or for all plants?

THE LATTER QUESTIONS ARE PROPOSED BY THE ART OF AGRICULTURE; but they are not susceptible of solution by means of the art.

When a farmer institutes experiments for the purpose of making a field fertile for plants which it would not formerly bear, the anticipation of success must be small, unless he is guided by scientific principles. Thousands of farmers try analogous experiments in various ways, and the results of these constitute a mass of experience, out of which a method of culture is finally formed; and this method suffices for a certain district. But the same method fails with a neighbouring district, or it may prove actually injurious.

What an immense amount of capital and power is lost in these experiments! What a very different and much more certain path does Science follow! It does not put us in danger of failure, and it gives us the best security of success.

If the causes of failure or the causes of sterility of a soil for one, two, or three plants be ascertained, the means of obviating the sterility follow as a matter of course.

The methods of cultivating soils vary with their geological characters. In basalt, grauwacke, porphyry, sandstone, limestone, &c., let us suppose that there are present, in different proportions, certain chemical compounds essential to the growth of plants, and which must therefore exist in fertile soils; then we are able to explain in a very simple manner the difference in the methods of culture; for it is obvious that the soils formed by the disintegration of the above rocks must vary in the proportion of their essential constituents, just as the rocks themselves vary.

Wheat, clover, and turnips require certain constituents from the soil; and hence they cannot flourish in a soil from which these are absent. Science enables us to recognise the necessary constituents, by the analysis of the ashes of the plants; and if we discover the absence of these ingredients from the soil, the cause of its sterility is obvious.

The means of obviating this sterility follows from a knowledge of its cause.

Empiricism ascribes all results to the art, that is, to the mechanical operations employed in cultivation, without inquiring the causes upon which their use depends. But a knowledge of these causes is of the highest importance; for such knowledge would prevent the lavish expenditure of capital and of power, and would enable us to use them in the most advantageous manner. Is it possible that the entrance of the ploughshare, or of the harrow into the earth—that the contact of iron with the soil—can act as a charm to impart fertility? No one can entertain such an opinion; and yet the causes of their action have not yet been inquired into, and much less have they been explained. It is quite certain that careful ploughing and breaking up of the soil, by producing the change and increase of its surface, exercise a very favourable influence on its fertility; but these mechanical operations are only means to an end.

In the effects produced by time, particularly in the case of fallow, or that period during which a field remains at rest, Science recognises certain chemical actions, which proceed continuously by means of the influence exercised by the constituents of the atmosphere upon the surface of the soil. By the action of the carbonic acid and oxygen in the air, aided by moisture and by rain-water, the power of dissolving in water is given to certain constituents of rocks, or of their debris, from which arable land is formed; these ingredients, in consequence of their solubility, become separated from the insoluble constituents.

These chemical actions serve to explain the effects produced by the hand of time, which destroys human structures, and converts gradually the hardest rocks into dust. It is by their influence that certain ingredients of arable land become fit for assimilation by plants; and the object of the mechanical operations of the farm is to obtain this result. Their action consists in accelerating the weathering or disintegration of the soil, and thus offers to a new generation of plants their necessary constituents, in a form fit for reception. The celerity of the disintegration of a solid body must be in proportion to its surface; for the more points which we expose to the action of the destructive agencies, the more rapidly will their effects be produced.

When a chemist subjects a mineral to analysis, in order to break up the compound, that is, to give solubility to its constituents, he is obliged to perform the very tedious and difficult task of reducing it to an impalpable powder. He separates the fine dust from the grosser particles by means of a fine sieve, or by elutriation, and exerts his utmost patience to obtain a fine powder; because he is aware that the breaking up of the mineral will be incomplete, and that all his operations will prove ineffectual, if he be at all careless in this preliminary operation.

The influence of an increased surface upon the weathering of a stone, or, in other words, on the changes which it suffers by the action of the constituents of the atmosphere, and by water, is very well pointed out in the interesting description given by Darwin of the gold mines at Yaquil, in Chili. The gold ores, after being reduced to a very fine powder in mills, are subjected to a process by which the particles of metal are separated from the lighter parts of the ore. The particles of stone are carried away by a stream of water; while those of gold fall to the bottom. The former are conducted into a tank, where they are permitted to deposit. As the tank fills gradually, they are removed from it, and are exposed in heaps to the action of the air and of moisture. From the nature of the elutriation to which it was subjected, the finely-divided ore can no longer contain any salts, or soluble ingredients. Whilst it lay at the bottom of the tank covered with water, and therefore excluded from air, it suffered no change; but when exposed to air, a powerful chemical action ensues in the heaps, and this action is recognised by the rich efflorescence of salts, which cover their surface, from the effects of disintegration. After the finely-divided ore has been exposed to the action of the weather for two or three years, during which time it hardens, it is again elutriated, and the processes of exposure and elutriation are repeated six or seven times, new quantities of gold being obtained each time, although in smaller proportions; this gold is liberated by the chemical process of weathering or of disintegration.

The same chemical actions as those now described proceed in our arable land, and it is to accelerate and increase these that we employ the mechanical operations of culture. We renew the surface of the soil, and endeavour to make every particle of it accessible to the action of carbonic acid and of oxygen. Thus we procure a new provision of soluble mineral substances, which are indispensable for the nourishment and luxuriance of a new generation of plants.

All cultivated plants require alkalies and alkaline earths, although each of them may use different proportions of the one or of the other: the cereals do not flourish in a soil deficient in silica in a soluble state.

Silicates, as they occur in nature, differ very materially in their tendency to suffer disintegration, and in the resistance which they offer to the action of atmospheric agents. The granite of Corsica and the felspar of Carlsbad crumble into dust in a space of time during which the polished granite of the Bergstrasse does not even lose its lustre.

There are certain kinds of soils so rich in silicates prone to disintegration, that every year, or every two years, a quantity of silicate of potash is rendered fit for assimilation sufficient for the formation of the leaves and stems of a whole crop of wheat. In Hungary there are large districts of land, on which, since the memory of man, corn and tobacco have been cultivated in alternate years, without the restoration of the mineral ingredients carried away in the corn and in the straw. There are other fields, on the contrary, which do not yield sufficient silicate of potash in two, three, or more years.

Fallow, in its most extended sense, means that period of culture during which a soil is exposed to the action of the weather, for the purpose of enriching it in certain soluble ingredients. In a more confined sense, the time of fallow may be limited to the intervals in the cultivation of cereal plants; for a magazine of soluble silicates and of alkalies is an essential condition to the existence of such plants. The cultivation of potatoes or of turnips during the interval will not impair the fertility of the field for the cereals which are to succeed, because the former plants do not require any of the silica necessary for the latter.

It follows from the preceding observations, that the mechanical operations in the field are the simplest and most economical means of rendering accessible to plants the nutritious matters in the soil.

But, it may be asked, are there no other means besides the mere mechanical operations, of liberating the ingredients of a soil, and of fitting them for reception by the organism of plants? There are such means, and one of the most simple and efficacious of them is the practice employed in England for the last century, of manuring soils with burnt lime.

In order to form a proper conception of the action of lime on soils, we must remember the processes employed by chemists to effect the speedy decomposition of a mineral, and to render soluble its ingredients. In order to dissolve finely-pulverised felspar in an acid, it would be necessary to expose it to continued digestion for weeks, or even for months. But when the felspar is mixed with lime, and is exposed to a moderately strong heat, the lime enters into chemical combination with the constituents of the felspar. A part of the alkali (potash) imprisoned in the felspar is now set at liberty, and a simple treatment of the felspar with acid, in the cold, now suffices to dissolve the lime and the other constituents of the mineral. The silica is dissolved by the acid to such an extent, that the whole assumes the consistence of a transparent jelly.

Most of the silicates of alumina, when mixed with slacked lime and kept in continued contact in a moist state, behave in a similar manner to felspar when heated with lime. When a mixture of common clay, or of pipe-clay, and water, is added to milk of lime, the whole becomes immediately thick on agitation. When they are left in contact for several months, it is found that the mixture gelatinises on the addition of an acid—a property which the mixture of clay and water did not possess, or only to a very small degree, before the contact with lime. The clay is broken up by the union of certain of its constituents with lime; and, what is still more remarkable, most of the alkalies contained in it are set at liberty. These beautiful observations were first made by Fuchs of Munich; and they have not only led to conclusions on the nature and properties of hydraulic limestones, but, what is far more important, they have explained the action of slacked lime upon soils, and they have thus furnished an invaluable means of liberating from the soil the alkalies which are indispensable to the existence of plants.

In October, the fields in Yorkshire and Lancashire have the appearance of being covered with snow. The soil for miles is seen covered either with lime previously slacked, or with lime that has slacked itself by exposure to air. During the moist months of winter, it exercises its beneficial influence on the stiff clayey soils.

According to the old theory of humus, we ought to suppose that burnt lime would exercise a very injurious influence on soils, by destroying the organic matter contained in them, and by thus rendering them unfit to supply a new vegetation with humus. But, on the contrary, it is found that lime heightens the fertility of a soil. The cereals require the alkalies and silicates liberated by the lime and rendered fit for assimilation by plants. If there be present decaying matter yielding to the plants carbonic acid, their development may be favoured by this means; but this is not necessary. For if we furnish to the soil ammonia, and to the cereals the phosphates essential to their growth, in the event of their being deficient, we furnish all the conditions necessary for a rich crop, as the atmosphere forms an inexhaustible magazine of carbonic acid.

In districts where fuel is cheap, an equally favourable influence is exerted on clayey soils by the system of burning.

It is not very long since that chemists observed the

remarkable changes which take place in the properties of clay when it is burned: these were first studied in the analysis of several silicates of alumina. Many of them, which are not at all attacked by acids in their natural state, acquire complete solubility when they are previously melted by heat. To this class of silicates belong pipe and potter's clay, loam, and the different varieties of clay occurring in soils. In the natural state of clay, it may be digested with concentrated sulphuric acid for hours, without dissolving in any appreciable quantity; but when the clay is slightly burnt (as is done, for example, in several alum works) it dissolves in acids with great ease, while the silica is separated in its gelatinous and soluble form. Common potter's clay forms generally very sterile soils, although it contains within it all the conditions for the luxuriant growth of plants; but the mere presence of these conditions does not suffice to render them useful to vegetation. The soil must be accessible to air, oxygen, and carbonic acid, for these are the principal conditions to favour the development of the roots. Its constituents must be contained in a state fit to be taken up by plants. Plastic clay is deficient in all these properties, but they are communicated to it by a gentle calcination.*

The great difference between burnt and unburnt clay may be observed in places where burnt bricks are used for building. In Flanders, where almost all the houses are constructed with burnt bricks, the surface of the walls, after exposure for a few days to the action of the weather, becomes covered with an efflorescence of salts. When these salts are washed away by the rain, a new efflorescence again appears; and in some cases, as in the gateway of the fortress at Lille, this may be observed, even though the walls have stood for centuries. The efflorescence consists of carbonates and of sulphates with alkaline bases—salts that are known to play a most important part in

* The author saw an example of this in the garden of Mr. Baker, at Hardwick Court, near Gloucester. The soil consisted of a stiff clay, and, from a state of complete sterility, had been made remarkably fertile, by simple burning. The operation, in this case, was carried on to a depth of three feet, —certainly not an economical, although a completely successful experiment.

the economy of vegetation. Lime exercises a striking effect upon these saline efflorescences, for it may be observed, that they first appear in those parts where the mortar and bricks come in contact.

It is obvious that mixtures of clay and lime contain all the conditions necessary for the liberation of the silicate of alumina, and for rendering soluble the alkaline silicates. Lime dissolved in water by means of carbonic acid acts upon clay in the same way that milk of lime does. This fact explains the favourable influence of marl upon most soils, marl being a clay rich in lime. Indeed there are certain marly soils surpassing in fertility, for all plants, soils of any other kind. Burnt marl must be in a very superior state for manure; and this remark applies to all substances of a similar composition,—to the hydraulic limestones, for example. By these the plants are furnished, not only with alkalies, but also with silica, in a state fit for reception. Many of the hydraulic limestones, or the natural cements, as they are called, after being mixed in their burnt state with water, yield to it, in a few hours, so much caustic alkali, that the water may be employed as a weak ley for the purposes of washing.

The ashes of brown coal and of mineral coal are used in many districts as excellent means of improving certain soils. Those ashes are to be preferred that gelatinise on the addition of an acid, or that become stony and hard after some time, like hydraulic cement, when mixed with lime and water.

The mechanical operations of the farm, fallow, the applications of lime, and the burning of clay, unite in elucidating the same scientific principle. They are the means of accelerating the disintegration of the alkaline silicates of alumina, and of supplying to plants their necessary constituents at the commencement of a new vegetation.

It must be distinctly understood, that the previous remarks apply only to those fields which are in a favourable physical state for the development of plants; for this, in conjunction with the other necessary conditions, has the greatest influence on fertility. A stiff, heavy clayey soil offers too much resistance to the spreading out and increase of the roots of a quick-growing summer plant. It is obvious that such a soil will be rendered more accessible to the roots, as well as to air and moisture, by a simple mixture with quarz or with sand, and this may often prove more effectual in improving it than the most diligent ploughing. When we supply to a soil easily penetrable by the roots of plants, as well by air and moisture, IN THE FORM OF ASHES, the constituents that we removed in the crops, the soil will retain all its original favourable physical state. In like manner, we can restore the original chemical composition to stiff, heavy clay soils; but it is better for such soils to restore the necessary ingredients IN THE FORM OF ANIMAL EXCREMENTS, than to do so, as in the former case, by means of ashes. By the improvement of the physical condition of the soil, its fertility is increased. In this respect excrements are of very various values, although they may contain the same chemical constituents; thus sheep's dung is close and heavy, while the dung of cows and of horses is light and porous.

In hot summers, accompanied by light and partial showers of rain, porous soils of no great fertility yield often better crops than richer stiff soils. The rain falling on the porous soil is immediately absorbed and reaches the roots, whilst that falling on the heavy soils is evaporated before it is enabled to penetrate them.

A soil destitute of cohesion, like quick-sand, is not fitted for the cultivation of plants in general. Finally, there are certain kinds of soils which ought, from their chemical composition, to be very fertile, but which, on the contrary, are sterile for many kinds of plants: such soils are those that consist of clay mixed with a large quantity of very fine sand. Such a soil converts itself into a kind of thick mud after a heavy fall of rain, and thus prevents all access of air, and it dries without much contraction.

If we were to apply, in all their extent, to porous, sandy, or calcareous soils, or to a soil of the nature mentioned above, the principles upon which depend the improvement of land by fallow, we could not hope to obtain favourable results. A soil of great porosity, through which water penetrates with great ease, and which does not yield sufficient hold to the roots of plants, and also a stiff soil, with its particles too finely divided, and of small fertility on account of its physical properties, cannot be benefited by the mechanical operations of the field; for these are intended to effect a still further reduction of the particles.

The physical conditions essential to the fertility of a soil are usually neglected in the calculations of the chemist, and thus render a mere chemical analysis of very subordinate value; for the existence of all the mineral means of nourishment in a soil does not necessarily indicate its value. But when the chemical is combined with the mechanical analysis* (for the latter of which Mr. Rham has described an equally simple and convenient instrument), then we are furnished with data upon which to form accurate conclusions.

* The estimation of the unequal quantities of mixed ingredients, such as of the coarse and fine sand, and of the clay and vegetable matters.

CHAPTER XI.

ON THE ROTATION OF CROPS.

It has been shown, by accurate examinations of animal bodies, that the blood, bones, hair, &c., as well as all the organs, contain a certain quantity of mineral substances, without the presence of which in the food, these tissues could not be formed.

Blood contains potash and soda in combination with phosphoric acid; the bile is rich in alkalies; the substance of the muscles contains a certain amount of sulphur; the blood globules contain iron; the principal ingredient of bones is phosphate of lime; nervous and cerebral substance contains phosphoric acid and alkaline phosphates; and the gastric juice contains free muriatic acid.

We know that the free muriatic acid of the gastric juice and that part of the soda of the bile is obtained from common salt; and we are enabled, by the mere exclusion of this material from food, to put an end to the digestive process and life of an animal.

When a young pigeon is fed upon grains of wheat in which phosphate of lime, the principal constituent of the bones, is deficient, and when it is prevented receiving this substance from other sources, its bones become thin and friable, and death ensues if the supply of this mineral substance is still prevented. (Choiset, Report to the Academy of Paris, June, 1842.) In like manner, if we exclude carbonate of lime from the food of fowls, they lay eggs without the hard exterior shell.

When a cow is fed upon an excess of roots, such as potatoes and turnips, the same thing must happen to it, as in the case of the pigeon cited above; for these roots contain phosphate of magnesia, and only traces of lime. Now, if we remove daily from the same cow a certain amount of phosphate of lime in its milk, without restoring this in the food, the lime will be obtained from its bones, which will thus lose gradually their strength and solidity, until they are no longer able to support the weight of the body. But if we give to the pigeon as food barley or peas, and to the cow barley-straw or clover, we will be able to sustain the health of the animals; for these materials abound in salts of lime.*

Man and animals receive the constituents of their blood and of their bodies from the vegetable world; and an Infinite Wisdom has so ordained, that the life and luxuriance of plants is strictly connected with the reception of the same mineral substances that are indispensable for the development of the animal organism; without the presence of the inorganic matters found in the ashes of plants, the formation of the germ, leaves, blossoms, or fruit could not be effected.

The amount of nutritive matters in the different kinds of cultivated plants is very unequal. The bulbous plants and roots approach each other much more nearly in their chemical constituents than the seeds; while the latter possess always an analogous composition.

Potatoes, for example, contain from 75 to 77 per cent. of water, and from 23 to 25 per cent. of solid matter. By means of a mechanical process, we may divide the latter into 18 or 19 parts of starch, and 3 or 4 parts of a fibre resembling starch. Both of these added together weigh nearly as much as the dry potato. The two per cent. not accounted for consists of salts, and of the substance containing sulphur and nitrogen, known under the name of albumen.

* The labourers in the mines of South America, whose daily labour (perhaps the most severe in the world), consists in carrying upon their shoulders a load of earth of from 180 to 200 lbs. weight, from a depth of 450 feet, subsist only upon bread and beans. They would prefer to confine themselves to bread, but their masters have found that they cannot work so much on this diet, and they, therefore, compel them, like horses, to eat beans. —(*Darwin's Journal of Researches.*) Beans are proportionally much richer in bone earth than bread.

Beet contains from 88 to 90 per cent. of water. Five-and-twenty parts of dry beet contain very nearly the same elements as 25 parts of dry potatoes. In the beet there are 18 or 19 parts of sugar and 3 or 4 parts of cellular tissue; the two per cent. not accounted for consist partly of salts, and the remainder of albumen.

Turnips contain from 90 to 92 parts of water. From 23 to 25 parts of dry turnips contain 18 to 19 parts pectin, with very little sugar, 3 or 4 parts cellular tissue, and 2 parts salts and albumen. Sugar and starch do not contain nitrogen; they exist in the plant in a free state, and are never combined with salts, or with alkaline bases. They are compounds formed from the carbon of the carbonic acid and the elements of water. In the potato, these assume the form of starch, and in the turnip the form of pectin.

In the seeds of cereals we find VEGETABLE FIBRIN, a constituent containing sulphur and nitrogen; in peas, beans, and lentils, we find CASEIN; and in the seeds of oily plants, ALBUMEN and a substance very analogous to casein.

Vegetable fibrin is accompanied by starch in the seeds of the cereals; the latter body occurs with casein in leguminous plants; but, in the oily seeds, its place is supplied by another body devoid of nitrogen, such as oil, butter, or a constituent resembling wax.

It is obvious that we must furnish to plants the peculiar conditions necessary for the development of these constituents, according to our object in cultivation. In order to procure sugar or starch, we must supply the plant with other materials than we would do were our object to obtain the ingredients containing sulphur and nitrogen.

In a hot summer, when the deficiency of moisture prevents the absorption of alkalies, we observe the leaves of the lime-tree, and of other trees, covered with a thick liquid containing a large quantity of sugar; the carbon of this sugar must, without doubt, be obtained from the carbonic acid of the air. The generation of the sugar takes place in the leaves; and all the constituents of the leaves, including the alkalies and alkaline earths, must participate in effecting its formation. Sugar does not exude from the leaves in moist seasons; and this leads us to conjecture, that the carbon which appeared as sugar in the former case would have been applied in the formation of other constituents of the tree, in the event of its having had a free and unimpeded circulation. When the soil is frozen in winter, there cannot be an absorption of alkalies by the roots; but notwithstanding this, it cannot be doubted that during the day the evergreens and the leaves of firs and pines must absorb continually from the air carbonic acid, which will be constantly decomposed by the action of the light. When circulation is unimpeded, the carbon of this carbonic acid may perhaps be converted into wood or into other constituents of the plant; but, in the absence of the conditions necessary for this conversion, it may now secrete resin, balsam, and volatile oils. In the generation of the sugar, or in that of resin and volatile oil in the firs and pines, all the constituents of the leaves must take part; and hence we cannot suppose that their alkalies, their lime, &c., are either accidental, or that they are unnecessary to the exercise of this vital function.

For the conversion of the carbon of carbonic acid into sugar, it is necessary that certain conditions exist in the plant itself, in addition to the external circumstances (such as heat and air).

We will furnish the conditions essential to the formation of starch, or of sugar, when we supply to the leaves—that is, to the organs destined for the absorption and assimilation of the carbonic acid—their necessary constituents.

The sap of such plants as are rich in sugar or in starch, and also the sap of most woody plants, contains much potash and soda, or alkaline earths. We cannot suppose that these are mere accidental ingredients; on the contrary, we must believe that they serve some purposes of the plants, and that they assist in the formation of certain of their constituents. It has already been mentioned, that they exist in the plants in a state of combination with certain organic acids. These acids are so far characteristic of certain genera, that they are never absent from them. Hence the organic acids themselves must assist in some of the vital functions. Now, when it is remembered that unripe fruits, such as grapes, are unfit to eat on account of their acidity; that these fruits possess the same power as the leaves of absorbing carbonic acid, and of giving off oxygen on exposure to light; and further, that the sugar increases on the diminution of the acid; we can scarcely avoid coming to the conclusion, that the carbon of the organic acid in the unripe fruit becomes a constituent of the sugar when it is ripe, and that, in consequence of the separation of oxygen and the union of the constituents of water, the acid passes into sugar.

The tartaric acid in grapes, the citric acid in cherries and in currants, the malic acid in summer apples, which ripen on the trees, form in these plants the intermediate members of the passage of carbonic acid into sugar; and when there is a deficiency of proper temperature, or of the action of solar light, the changes necessary for the conversion into sugar are not furnished, and the acids remain.

In the fruit of the mountain ash, malic acid succeeds the tartaric acid at first present, or in other words, an acid poor in oxygen succeeds one rich in that element; afterwards the malic acid in the berries disappears almost entirely, and in its place are found gum and mucilage, neither of which formerly existed in them; and with the same reason that we consider that the carbon of the tartaric acid forms a constituent of the succeeding malic acid—and this few would be inclined to dispute—we suppose that the carbon of the acids passes over into the sugar which succeeds on their disappearance.

It surely cannot be supposed that a plant assimilates carbonic acid, and that this carbonic acid is converted in the organism of the plant into tartaric, racemic and citric acids, merely for the purpose of being reconverted into carbonic acid.

If then the view be confirmed, that the organic acids in

cultivated plants aid in the formation of sugar, it must be admitted, that they are of equal importance in the production of all other non-azotised ingredients similarly composed. The formation of starch, of pectin, and of gum, do not take place immediately, that is, they do not arise at once from the union of the carbon of the carbonic acid with the constituents of water; but a gradual conversion takes place, in consequence of the production of compounds that are always poorer in oxygen, and always richer in hydrogen. We cannot suppose that oil of turpentine could be formed without the existence of analogous intermediate members of the series.

Now, if the organic compounds rich in oxygen, viz. the ACIDS, be the means of producing the compounds poorer in this element, such as SUGAR, STARCH, &c., then the alkalies and alkaline bases must be looked upon as the conditions essential for the formation of these non-azotised constituents, because the acids existing in cultivated plants are generally in the form of salts and are rarely free. An organic acid may perhaps be formed, without the presence of these bases, but, in the absence of an alkali, or of a body possessing an analogous action, sugar, starch, gum, and pectin cannot form in the organism of a plant. Sugar is not formed in those fruits and seeds in which the organic acids are free, that is, in which they do not exist as salts, as, for instance, citric acid in the lemon, or oxalic acid in the chick-pea. It is only in plants containing the acids combined with bases in the form of soluble salts, that sugar, gum, and starch, are produced.

It is a matter of little consequence what value is attached to the opinion now given of the part taken by alkaline bases in the process of vegetable life. But the following facts are of the greatest significance and value to agriculture, namely, that the newly-developed sprouts, leaves, and buds; * or in

* 1000 parts of Firwood gave 3·28 parts of Ashes.
1000 ,, Fir-leaves ,, 62·25 ,,
The ashes of the leaves of the fir correspond to more than 20 times those

other words, those parts of the plants possessing the greatest intensity of assimilation, contain the greatest proportion of alkaline bases, and that the plants richest in sugar and in starch are no less distinguished for their quantity of alkaline bases and of organic acids.

As we find sugar and starch accompanied by salts of an organic acid; and as experience proves that a deficiency of alkalies causes a deficient formation of woody fibre, sugar, and starch; and that, on the contrary, a luxuriant growth is the consequence of their abundant supply; it is obvious that the object of culture, viz., a maximum of crops, cannot be obtained, unless the alkalies necessary for the transformation of carbonic acid into starch and sugar are supplied in abundant quantity, and in a form fit for assimilation by plants.*

---

in the wood freed from its bark. 100 parts of the former contain:—(HERTWIG.)

Alkaline carbonates, Common salt	10·72	12·70 salts soluble in water.
Sulphate of potash	1·95	
Silicate of potash	3·90	
Carbonate of lime	63·32	86·30 compounds insoluble in water.
,, magnesia	1·86	
Phosphate of magnesia, ,, lime	6·35	
Basic per phosphate of iron	0·88	
Basic phosphate of alumina	0·71	
Silica	10·31	

* The acids—malic, tartaric, citric, oxalic, &c.—are generated in the organism of plants, and their carbon must be derived from carbonic acid.

In plants these acids are found combined with potash, lime, and magnesia, in the form of salts, the smallest particles of which, when left to themselves, follow their own attractions; this is indicated by their crystallisation.

There is no doubt that these compounds do not possess organic life, because the active power observed in them is not vitality, but cohesion. The same must be the case with sugar, which crystallises in a similar manner.

We must presume that the smallest particles of the products formed from carbonic acid are subject to the powers acting upon them in the living plant, in the same way that a particle of carbonic acid is; that, therefore, the carbon of oxalic acid, tartaric acid, &c., must possess the power of passing into a constituent of an organ endowed with life.

The conversion of organic acids into organs may be followed with ease.

If we suppose that 12 equivalents of carbonic acid, in the presence of a base, and by the action of light, loses the fourth part of its oxygen, in consequence of the action of vitality upon its elements, then oxalic acid would

Every part and constituent of the body is obtained from plants. By the organism of the plants, are formed those compounds which serve for the formation of the blood; there

---

be produced. In its anhydrous state, we cannot conceive it to be formed from carbonic acid in any other way.

$$C_{12}\ O_{24} - O_{6} = C_{12}\ O_{18} = 6 \text{ Eq. anhydrous oxalic acid.}$$

Oxalic acid does not exist in an anhydrous state. Hydrated oxalic acid contains one equivalent of water; the oxalates of potash, lime, and magnesia also contain water. Hydrated oxalic acid consists of—

$$C_{12}\ O_{18} + H_{6}\ O_{6} = C_{12}\ H_{6}\ O_{24} = 6 \text{ Eq. hydrated oxalic acid.}$$

From this it may be observed that carbonic acid and hydrated oxalic acid contain the same quantity of oxygen. We can, therefore, suppose that hydrated oxalic acid has been formed from carbonic acid, to which a certain amount of hydrogen has been added.

By the continued action of the same agents a new quantity of oxygen might become separated from carbonic acid, in which case tartaric acid or malic acid would result. By the separation of 9 equivalents of oxygen, tartaric acid would be produced; the separation of 12 equivalents would produce malic acid.

Hydrated oxalic acid $C_{12}\ H_{6}\ O_{24} - O_{9} = C_{12}\ H_{6}\ O_{15} = 3$ Eq. tartaric acid.
,, ,, $C_{12}\ H_{6}\ O_{24} - O_{12} = C_{12}\ H_{6}\ O_{12} = 3$ Eq. malic acid.

By the simple separation of water from the elements of malic acid, citric acid is produced; we know that we can produce, by means of heat, aconitic acid from citric acid, and fumaric acid, or maleic acid, from malic acid.

Malic acid $C_{12}\ H_{6}\ O_{12} - HO = C_{12}\ H_{5}\ O_{11} = 3$ Eq. citric acid.
,, $C_{12}\ H_{6}\ O_{12} - 3HO = C_{12}\ H_{3}\ O_{9} = 3$ Eq. fumaric acid.

Now we can view tartaric, citric, and malic acids as compounds of oxalic acid with sugar, with gum, with woody fibre, or with the elements of these:

Tartaric Acid. Oxalic Acid. Dry Sugar of Grapes.

$$2\ (C_{12}\ H_{6}\ O_{15}) = C_{12}\ O_{18} + C_{12}\ H_{12}\ O_{12}$$

In such a manner, therefore, that the addition of new quantities of hydrogen would enable all these acids to aid in the formation of sugar, starch, and gum. When this conversion is effected, the alkalies in union with the acids must of course be liberated, and they will thus be rendered capable of playing anew the same part. According to this view, it is quite conceivable that one equivalent of an alkali may enable 10, 20, or 100 equivalents of carbon to pass into constituents of a plant; but the time necessary to effect the transformation will vary according to the amount of base present.

If a perennial evergreen, by the help of a certain quantity of alkali, is able to assimilate a certain amount of carbon during the whole year, it will be necessary to convey to a summer plant four times the quantity of alkali, in order to enable it to assimilate the same amount of carbon in one-fourth the time.

Gay-Lussac first observed that by the contact of an alkali, at a high temperature, with tartaric, citric, and oxalic acids, or sugar, woody fibre, &c., these substances were reconverted into carbonic acid.

This mode of decomposition is quite the reverse of that which takes place can be no doubt that the nutritive parts of plants must contain all the constituents of the blood, and not merely one or two of them.

It cannot be supposed that blood will be formed in the body of an animal, or milk in that of a cow, if their food fail in even one of the constituents necessary for the sustenance of the vital functions. The compounds containing nitrogen and sulphur, as well as the alkalies and phosphates, are constituents of the blood; but the conversion of the former into blood cannot be conceived without the presence and co-operation of the latter.

According to this view, the power of any part of a plant to support the life of an animal, and to increase its blood and flesh, is in exact proportion to its amount of the organic constituents of the blood, and of the materials necessary for their conversion into blood—viz., of alkalies, phosphates, and chlorides (common salt or chloride of potassium).

It is highly worthy of observation, and of great significance to agriculture, that the vegetable compounds containing sulphur and nitrogen, which we have designated as the organic constituents of the blood, are always accompanied, in the parts of the plants where they occur, with alkalies and with phosphates. The juice of potatoes and of beet contains vegetable albumen accompanied by salts of alkaline bases, and by soluble phosphate of magnesia; in the seeds of cereals and

---

in plants. In the latter the elements of water unite with the compound of carbon (carbonic acid); and oxalic acid, tartaric acid, &c., are thus produced, IN CONSEQUENCE OF A SEPARATION OF OXYGEN. But in the chemical process referred to, the elements of the water unite with the elements of oxalic and tartaric acids, &c., and they are reconverted into carbonic acid, IN CONSEQUENCE OF A SEPARATION OF HYDROGEN.

Without the development of any gas, tartaric and citric acids, in contact with alkali, even at a temperature of 400° F., are decomposed into oxalic and acetic acids. Anhydrous acetic acid contains carbon and the constituents of water, in exactly the same relative proportions as woody fibre (PELIGOT), which also yields acetic acid under similar circumstances.

These methods of decomposition have induced a distinguished French chemist to assume the existence of ready-formed oxalic acid in tartaric acid; certainly its elements are present, beside those of a second body, which, like sugar, gum, and woody fibre, may be viewed as a compound of carbon with water.

of peas, beans and lentils, there are alkaline phosphates and earthy salts.

The seeds and fruits, which are richest in the organic constituents of the blood, contain also the inorganic, such as the phosphates, in large quantity; other parts of plants, as the potato, and the various roots, which are proportionally so poor in the former ingredients, contain a much smaller quantity of the latter.

The contemporaneous occurrence of both these classes of compounds is so constant, that it would be difficult to trace a case of more intimate connection. It is extremely probable that the origin and formation of the organic constituents of the blood in the organism of plants is closely connected with the presence of phosphates. It must be supposed that the organic constituents of the blood will not be formed in a condition adapted for their conversion into blood, without the presence of alkalies and of phosphates, which are found constantly to accompany them; and this will be the case, even although carbonic acid, ammonia, and sulphates as a source of sulphur, be presented to them in the most abundant quantity. But, even on the assumption that they could be generated in the organism of the plant, without the action of these substances, we cannot suppose that they could be converted into blood and flesh in the body of the animals, when the mineral constituents of the blood were absent from the vegetable given as food.

But independently of these views, a rational farmer must endeavour to effect the purpose desired, and in doing so he must act exactly as if the presence of the inorganic constituents of blood (the alkalies and phosphates) were indispensable for the production of the organic constituents; for he must furnish to the plants all the materials necessary for the formation of the stem, leaves, and seeds. If he is desirous of making his land yield a maximum of blood and flesh, he must furnish to it in abundant quantity those constituents which the atmosphere cannot yield.*

* When fresh arterial blood is evaporated to dryness and incinerated, ashes are obtained which yield to water salts of an alkaline reaction, but not

Starch, sugar, and gum contain carbon and the elements of water, but they are never combined with alkalies, nor do they contain phosphates. We can suppose that two specimens of the same plant, when supplied with the same amount of mineral food, may yet form very unequal quantities of sugar and of starch; and that two equal surfaces of land prepared in exactly the same manner may bear two samples of barley, the one of which may yield half or double the weight of the seeds that the other does. But the excess of weight must depend upon the amount of unnitrogenous ingredients, and not on the constituents containing sulphur and nitrogen; for, if the same quantity of the inorganic constituents of blood be supplied to the soil, and if they enter into the plants, a corresponding quantity of the organic constituents of blood must

---

any alkaline carbonates, for no effervescence is occasioned by the addition of an acid. These ashes consist of variable quantities of:—

Phosphates of the alkalies,
Phosphate of lime,
Phosphate of magnesia,
Basic perphosphate of iron,
Common salt,
Sulphates of the alkalies.

The ashes of seeds contain:—

	Red Wheat. Fresenius.	White Wheat. Will.	Rye. Fresenius.	Peas. Will.	Beans (*vicia faba*). Buchner.
Phosphate of potash	36·51	52·98	52·91	52·78 }	68·59
Phosphate of soda	32·13	0·00	9·27	5·67 }	
Phosphate of lime	3·35	5·06	5·21	10·77	9·35
Phosphate of magnesia	19·61	32·96	26·91	13·78	19·11
Perphosphate of iron	3.04	0·67	1·88	2·46	
Sulphate of potash }	Traces.		2·98	9·09 }	1·84
Common salt }				3·96 }	
Silicate of potash			0·34		1·11
Silica	0·15	0·30			
Coal }	4·99	8·03	0·50		
Sand }					

In the above analyses, the phosphates of the alkalies in the peas and beans are contained and calculated as tribasic salts, those in the seeds of the cereals, as bibasic. The ashes of the seeds could not effervesce with acids, because they do not contain an alkaline carbonate; in this respect they are similar to the ashes of blood; and it may be observed that the salts in both are quite the same. If the ashes either of blood or of the seeds be exposed to air, they absorb carbonic acid and moisture; the tribasic phosphate becomes bibasic, and the third atom of alkali is converted into a carbonate.

be formed in the seeds, so that one cannot contain more than the other. A difference in the result can happen only when the one plant receives a less supply of nitrogen than the other, in a given time; for when there is a deficiency of ammonia, a corresponding quantity of the inorganic constituents of the blood is left unemployed.

When two species of plants are cultivated on a field of the same nature throughout, that species which generates the greatest quantity of the organic constituents of the blood (compounds containing sulphur and nitrogen), will remove from the soil the greatest amount of inorganic constituents (phosphates).

The one plant will exhaust a soil of these ingredients, but it may still remain in a good condition for a second kind of plant requiring a smaller quantity of phosphates, and may even be fertile for a third kind.

Hence it happens that the greater development of certain parts of plants, such as the seeds, which contain much more of the organic constituents of the blood than any of the other parts, exhaust and remove from the soil a much greater amount of phosphates than would be done by the culture of herbaceous plants, tubers, or roots, these being proportionally much poorer in the above ingredients. It is further evident, that two plants growing together on the same soil will share its ingredients between them, if they both require in equal periods equal quantities of the same constituents. The ingredients taken up into the organism of one of the plants cannot be used by the other.

If a given space of a soil (in surface and in depth) contains only a sufficient quantity of inorganic ingredients for the perfect development of ten plants, twenty specimens of the same plant, cultivated on this surface, could only obtain half their proper maturity; in such a case, there must be a difference in the number of their leaves, in the strength of their stems, and in the number of their seeds.

Two plants of the same kind growing in close vicinity must prove prejudicial to each other, if they find in the soil, or in the atmosphere surrounding them, less of the means of nourishment than they require for their perfect development. There is no plant more injurious to wheat than wheat itself, none more hurtful to the potato than another potato. Hence we actually find that the cultivated plants on the borders of a field are much more luxuriant, not only in strength, but in the number and richness of their seeds or tubers, than plants growing in the middle of the same field.

The same results must ensue in exactly a similar manner, when we cultivate on a soil the same plants for successive years, instead of, as in the former case, growing them too closely together. Let us assume that a certain soil contains a quantity of silicates and of phosphates sufficient for 1000 crops of wheat, then, after 1000 years, it must become sterile for this plant. If we were to remove the surface soil and bring up the subsoil to the surface, making what was formerly surface soil now the subsoil, we would procure a surface much less exhausted than the former, and this might suffice to supply a new series of crops, but its state of fertility would also have a limit.

A soil will naturally reach its point of exhaustion sooner the less rich it is in the mineral ingredients necessary as food for plants. But it is obvious, that we can restore the soil to its original state of fertility, by bringing it back to its former composition; that is, by returning to it the constituents removed by the various crops of plants.

Two plants may be cultivated side by side, or successively when they require unequal quantities of the same constituents at different times; they will grow luxuriantly without mutual injury, if they require for their developement DIFFERENT ingredients of the soil.

The experiments of Saussure, and of many other philosophers, have shown that the seeds of beans, of *Phaseolus vulgaris*, of peas, and of garden cresses, germinate and even grow to a certain extent in moist sand or moistened horse-hair; but when the mineral substances contained in the seeds no longer suffice for their further growth, then the plants begin to droop; they may even perhaps blossom, but they never bear seeds.

Wiegmann and Polstorf allowed plants of various kinds to vegetate in a white sand previously treated with aqua regia, and freed from the acid by careful washing*. Barley and oats growing in this sand, on being properly treated with water free from ammonia, reached a height of 1½ foot; they blossomed, but did not come to seed, and withered after blossoming. *Vetches* reached a height of ten inches, blossomed, and put out pods, but they did not contain any seeds.

Tobacco sowed in the sand, developed itself at first in the usual way, but from June to October the plants reached the height only of five inches; they had only four leaves and no stem.

The analyses of the ashes of these plants, and also the analyses of their seeds, proved that although this sterile sand contained such a small quantity of potash and of soluble constituents, still it had yielded a certain quantity of these, and on this quantity the growth of the leaves and of the stem depended: but it was impossible that the plants could come to seed, because the constituents necessary for the formation of the seeds were entirely absent.

Phosphoric acid was detected in the ashes of most of the plants growing in this sand, but its quantity corresponded only to that introduced to the soil in the seeds themselves. In the ashes of the tobacco plant, which it is known contains scarcely an appreciable quantity of phosphoric acid, not a trace of that acid could be detected.

What theory distinctly indicated as the cause of the sterility of this sand, the experiments of Wiegmann and Polstorf completely established. They took the same sand and prepared an artificial soil by the addition of salts manufactured in a laboratory (see Appendix); they then sowed in this soil the same plants, and saw that they flourished in the

* This sand contained in 1000 parts:—(PREISSCHRIFT, p. 28.)

Silica	979·00
Potash	3·20
Alumina	8·76
Peroxide of iron	3·15
Lime	4·84
Magnesia	0·09

most luxuriant manner. The tobacco became three feet in height, and put forth many leaves; on the 25th of June it began to blossom, and on the 10th of August obtained seeds, from which on the 8th of September ripe seed capsules with complete seeds were taken. In exactly the same way, barley, oats, buck-wheat, and clover grew luxuriantly, blossomed and yielded ripe and perfect seeds.

It is quite certain, that the growth of these plants in the formerly sterile sand, depended upon the salts added to it. An equal fertility for all plants is given to this artificial soil by the addition of certain substances which are absolutely necessary for the life of the plants, because they are present in the matured plant, and in its stem, leaves, and seeds.

Thus we are in a position to give to the most sterile soil a state of the greatest fertility for all plants, if we furnish to it the constituents which they require for their growth. It would, indeed, neither repay the labour nor the expense to render fertile on those principles an absolutely sterile soil. But in our ordinary arable soils, which contain already many of these constituents, it suffices to supply the absent ingredients, or to increase those which are in deficient quantity. At the same time, by the art of farming, the soil must be put into a proper physical state, by which it is rendered accessible to moisture and to rain, and is fitted to enable the plants to appropriate these ingredients of the soil.

Different genera of plants require for their growth and perfect maturity, either the same inorganic means of nourishment, although in unequal quantities and at different times, or they require different mineral ingredients. It is owing to the difference of the food necessary for the growth of plants, and which must be furnished by the soil, that different kinds of plants exert mutual injury when growing together, and that others, on the contrary, grow together with great luxuriance.

Very little difference is observed in the composition of the ashes of the same plants, even although they have grown on different soils. Silica and potash form invariable constituents in the straw of the Gramineæ; and, in their seeds, there is always present phosphate of potash and phosphate of magnesia. A large quantity of lime occurs in the straw of peas and in clover. We know, further, that in certain kinds of plants, the potash is replaced by soda, and the lime by magnesia.

It has been shown by the experiments of Boussingault,* that the five following crops grown in succession on an equal surface of the same field once manured, removed from the soil:—

			Ingredients of the soil.
1 Year crop of	Potatoes (tubers without herb)	. .	246·8 lbs.
2 " "	Wheat (straw and corn)	. . . .	371·0 "
3 " "	Clover	. . . .	620·0 "
4 " "	Wheat †	. . . .	488·0 "
	Fallow Turnips	. . . . .	108·8 "
5 " "	Oats (corn and straw)	. . . .	215·0 "
By a crop of	Beetroot ‡ (roots without leaves)	. .	399·6 "
" "	Peas (peas and straw)	. . . .	618·0 "
" "	Rye	. . . .	284·6 "
" "	Helianthus tuberosus	. . . .	660·0 "

These numbers express the quantities of inorganic substances removed from the same soil by different plants, and carried away with the crops. They, therefore, prove that different plants take up into their organism, unequal weights of these ingredients of the soil. It is shown by a further consideration of the composition of these ashes, that they differ essentially from each other with respect to their quality.

One thousand parts of beet, turnips, or potatoes, yield by incineration 90 parts of ashes; the latter are easily fusible, and contain a large quantity of carbonate of potash, and of salts with alkaline bases. Of these 90 parts, 75 parts are soluble in cold water.

---

* Annales de Chimie et de Physique, t. i. p. 242.

† On a system of alternate crops.

‡ In the above five-yearly rotation, wheat was introduced twice. In the second year the crop of wheat removed from the soil 371 lbs. and in the fourth year 458 lbs. of inorganic substances. This difference depends upon the unequal quantities of straw and corn in the crops of the two years. The weight of the straw and corn of one year was 8790 lbs., in the other year it amounted to 10,858 lbs. The relative proportion of their ashes is exactly the same.

Two thousand parts of dry fern yield also 90 parts of ashes; but of these 90 parts none, or only a trace, is soluble in water (BERTHIER).

The ashes of wheat, barley, pea and bean straw, differ in like manner in their composition. Equal quantities of their ashes contain very unequal amounts of ingredients soluble in water. There are certain ashes of plants wholly soluble in water; others are only partially soluble, while certain kinds yield only traces of soluble ingredients.

When the parts of the ashes INSOLUBLE in water are treated with an acid (muriatic acid), this residue, in the case of many plants, is quite soluble in the acids (as for instance the ashes of beet, turnips, and potatoes); with other plants, only half the residue dissolves, the other half resisting the solvent action of the acid; while in the case of certain plants only a third or even less of the residue is taken up by the acid.

The parts of the ashes soluble in cold water consist entirely of SALTS WITH ALKALINE BASES (POTASH AND SODA). The ingredients soluble in acids are SALTS OF LIME AND MAGNESIA; and the residue insoluble in acids consists of SILICA.

These ingredients being so different in their behaviour to water and to acids, afford us a means of classing the cultivated plants according to their unequal quantity of these constituents. Thus POTASH PLANTS are those the ashes of which contain more than half their weight of soluble alkaline salts; we may designate as LIME PLANTS, and as SILICA PLANTS, those in which lime and silica respectively predominate. The ingredients thus indicated are those which form the distinguishing characteristics of the plants which require an abundant supply of them for their growth.

The POTASH PLANTS include the chenopodia, arrach, wormwood, &c.; and amongst cultivated plants, the beet, mangel-wurzel, turnip, and maize. The LIME PLANTS comprehend the lichens (containing oxalate of lime), the cactus (containing crystallised tartrate of lime), clover, beans, peas, and tobacco. SILICA PLANTS include wheat, oats, rye, and barley.*

		Salts of Potash and Soda.	Salts of Lime and Magnesia.	Silica.
Silica Plants.	Oatstraw with seeds (1)	34.00	4.00	62.00
	Wheat-straw (2)	22·00	7·20	61·05
	Barley-straw with seeds (1)	19.00	25·70	55·03
	Rye-straw (3)	18·65	16·52	63·89
Lime Plants.	Tobacco (Havannah), (4)	24·34	67·44	8·30
	,, (Dutch), (4)	23·07	62·23	15·25
	,, (grown in an artificial soil), (1)	29·00	59·00	12·00
	Pea-straw (4)	27·82	63·74	7·81
	Potato-herb (5)	4·20	59·40	36·40
	Meadow-clover (1)	39·20	56·00	4·90
Potash Plants.	Maize-straw (2)	71·00	6·50	18·00
	Turnips	81·60	18·40	
	Beetroot (6)	88·00	12·00	
	Potatoes (tubers), (6)	85·81	14·19	
	Helianthus tuberosus (7)	84·30	15·70 †	

This classification, however, is obviously only a very general one, and permits division into a great number of subordinate classes; particularly with respect to those plants in which the alkalies may be replaced by lime and magnesia. As far as we are authorised to judge by our present knowledge, a substitution of soda for potash takes place in our cultivated plants; but it has not yet been observed, that in these plants the alkalies can be replaced by lime.

The potato plant belongs to the LIME PLANTS, as far as

* The ashes of good meadow hay (consisting of a mixture of the ashes of potash, lime, and silica plants,) gave in 100 parts—(HAIDLEN):

Silica	60·1
Phosphate of lime	16·1
Phosphate of peroxide of iron	5·0
Lime	2.7
Magnesia	8·6
Gypsum	1·2
Sulphate of potash	2·2
Chloride of potassium	1·3
Carbonate of soda	2·0
Loss	0·8

† In the above analysis, the figures represent the chemists as under:—

(1) Wiegmann and Polstorf.
(2) Saussure.
(3) Fresenius.
(4) Hertwig.
(5) Berthier and Braconnot.
(6) Hruschauer.
(7) Braconnot.

regards the ingredients of its leaves, but its tubers (which contain only traces of lime) belong to the class of POTASH PLANTS. With reference to the siliceous plants, this difference of their parts is very marked.

Barley must be viewed as a lime plant, when compared with oats or with wheat, in reference to their ingredients soluble in muriatic acid; but it would be considered as a siliceous plant, if viewed only in reference to its amount of silica. Beet-root contains phosphate of magnesia, and only traces of lime; while the turnip contains phosphate of lime and only traces of magnesia.

When we take into consideration the quantity of ashes, and their known composition, we are enabled to calculate with ease, not only the particular ingredients removed from a soil, but also the degree in which it is exhausted of these by certain species of plants belonging to the POTASH, LIME, or SILICEOUS plants. This will be rendered obvious by the following examples.

A soil, consisting of four Hessian acres, has removed from it by a crop of—

		Salts of potash and soda. lbs.	lbs.	Salts of lime, magnesia, and peroxide of iron. lbs.	lbs.	Silica. lbs.
Wheat	In straw .	95·31	130·51	34·75	67·55	260·05
	In corn .	35·20		32·80		
Peas .	In straw	154·40	198·42	354·80	371·46	46·60
	In corn	44·02		16·68		
Rye .	In straw .	40·73	82·78	36·00	57·82	139·77
	In corn .	42·05		21·82		
Beet-root, without leaves . .			361·00		37·84	
Helianthus tuberosus . . .			556·00		104·00	

The same surface is deprived by these crops of the following quantity of phosphates *—

* In the above numbers we have not an exact, but an approximate, proportion of the ingredients of the soil removed by the various crops. The analyses of the ashes have been used, as far as they are already made and known. The analysis of the ashes of the seeds and of the straw of wheat is by Saussure; that of pea-straw by Hertwig; that of peas by Dr. Will; that of the ashes of the straw and seeds of rye by Dr. Fresenius, of the beet-root by Hruschauer, and of Helianthus tuberosus by Braconnot. Exact and trustworthy results can only be obtained by estimating the ashes of the crops grown on a given surface, and by subjecting these ashes to analysis; and not as in the above cases, in which the analyses were made upon the ashes of plants grown under

Peas.*	Wheat.	Rye.	Helianthus Tuberosus.	Turnips.†
117	112·43	77·05	122	37·84

According to the preceding views, plants must obtain from the soil certain constituents, in order to enable them to reach perfect maturity—that is, to enable them to bear blossoms and fruit. The growth of a plant is very limited in pure water, in pure silica, or in a soil from which these ingredients are absent. If there be not present in the soil alkalies, lime, and magnesia, the stem, leaves, and blossoms of the plants can only be formed in proportion to the quantity of these substances existing as a provision in the seed. When phosphates are wanting, the seeds cannot be formed.

The more quickly a plant grows, the more rapidly do its leaves increase in number and in size, and therefore the supply of alkaline bases must be greater in a given time.

As all plants remove from the soil certain constituents, it is quite obvious that none of them can render it either richer or more fertile for a plant of another kind. If we convert into arable land a soil which has grown for centuries wood, or a vegetation which has not changed, and if we spread over this soil the ashes of the wood and of the bushes, we have added to that contained in the soil a new provision of alkaline bases, and of phosphates, which may suffice for a hundred or more crops of certain plants. If the soil contains silicates

different circumstances and in different situations. Boussingault, for example, obtained from pea-straw (from a crop heavily manured) 11·2 per cent. of ashes; Saussure obtained only 8 per cent. (in straw with seeds), and Hertwig only 5 per cent. These numbers change the absolute quantities, but have little or no influence on the relative proportions.

The analyses by Sprengel of the ashes of various plants are quite inexact, and do not merit the slightest confidence. The ashes of the seeds of wheat, of peas, of beans, rye, &c., consist of phosphates, without any mixture of carbonic acid; these ashes do not contain silica. But Sprengel finds in peas 18 per cent. and in rye 15 per cent. of silica. The ashes of the seeds of rye contain 48 per cent., those of peas 34·23 per cent., of anhydrous phosphoric acid; but Sprengel finds in peas only 4 per cent., in rye 8 per cent., of phosphoric acid. It is worthy of observation, that all the bases in the ashes of peas are present as tribasic phosphates, while in the ashes of rye and of wheat, they exist as bibasic salts.

* Heavily manured. † Heavily manured.

susceptible of desintegration, there will also be present in it soluble silicate of potash or soda, which is necessary for the rendering mature the stem of the siliceous plants; and, with the phosphates already present, we have in such a soil all the conditions necessary to sustain uninterrupted crops of corn for a series of years.

If this soil be either deficient or wanting in the silicates, but yet contain an abundant quantity of salts of lime and of phosphates, we will be enabled to obtain from it, for a number of years, successive crops of tobacco, peas, and beans.

But, if none of the ingredients furnished to these plants be again returned to the soil, a time must come when it can no longer furnish these constituents to a new vegetation; when it must become completely exhausted, and be at last quite sterile, even for weeds.

This state of sterility will take place earlier for one kind of plant than for another, according to the unequal quantity of the different ingredients of the soil. If the soil is poor in phosphates, but rich in silicates, it will be exhausted sooner by the cultivation of wheat than by that of oats or of barley, because a greater quantity of phosphates is removed in the seeds and straw of one crop of wheat, than would be removed in three or four crops of barley or of oats.* But if this soil be deficient in lime, the barley will grow upon it very imperfectly.

It is owing to the deficiency of these salts, so indispensable to the formation of the seeds, that it happens, however abundant may be the quantity of silicates, that in one year we may obtain nine times, in a second thrice, in a third twice as much corn as may grow on the same soil in another year.

In a soil rich in alkaline silicates, but containing only a limited supply of phosphates, the period of its exhaustion for these salts will be delayed if we alternate with the wheat plants which we cut before they have come to seed; or, what

* The weight of the ashes of a crop of the seeds of wheat is to that in a crop of oats as 34 : 42·6, the phosphates contained in them as 26 : 10; the phosphates of the straw not being included in the calculation.

is the same thing, with plants that remove from the soil only a small quantity of phosphates. If we cultivate on this soil peas or beans, these plants will leave, after the removal of the crop, a quantity of silica in a soluble state sufficient for a succeeding crop of wheat; but they will exhaust the soil of phosphates quite as much as wheat itself, because the seeds of both require for their maturity nearly an equal quantity of these salts.

We are enabled to delay the period of exhaustion of a soil of phosphates by adopting a rotation, in which potatoes, tobacco, or clover, are made to alternate with a white crop. The seeds of the plants now named are small, and contain proportionally only minute quantities of phosphates; their roots and leaves, also, do not require much of these salts for their maturity. But it must be remembered, at the same time, that each of these have rendered the soil poorer, by a certain quantity of phosphates. By the rotation adopted, we have deferred the period of exhaustion, and have obtained in the crops a greater weight of sugar, starch, &c., but we have not acquired any larger quantity of the constituents of the blood, or of the only substances which can be considered as properly the nutritious parts of plants. When the soil is deficient in salts of lime, tobacco, clover, and peas will not flourish; whilst, under the same conditions, the growth of beet-root or turnips will not be impeded, if the soil, at the same time, contain a proper quantity of alkalies.

When a soil contains silicates not prone to disintegrate, it may be able, in its natural state, to liberate by the influence of the atmosphere, in three or four years, only as much silica as suffices for one crop of wheat. In this case, such a crop can only be grown on it in a three or four years' rotation, assuming that the phosphates necessary for the formation of the seeds exist in the soil in sufficient quantity. But we can shorten this period by working well the soil, and by increasing its surface, so as to make it more accessible to the action of the air and moisture, in order to disintegrate the soil, and to procure a greater provision of soluble silicates.

The decomposition of the silicates may also be accelerated by the use of burnt lime; but it is certain that, although all these means may enable us to ensure rich crops for a certain period, they induce, at the same time, an earlier exhaustion of the soil, and impair its natural state of fertility.

If the proportion of alkali and of silica liberated from the soil in the course of three or four years be sufficient only for one crop of wheat, we cannot in the interval, without injury to this crop, cultivate on the same soil any other plant; because the alkali necessary for the growth of the latter cannot be applied to the use of the wheat.

By examining the known proportion of alkali and of silica liberated by the disintegration of the silicates in their conversion into clay, and by the weathering of the latter itself,* we find that, for a given quantity of silica rendered soluble, a much larger amount of alkali is liberated than corresponds to the proportion in which both are taken up into the straw of the cereals.

During the time of fallow, which in this case must elapse before two crops of wheat can be obtained, we may employ the excess of alkalies in the culture of other plants requiring salts with alkaline bases without silica. Between these crops, therefore, we may grow mangel wurzel, or even potatoes, if we remove only the tubers of the latter, and allow the plant itself, which contains much silica, to remain on the field.

In the preceding remarks, we have considered the changes in the nature and composition of a field on which a rotation of cultivated plants has been carried on for a series of years. If this field contain an ordinary proportion of alkaline silicates, clay, lime, and magnesia, it will possess an almost inexhaustible provision of alkalies, alkaline earths, and silica; with this difference, however, that they are not all in a fit state to be used by the plant at the same time. By the mechanical operations of the farm, and by chemical means

* One equivalent of silica is liberated for every equivalent of potash separated from the constituents of an equivalent of felspar. In the straw of wheat, oats, and rye, for 10 eq. of silica, there is only 1, or, at the most, 2 eq in combination with alkalies.

(by the use of lime, &c.), we may shorten the time in which these obtain a form fitted for the vital functions of the plant; but these matters do not suffice for its complete maturity.

When phosphates and sulphates are absent from the soil, the plants growing on it cannot form seeds, because all seeds, without exception, contain compounds in which phosphoric acid and sulphur are invariable constituents. Although all the other ingredients of plants be present in superabundance, the soil will become completely sterile, when the period arrives at which it can no longer furnish phosphates or sulphates to a new vegetation.

We must suppose that, for the formation of the stem and herb, for the fixation of carbon, and for the production of sugar, starch, and woody fibre, a certain amount of alkalies (in the case of the potash-plants), or an equivalent of lime (in the case of the lime plants) is necessary. But we must bear in mind, at the same time, that the constituents of blood can be formed in the organism of the plant only in quantity corresponding to that of the phosphates, however abundantly ammonia or carbonic acid may be supplied. The production of the constituents of the juice containing sulphur and nitrogen is inseparably connected with the presence of these salts.

Every soil upon which a weed attains maturity is fitted for culture if that weed yields, on incineration, alkaline ashes.

The alkalies of these ashes arise from silicates, so that in addition to the alkalies, soluble silica must exist in the same soil. Such a soil may contain a quantity of phosphates of lime and magnesia sufficient for potatoes and turnips, without on that account being rich enough for crops of wheat.

These considerations must show the great importance which ought to be attached to phosphates in the practice of agriculture. These salts are present in the soil only in small quantity, and therefore the greater attention should be paid to prevent its exhaustion.

In the limited but enormous extent of the ocean, whole worlds of plants and animals succeed each other. A generation of these animals obtain all their elements from plants, and the constituents of the organs of the animal after death assume their original form, and serve for the nourishment of a new generation of animals.

The oxygen employed by the marine animals in the process of respiration, and removed from the air, dissolved in the water, (this air contains from 32 to 33 volume per cent. of oxygen, while atmospheric air contains only 21 per cent.,) is restored again to the water, by the vital processes of marine plants. In the products of the putrefaction of the carcasses of the dead animals, their carbon is converted into carbonic acid, their hydrogen into water, and their nitrogen assumes again the form of ammonia.

Thus, we observe that in the sea, a perpetual circulation takes place, without the accession or removal of an element, and this circulation is unlimited in its duration, although it may be in its extent, by the finite quantity of nourishment contained in plants in a limited space.

With respect to marine plants, there cannot be any discussion as to their receiving food by their roots in the form of humus. What nourishment indeed can the thick roots of the giant sea-weed draw from a naked rock, the surface of which does not suffer the slightest change—a plant which reaches a height of 360 feet (Cook), and one of which, with its leaves and twigs, affords nourishment to thousands of marine animals. These plants require obviously only a fastening point in order to prevent a change of place, or an arrangement by which their small specific weight is compensated; they live in a medium which conveys the necessary nourishment to all their parts. Sea-water does not only contain carbonic acid and ammonia, but also phosphates, and earthy and alkaline carbonates, salts invariably found in the ashes of marine plants, and indispensable for their growth.

All our knowledge tends to prove that the conditions necessary for the existence and duration of marine plants are the same as those upon which the existence of terrestrial plants depends.

But terrestrial plants do not live like marine plants, in a medium containing all their elements, and surrounding every part of their organism; but their existence depends upon two media, the one of which, the SOIL, contains constituents which are absent from the other, the ATMOSPHERE.

How is it possible, we may well ask, that there ever could have been a doubt as to the part which the constituents of the soil took in the growth of the vegetable world? Yet, there was a time when it was considered that the mineral constituents of plants were not necessary and essential to their existence!

The same circle exists on the surface of the earth as in the sea; there is an unceasing change—a perpetual destruction and re-establishment of equilibrium. Practice in agriculture has taught us that the amount of vegetable matters on a given surface increases with the supply of certain substances, WHICH WERE ORIGINAL CONSTITUENTS OF THE SAME SURFACE OF THE SOIL, and had been removed from it by means of plants. The excrements of men and of animals arise from plants; they are exactly the materials which, during the life of the animal, or after its death, obtain again the same form that they possessed as constituents of the soil.

We know that the atmosphere does not contain these materials, and that it does not replace them; we know further that, by their removal from the soil, an inequality of production is occasioned, and, finally, even a want of fertility; but that, by the restoration of these materials, the fertility may be sustained, and even increased.

After so many striking proofs respecting the origin of the constituents of plants and of animals, and of the use of alkalies, of phosphates, and of lime, can we entertain the slightest doubt of the principles upon which a rational system of agriculture must depend?

Does the art of farming, then, depend upon anything else than the restoration of a disturbed equilibrium?

Is it conceivable that a rich fertile land, with a flourishing trade, which has for centuries exported the products of its soil in the form of cattle and of corn, can retain its fertility, if the same trade do not restore to its land, in the form of manure, the constituents abstracted from it, and which cannot be replaced by the atmosphere? In such a case, would not the same fate await this land as that which befel Virginia, upon the soil of which wheat and tobacco can no longer be cultivated?

In the large towns of England, the products of English as well as of foreign agriculture are consumed; and to supply this great consumption, the constituents of the soil necessary to the plants are removed with them, from an immense surface of land, to which they are not again returned. The domestic arrangements peculiar to the English render it difficult, perhaps even impossible, to collect the immense quantity of phosphates (the most important ingredients of the soil, although present in it in small quantity) which are daily sent into the rivers in the form of urine and of solid excrements. We have seen upon the fields of England exhausted of their phosphates, the most beneficial effects produced upon the crops by the introduction of bones (phosphate of lime) from the Continent; in some cases, the crops on the soil were doubled by the use of this manure, as if by a charm.

But if this exportation of bones be continued on the same scale as at present, the German soil will become gradually exhausted; and the loss will be perceived to be greater than at first is apparent, when it is considered that a single pound weight of bones contains as much phosphoric acid as a whole hundred weight of corn.

Thousands of hundred weights of phosphates flow annually into the sea with the Thames, and with other of the British rivers.

Thousands of hundred weights of the same materials, arising from the sea, annually flow back again into that land in the form of guano.

In the alchemistical era, the imperfect knowledge of the properties of matter gave rise to the supposition, that metals, uch as gold, could be developed by seeds. Crystalline forms, and the ramifications which they assume, appeared to alchemists to be the leaves and branches of metallic plants, and their great endeavours were to find an earth fitted for the peculiar growth and development of their seeds. Without there being any apparent nourishment given to a seed, it was seen to grow to a plant, which put forth blossoms and seeds. This led to the belief, that could the seeds of metals be procured, similar hopes of their reaching maturity might be entertained.

Such ideas could only belong to a time when scarcely anything was known of the nature of the atmosphere, and when there was not a conception of the part taken by the earth and air in the vital processes of plants.

The chemistry of the present day exhibits the elements of water; it can even prepare this water with all its properties from their elements; but it cannot manufacture these elements —it can only prepare them from water. The newly-formed water, which has been artificially prepared, was water previously to the separation of its elements.

Many of our farmers, like to these alchemists of old, in searching after the philosopher's stone, look now to find the wonderful seeds; for they expect that their land should bear a hundred-fold, without supplying to it any food, even although this land is scarcely rich enough to bear its own indigenous plants!

The experience of centuries or of thousands of years is not sufficient to protect them from the new fallacies which are constantly arising; the power of resisting the effects of credulity or superstition can only be obtained from a knowledge of true scientific principles.

In the first stage of the philosophy of nature, it was supposed that the organic kingdom was developed from water alone; then it was considered that both water and air were necessary; and now we know, with the greatest certainty, that the soil furnishes other important conditions, which must be added to the former, otherwise plants will not obtain the power of propagating and of multiplying themselves.

The quantity of food for plants in the atmosphere is limited, but still it must be sufficient to cover the surface of the earth with a rich vegetation.

In the tropics, and in those regions of the earth where a favourable soil, moisture, light, and an elevated temperature—the usual conditions of fertility—are combined, the vegetation is scarcely confined by the space on which it grows; there, when the soil is deficient, the bark and branches of dead plants soon form soil for succeeding ones. It is obvious, therefore, that there is no deficiency of atmospheric food for the plants of these regions, and there can be none for our own cultivated plants.

The constant movement to which the atmosphere is subjected, causes an equal distribution of the gaseous food necessary for the growth of plants, so that the tropics do not contain more of it than the cold zones; and yet, how different appears to be the power of production of equal surfaces of land in these regions!

All plants of tropical regions, such as the sugar-cane, the palms bearing wax and oil, contain, in comparison with our own cultivated plants, only a small quantity of the constituents of blood necessary for the nourishment of animals. The produce in tubers, of an acre of potatoes, growing, as in Chili, to the height of a tall bush, would scarcely suffice to prolong the life of an Irish family for a day (Darwin). The nutritious plants which are the objects of culture, are only a means for generating the necessary constituents of the blood. If the ingredients of the soil indispensable to their formation be absent from it, the constituents of the blood cannot be formed in the plants, although it is possible that wood, sugar, or starch, might be produced under such circumstances. If we desire to produce from a given surface more of these constituents of the blood, than the plants growing on it could receive from the atmosphere or from the soil in their natural wild and and normal condition, we must procure an artificial atmosphere, and we must add to the soil the ingredients in which it is deficient.

Very unequal quantities of nourishment must be furnished to different plants in a given time, in order to procure a free and unimpeded growth. On arid sands, simple calcareous soils, or on naked rocks, few plants flourish, and those that do are generally perennial. These, growing slowly, require only small quantities of mineral ingredients, so that soils sterile for other kinds of plants are still able to furnish to them mineral ingredients in sufficient quantity. The annuals, particularly summer plants, reach complete maturity in comparatively a short time, so that they do not flourish on a soil poor in the mineral substances necessary for their growth.

The food contained in the atmosphere does not suffice to enable these plants to obtain their maximum of size in the short period of their life. If the object of culture is to be attained, there must be present in the soil itself an artificial atmosphere of carbonic acid and ammonia, and this excess of nourishment, which the leaves cannot get, must be conveyed to corresponding organs existing in the soil.

But the ammonia with carbonic acid does not suffice to enable itself to become a constituent of a plant fit for the food of animals. Albumen cannot be formed without alkalies, and vegetable fibrin and casein cannot be produced without the aid of phosphoric acid and of earthy salts. We know that phosphoric acid is indispensable for the production of the seeds of our cereals and culinary vegetables, although the same acid is found in large quantity in an excrementitious form in the bark of woody plants.

How very different, in comparison with summer plants, are the characters of evergreens, of mosses, ferns, and pines! During every part of the day, both in summer and in winter, they absorb by their leaves carbonic acid, which the sterile soils cannot yield; their fleshy leather-like leaves retain with great tenacity the water absorbed, and lose very little of it by evaporation, in comparison with other plants. And yet how very small is the quantity of mineral substances which they abstract from the soil during the whole year of almost perpetual growth, when we compare it with the quantity removed from the soil in three months by a crop of wheat of equal weight!

It follows, then, from the preceding observations, that the advantage of the alternate system of husbandry consists in the fact that the cultivated plants abstract from the soil unequal quantities of certain nutritious matters.

A fertile soil must contain in sufficient quantity, and in a form adapted for assimilation, all the inorganic materials indispensable for the growth of plants.

A field artificially prepared for culture, contains a certain amount of these ingredients, and also of ammoniacal salts and decaying vegetable matter. The system of rotation adopted on such a field is, that a potash-plant (turnips or potatoes) is succeeded by a silica plant, and the latter is followed by a lime-plant. All these plants require phosphates and alkalies—the potash-plant requiring the largest quantity of the latter and the smallest quantity of the former. The silica plants require, in addition to the soluble silica left by the potash plants, a considerable amount of phosphates; and the succeeding lime plants (peas or clover) are capable of exhausting the soil of this important ingredient to such an extent, that there is only sufficient left to enable a crop of oats or of rye to form their seeds.

The number of crops which may be obtained from the soil depends upon the quantity of the phosphates, of the alkalies, or of lime, and the salts of magnesia existing in it.

The existing provision may suffice for two successive crops of a potash or of a lime plant, or for three or for more crops of a silica plant, or it may suffice for five or seven crops of all taken together; but after this time, all the mineral substances removed from the field in the form of fruit, herbs, or straw, must again be returned to it; the equilibrium must be restored, if we desire to retain the field in its original state of fertility.

This is effected by means of MANURE. It may be assumed that the soil receives again, in the roots and stubble of the cereals, or in the fallen leaves of trees, as much carbon as its humus yielded in the form of carbonic acid at the commencement of a new vegetation; in like manner, the herb of the potato and the roots of the clover remain in the soil. The remains of these plants enter into decay during winter, and thus furnish to the young plant a new source of carbonic acid. The soil, therefore, is not exhausted of its humus by the cultivation of these plants.

It may also be deduced from theoretical considerations, that the soil receives during the life of the plants as much, or perhaps even more, of carbonaceous materials as it yields to them; and that the soil is enriched in these matters by the process of secretion proceeding at the surface of the fibres of the root; the matters thus secreted enter into decay during winter, and pass into humus.

Physiologists differ in their opinions with regard to the processes of secretion and excretion by plants; some affirm that these processes do exist, while others deny their existence; so that, at this moment, the opinions are divided on the subject. But still, no one denies that the oxygen separated by the leaves and green parts of plants ought to be considered as an excrement. In the act of vital activity, the plants assimilate the carbon of carbonic acid, and the hydrogen of water, making them constituents of their organs, while they separate the oxygen with which these elements were combined.

In the blossoms we find volatile oils, compounds rich in carbon and in hydrogen, and which are not further employed in any of the vital processes: out of the bark we see flowing resin, balsam, and gum; and out of the leaves, sugar and mucilaginous substances.

Oxygen is not separated from the surface of the bark, roots, or other parts that are not green; but, on the contrary, these parts separate substances rich in carbon which have been generated by the vital processes of the plant, but have not experienced any further change.

When we compare the barks of the fir,* pine, beech, or oak, with their sap and wood, we find that they differ essentially from each other, both in their composition and characters.

True wood yields only one-fourth to two per cent. of ashes, while the bark of the oak, fir, willow, and beech, give 6, 10, to 15 times more. The ashes of wood and of the bark have a very different composition. The inorganic ingredients of the bark are obviously inorganic substances expelled by the living organism.

The same reasoning holds good in the case of the organic substances as it does in the case of the bark. The bark of the cork-tree contains nearly half its weight of fatty, or of fatty substances, which we also find present, although in smaller proportion, in the bark of firs and pines. The solid material (insoluble in alcohol or ether) of these barks is entirely different from woody fibre. The barks of firs and pines are completely soluble in potash leys, forming a liquid of a dark brown colour, which yields, on the addition of an acid, a precipitate strongly resembling the substance called humic acid. But wood is not attacked by potash ley.

These barks are in so far true excrements, that they arise from living plants, and play no further part in their vital

---

		* Ashes of Wood of the Fir. HERTWIG.	Ashes of the Bark of the Fir. HERTWIG.	
		1000 wood gave 3·28 ashes.	1000 bark gave 17·85 ashes.	
Soluble Salts 18·72.	Carbonate of soda	7·42		
	Carbonate of potash	11·30	Soluble salts 2·95	
	Chloride of sodium	Traces		
	Sulphate of potash	Traces		
Insoluble Salts.	Carbonate of lime	50·94	64·98	Insoluble Salts 97·05.
	Magnesia	5·60	0·93	
	Phosphate of lime	3·43	5·03	
	,, magnesia	2·90	4·18	
	,, manganese	Traces		
	,, peroxide of iron	1·04	1·04	
	,, alumina	1·75	2·42	
	Silica	13·37	17·28	
	Loss	2·26	1·79	
		100·00	100·00	

functions; they may even be removed from them, without thereby endangering their existence. It is known that certain trees throw off annually their barks: this circumstance, viewed in its proper light, shows that, during the formation of certain products formed by the vital processes, materials arise which are incapable of experiencing a further change.

There is every reason to believe that this separation takes place over the whole surface; it is observed not only on the stem, but also on the smallest twigs; and hence we must conclude that the same excretory process goes on in the roots.

When a branch of a willow is allowed to vegetate in rain water, the latter assumes gradually a dark-brown colour. The same phenomenon is observed in bulbous plants (such as hyacinths) allowed to grow in pure water. It therefore cannot be denied that excrements are actually separated by plants, although it is very possible that they do not all separate them in the same degree.

It is generally admitted by farmers, as the result of experience, that a soil is enriched in organic matters by the culture of perennial plants, such as sainfoin and lucerne, which are distinguished for the extensive ramification of their roots and strong growth of their leaves; the cause of their enriching power will perhaps be explained from the above remarks.

We cannot effect the formation of ammonia on our cultivated land, but it is in our power to obtain an artificial production of humus: this must be viewed as one of the objects of a system of rotation, and as a second cause of the advantage arising from it.

By sowing a field with a fallow crop, such as clover, rye, lupins, buck-wheat, &c., and by ploughing and incorporating with the soil, the plants, when they have nearly come to blossom, we procure to the young plants of a new crop sown on the field a maximum of nourishment and an atmosphere of carbonic acid, in consequence of the decay of the preceding crops. All the nitrogen drawn from the atmosphere by the preceding plants, and all the alkalies and phosphates received from the soil, now serve to cause a luxuriant growth to the plants which succeed.

CHAPTER XII.

ON MANURE.

In order to obtain clear ideas on the value and action of animal excrements, it is most important to bear in mind their origin. It is well known, that, when a man or an animal is deprived of food, he becomes emaciated, and his body diminishes in weight from day to day. This emaciation becomes visible after a few days; and in the case of persons who are starved to death, their fat and muscular substance disappears, their body becomes empty of blood, and at last nothing remains except skin and bones.

On the other hand, the weight of the body does not alter, even though supplied with sufficient food; for in the body of a healthy man there is neither a marked increase nor diminution of weight from one twenty-four hours to another.

These phenomena prove with certainty that a change proceeds in the organism of an animal during every moment of its life; and a part of the living substance of the body passes out from it in a state more or less changed. The weight of the body, therefore, would decrease continually, if the parts separated or changed were not again prepared and replaced.

The restoration and replacement of the original weight is effected by means of FOOD.

A man or an animal consumes daily a certain number of ounces or of pounds of bread, flesh, or other nutritive substances, so that in a year he consumes an amount of food of a much greater weight than that of his own body. He takes in the food a certain quantity of carbon, hydrogen, nitrogen, and sulphur, as well as a very considerable quantity of mineral ingredients, which we have learned to know as the ashes of food.

What, it may be asked, has become of all these constituents of the food, to what purposes have they been applied, and in what form have they been expelled from the body? Carbon and hydrogen have been furnished to the body, and yet the weight of the body, with respect to these elements, has not increased; the body has received in the food a quantity of alkalies and of phosphates, but still the amount of these substances in our body has not been rendered greater.

These questions are easily solved, when it is considered that the food does not supply the only conditions necessary for the support of the vital processes, for there are other conditions which distinguish animals essentially from plants.

The life of an animal is essentially connected with a continual introduction into its system of the oxygen contained in air. Without air and oxygen, animals cannot exist. In the process of respiration, a certain quantity of oxygen is introduced into the blood by means of the lungs. The air which we respire contains this oxygen, and yields it to the constituents of the blood; the blood of an adult man removes from the air, at each respiration, about two cubic inches of oxygen. A man consumes, in 24 hours, from 10 to 14 ounces of oxygen—in a year, hundreds of pounds; what then becomes of this oxygen? We take into our bodies pounds weight of food and pounds weight of oxygen, and nevertheless the weight of our body either does not increase to any sensible extent, or it does so in a much smaller proportion than corresponds to the food; in certain individuals (in old age), it experiences a continued reduction.

It must be obvious, that this phenomenon is explicable only on the supposition that the oxygen and the constituents of the food exercise on each other in the organism a certain action, in consequence of which both disappear from the body. This is actually the case; for none of the oxygen respired as a gas into the body remains in it; it is separated in the form of carbonic acid and water. The carbon and hydrogen, which have combined with the oxygen, are derived from the organism; but, as these elements of the body are obtained from the food, it may be said, that, in their final form, all the elements of food capable of uniting with oxygen are converted, in the living body, to oxygenised compounds, or, what expresses the same thing, they enter into combustion.

When bread, flesh, potatoes, hay, or oats, are burned in a common fire-place, with an ordinary draft, but perfectly exposed to the entrance of the air, the carbon of these substances is converted into carbonic acid, their hydrogen into water, their nitrogen is set at liberty in the form of ammonia, and their sulphur assumes the form of sulphuric acid, so that at last nothing remains except the mineral ingredients of these substances in the form of ashes. In the form of volatile products, we obtain carbonic acid, carbonate of ammonia, and water, and besides these, (if the combustion be incomplete,) smoke or soot; in the incombustible residue we obtain all the salts contained in the food.

When water is poured over these ashes, the alkalies dissolve, and also the soluble phosphates, common salt, and sulphates; the residue insoluble in water contains salts of lime and magnesia, and silica, if the substance burned contained the latter substance.

Exactly the same process ensues in the bodies of animals. Through the skin, and by means of the lungs, the carbon and hydrogen of the food is expelled in their final form of carbonic acid and water; all the nitrogen of the food is collected in the urinary bladder in the form of urea, which by the simple union of the elements of water is converted into carbonate of ammonia.

When the body regains its original weight, exactly as much carbon, hydrogen, and nitrogen, as it took in the food must have been expelled from it. It is only in youth, and in the process of fattening, that an increase in weight takes place, and that, therefore, part of the constituents of the food remains in the body: in old age, on the contrary, the weight decreases, that is, more is separated from the body than enters into it.

The nitrogen of the food is, therefore, daily expelled by the urine in the form of urea and of compounds of ammonia. The fæces contain the unburned substances of the food, such as the woody fibre, chlorophyl,* and wax, which have suffered no change in the organism; the carbon, hydrogen, and nitrogen of these substances are very small in quantity, in comparison with that in the food. The mixture of these indigestible materials with the secretions may be compared to the smoke and soot produced when food is imperfectly burned in a common fire-place.

It has been shown, by an examination of fæces and of urine, that the mineral ingredients of the food, the alkalies, salts, and silica, are eliminated in these excrements. Urine contains all the soluble mineral substances of the food, while the fæces contain the ingredients insoluble in water. As the food is burned in the body just as it would be in a fire-place, the urine may be said to contain the soluble salts of the ashes, and the fæces the insoluble salts.†

* Chlorophyl is the green colouring matter of the leaves and other parts of plants.

† Horse's Urine. VAUQUELIN.

Carbonate of lime	11
" soda	9
Hippurate of soda	24
Chloride of potassium	9
Urea	7
Water	940
	1000

Horse's Dung. JACKSON.

Phosphate of lime	5·0
Carbonate of lime	18·75
Phosphate of magnesia	36·25
Silica	40·
	100·00

Cow's Urine. BRANDE.

Chloride of potassium and Sal-ammoniac	15
Sulphate of potash	6
Carbonate of potash	4
" lime	3
Urea	4
Water	650
	682

Cow's Dung. HAIDLEN.

Phosphate of lime	10 9
" magnesia	10·
Perphosphate of iron	8·5
Lime	1·5
Gypsum	3·1
Chloride of potassium, copper	Traces.
Silica	63·7
Loss	1·3
	100·0

These analyses show, as nearly as can be expected from experiments of this kind, that all the constituents of the ashes of the food are again obtained, without alteration, in the solid and liquid excrements of the horse and cow. The action produced upon our fields, by the liquid and solid excrements of animals, ceases to be mysterious or enigmatical, as soon as we have attained a knowledge of their mode of origin.

The mineral ingredients of food have been obtained from our fields, having been removed from them in the form of seeds, of roots, and of herbs. In the vital processes of animals, the combustible elements of the food are converted into compounds of oxygen, while the urine and the fæces contain the constituents of the soil abstracted from our fields; so that, by incorporating these excrements with our land, we restore it to its original state of fertility. If they are given to a field deficient in ingredients necessary for the growth of plants, it will be rendered fertile for all kinds of crops.

A part of the crop taken from a field is used in feeding and fattening animals, which are afterwards consumed by man. Another part is used directly in the form of potatoes, meal, or vegetables; while a third part, consisting of the remnants of plants, are employed as litter in the form of straw, &c. It is evident that all the constituents of the field removed from it in the form of animals, corn, and fruit, may again be obtained in the liquid and solid excrements of man, and in the bones and blood of the slaughtered animals. It altogether depends upon us to keep our fields in a constant state of composition and fertility by the careful collection of

A HORSE CONSUMES OF INGREDIENTS OF THE SOIL *—	Ounces of Ashes.		THE EXCREMENTS OF THE HORSE RESTORE—	Ounces of Ashes.	
In 15lbs hay	18·61	21·49	In urine	3·51	21·87
In 4·54 oats	2·46		In fæces	18·36	
In its drink	0·42				

A COW—

In 30lbs. of potatoes	6·67	28·47	In urine	12·29	30·45
In hay	20·20		In fæces	16·36	
In its drink	1·6		In milk	1·80	

* Boussingault, Annales de Chimie et de Physique, lxxi.

N

these substances. We are able to calculate how much of the ingredients of the soil are removed by a sheep, by an ox, or in the milk of a cow,* or how much we convey from it in a bushel of barley, wheat, or potatoes. From the known composition of the excrements of man, we are also able to calculate how much of them it is necessary to supply to a field to compensate for the loss that it has sustained.

It is certainly the case, that we could dispense with the excrements of man and animals, if we were able to obtain from other sources the ingredients on which depends all their value for agriculture. It is a matter of no consequence whether we obtain ammonia in the form of urine, or in that of a salt from the products of the distillation of coal; or whether we obtain phosphate of lime in the form of bones, or as the mineral apatite. The principal object of agriculture is to restore to our land the substances removed from it, and which the atmosphere cannot yield, in whatever way the restoration can be most conveniently effected. If the restoration be imperfect, the fertility of our fields, or of the whole country, will be impaired; but if, on the contrary, we add more than we take away, the fertility will be increased.

The importation of urine or of solid excrements from a foreign land, is quite equivalent to the importation of corn and cattle. All these matters, in a certain time, assume the form of corn, flesh, and bones; they pass into the bodies of men, and again assume the same form which they originally possessed. The only true loss that we experience, and that

* 1000 parts of milk yielded, by incineration—

I.	67·7	Residue.
II.	49·0	,,

100 parts of the ashes of the milk consisted of:—

	I.	II.
Phosphate of lime	47·14	50·81
,, magnesia	8·57	9·45
Perphosphate of iron	1·43	1·04
Chloride of potassium	29·38	27·03
Common salt	4·89	5·03
Soda	8·57	6·64
	99·98	100·00

we cannot prevent, on account of the habits of our times, is the loss of the phosphates, which man carries in his bones to the grave. The enormous quantity of food, which man consumes during the 60 years of his life, and every constituent of it that was derived from our fields, may again be obtained and restored to them. It is quite certain, that it is only in the bodies of our youth, and in those of growing animals, that a certain quantity of phosphate of lime is retained in the bones, and of alkaline phosphates in the blood. With the exception of this extremely small proportion, in comparison with the actual quantity existing in the food, all the salts with alkaline bases, and all the phosphates of lime and magnesia, which animals daily consume in their food,—in fact, therefore, all the inorganic ingredients of the food,—are again obtained in the solid and liquid excrements. Without even instituting an analysis of these excrements, we can with ease indicate their quantity and their nature, and we can estimate their composition. We furnish to a horse daily 4½ lbs. of oats and 15 lbs. of hay; the oats yield 4 per cent., the hay 9 per cent. of ashes; and from these data we calculate, that the daily excrements of the horse must contain 21 ounces of inorganic materials, which have been obtained from our fields. The analyses of the ashes of hay and of oats inform us in per centage how much silica, alkalies, and phosphates are contained in them.*

* The ashes of oats contain, according to SAUSSURE—

	In 100 parts.
Soluble salts with alkaline bases	16
Phosphate of lime	24
Silica	60

The ashes of hay contain, according to HAIDLEN—

	In 100 parts.
Phosphate of lime	16·1
Perphosphate of iron	5·0
Lime	2·7
Magnesia	8·6
Sulphate of soda	1·2
,, potash	2·2
Chloride of potassium	1·3
Carbonate of soda	1·1
,, potash	0·9
Silica	60·6
Loss	0·8

The nature of the fixed ingredients in the excrements varies according to the food. If we feed a cow on mangel-wurzel, or potatoes, without hay or barley straw, its solid excrements will not contain silica, but they will contain phosphates of lime and magnesia, and the liquid excrements will contain carbonates of potash and soda, and also compounds of these bases with inorganic acids. If the fodder or food yield, on incineration, ashes containing soluble alkaline phosphates, (such as bread, meal, all kinds of seeds and flesh), we obtain from the animal fed upon these, urine in which the alkaline phosphates exist. But if the ashes of the food (such as hay, turnips, and potatoes), do not contain any soluble phosphate of the alkalies, but only insoluble earthy phosphates, then, the urine is free from the alkaline phosphates, and the fæces are found to contain the earthy phosphates. The urine of men and of animals subsisting upon flesh and grain contains alkaline phosphates; while that of animals living wholly upon grass is destitute of these salts. The analyses of human excrements,* those of birds living upon fish

* According to the analysis of Berzelius, 1000 parts of human urine contain—

	1000 parts of Urine.	1000 parts of the residue.
Urea	30·10	44·39
Free lactic acid, lactate of ammonia and animal matters not separable from them	17·14	25·58
Uric acid	1·00	1·49
Mucus of the bladder	0·32	0·48
Sulphate of potash	3·71	5·54
,, soda	3·16	4·72
Phosphate of soda	2·94	4·39
,, ammonia	1·65	2·46
Chloride of sodium	4·45	6·64
Muriate of ammonia	1·50	2·23
Phosphates of magnesia and lime	1·00	1·49
Silica	0·03	0·05
Water	933.00	
	1000·00	100·00

1000 parts of human fæces yielded 150 parts of ashes, which consisted of—(Berzelius) :—

Phosphate of lime „ magnesia Traces of gypsum	100

(guano),* and of the excrements of the horse and of the cow yield conclusive proof of the nature of the salts contained in them.

In the solid and liquid excrements of man and of animals, WE RESTORE TO OUR FIELDS THE ASHES OF THE PLANTS which served to nourish these animals. These ashes consist of certain soluble salts and insoluble earths, which a fertile soil must yield, for they are indispensable to the growth of cultivated plants.

It cannot admit of a doubt, that, by introducing these excrements to the soil, we give to it the power of affording food to a new crop, or, in other words, we reinstate the equilibrium which had been disturbed. Now that we know

Brought forward—Phosphate, &c. &c.	100
Sulphate of soda, ,, potash, Phosphate of soda	8
Carbonate of soda	8
Silica	16
Carbonaceous residue and loss	18
	150

*	Guano (a specimen from Liverpool). BARTELS.	Guano from Lima. VÖLCKEL.	Dung of the Nightingale. BRACONNOT.
Muriate of ammonia	6·500	4·2	0·2
Oxalate of ammonia	13·351	10·6	
Urate of ammonia	3·244	9·0	52·7 with potash.
Phosphate of ammonia	6·250	6·0	0·8 with potash.
Waxy substance	0·600		
Sulphate of potash	4·227	5·5	3·3
,, soda	1·119	3·8	
Phosphate of soda	5·291		
,, magnesia and ammonia	4·196	2·6	0·2
Chloride of sodium	0·100		0·8
Phosphate of lime	9·940	14·3	4·3
Oxalate of lime	16·360	7·0	
Alumina	0·104		
Residue insoluble in nitric acid	5·800	4·7	
Loss consisting of water, ammonia, and organic matters not estimated	22·718	32·3	37·7
			100·

that the constituents of the food pass over into the urine and excrements of the animal fed upon it, we can with great ease determine the different value of various kinds of manure. THE SOLID AND LIQUID EXCREMENTS OF AN ANIMAL ARE OF THE HIGHEST VALUE AS MANURE FOR THOSE PLANTS WHICH FURNISHED FOOD TO THE ANIMAL. The dung of pigs fed upon peas and potatoes, is in the highest degree adapted as a manure for fields growing peas and potatoes. We feed a cow upon hay and turnips, and we obtain a manure containing all the mineral constituents of grass and of turnips; this manure ought to be preferred, as being more suitable for turnips, than that procured from any other source. The dung of pigeons contains the mineral ingredients of the grains; that of the rabbit, the constituents of culinary vegetables; the liquid and solid excrements of man contain in very great quantity the mineral substances of all seeds.

According to the above view, a knowledge of the constituents of the ashes of food and of fodder, gives us an exact indication of the ingredients of the soil contained in the liquid and solid excrements of men and of animals.

If we know the quantity of the food, and the composition of its ashes, we know also with certainty how much soluble salts will be contained in the urine, and how much of the insoluble salts will exist in the fæces. It would, therefore, be superfluous and useless to state here a greater number of analyses of excrements, because these analyses must differ from each other, quite as much as the variation in composition of the ashes of the food on which the animal was fed.

Common stable manure is a mixture of solid excrements with urine, which gradually enters into putrefaction in the dunghill. In consequence of the putrefaction of the urine, all the urea contained in it is converted into volatile carbonate of ammonia. A large portion of the organic ingredients of the manure enter into decay and assume a gaseous condition, by the action of the air, with the continued evolution of heat. The weight of these ingredients diminishes while the relative proportion of the fixed mineral substances increases. If all the decaying matters entered into union with oxygen, the result of course would be, that those not susceptible of decay, or, in other words, the ashes, would remain behind. The following analyses will illustrate the meaning of this remark:—

100 parts fresh Cow-dung—

Water		85·900
Combustible substances .	12·352	14·100
Ashes	1·748	
		100·000

100 parts Stable Manure ½ year old. *

Water		79·3
Combustible substances .	14·04	20·7
Ashes	6·66	
		100·0

Now that we know that the proportion of the mineral food of plants increases with the age of the dung, that old dung may contain 4 to 6 times more of it than fresh dung, an explanation is furnished of the relatively greater action of the former, and of the preference accorded to it by farmers of experience.

It has been mentioned in the preceding part of the chapter, that animal excrements may be replaced in agriculture, by other materials containing their constituents. Now, as the principal action of the former depends upon their amount of mineral food so necessary for the growth of cultivated plants, it follows, that we might manure with the mineral food of wild plants, or in other words, WITH THEIR ASHES; for, these plants are governed by the same laws, in their nutrition and growth, as cultivated plants themselves. Thus, these ashes might be substituted for animal excrements; and if a proper selection were made of them, we might again furnish our fields with all the constituents

* Annales de Chimie et de Physique, iii. Série, 237.

removed from them by crops of cultivated plants. The vast importance of ashes as a manure is recognised by many farmers. In the vicinity of Marburg, and in the Wetterau, such a high value is attached to this costly material, as a manure, that the farmers do not object to send for it to a distance of 18 or 24 miles. The importance of this manure will be more obvious, when it is considered, that wood-ashes lixiviated with cold water contain silicate of potash, in exactly the same proportion as straw ($10\ Si\,O_3,+KO$); and that, in addition to this salt, it contains considerable quantities of phosphates.

Different kinds of wood-ashes possess very unequal value as manure. Thus, the ashes of the oak are of the smallest, those of the beech of the greatest value. Wood-ashes from oak contain 4 to 5 per cent. of phosphates; those from the beech contain the fifth part of their weight of these salts. The quantity of phosphates in the ashes of firs and pines amounts to from 9 to 15 per cent.; the ashes of the poplar contain $16\frac{3}{4}$ per cent., and those of the hazel-nut tree 12 per cent.*

With every hundred pounds of lixiviated ashes of the beech, we furnish to the soil as much phosphates as are contained in 460 lbs. of fresh human excrements.

According to the analysis of Saussure, 100 parts of the ashes of grains of wheat contain 32 parts soluble and 44·5 parts insoluble, or altogether 76·5 parts of soluble and insoluble phosphates. The ashes of wheat straw contain in all 11·5 per cent. of phosphates. Thus, with every 100 lbs. of the ashes of beech, we furnish to the field phosphoric acid sufficient for the production of 4000 lbs. of straw (calculating its ashes at 4 per cent., according to Saussure), or for 2000 lbs. of the grains of wheat (calculating their ashes at 1·3 per cent. —Saussure).

The dry fruit of the horse-chesnut (*Æsculus hippocastanum*) yields 34 per cent. of ashes, possessing a similar composition

* Ashes of pines from Norway contain the minimum of phosphates—viz., 0·9 per cent. (Berthier.)

to the ashes of maize, and of the grain of certain kinds of wheat.*

The importance of manuring with bones must be obvious to all. The bones of man, and of animals in general, have their origin from apatite (phosphate of lime), which is never absent from fertile land. The bone earth passes from the soil into hay, straw, and other kinds of food, which are afterwards consumed by animals. Now, when we consider that bones contain 55 per cent. of the phosphates of lime and magnesia (Berzelius), and if we assume that hay contains the same quantity of these salts as wheat straw, then it follows that 8 lbs. of bones contain as much phosphate of lime as 1000 lbs. of hay or of wheat straw, and 20 lbs. as much phosphoric acid as 1000 lbs. of the grain of wheat or of oats. These numbers are not absolutely correct, but they give a very fair approximation of the quantity of phosphates yielded annually by a soil to these plants. By manuring an acre of land with 60 lbs. of fresh bones, we furnish sufficient manure to supply three crops (mangel-wurzel, wheat, and rye) with phosphates. But the form in which they are restored to a soil does not appear to be a matter of indifference. For the more finely the bones are reduced to powder, and the more intimately that they are mixed with the soil, the more easily are they assimilated. The most easy and practical method of effecting their division is to pour over the bones, in a state of fine powder, half their weight of sulphuric acid diluted with three or four parts of water, and after they have been digested for some time, to add about one hundred parts of water, and to sprinkle this acid mixture (phosphates of lime and mag-

* Ashes of the fruit of the horse-chesnut (SAUSSURE):—

Potash	51
Alkaline phosphates	28
Chloride of potassium and sulphate of potash	3
Earthy phosphates	12
Silica	0·5
Metallic oxides	0·25
Loss	5·25
	100·00

nesia) before the plough. In a few seconds, the free acids unite with the bases contained in the earth, and a neutral salt is formed in a state of very fine division. Experiments instituted on a soil formed from grauwacke, for the purpose of ascertaining the action of the manure thus prepared, have distinctly shown that neither corn nor kitchen-garden plants suffer injurious effects in consequence; but that, on the contrary, they thrive with much more vigour.*

In the manufactories of glue, many hundred tons of a solution of phosphates in muriatic acid are yearly thrown away as being useless. It would be important to ascertain how far this solution might be substituted for bones. The free acid would combine with the alkalies in the soil, especially with lime, and a soluble salt would thus be produced, which is known to possess a favourable action on the growth of plants. This salt (muriate of lime, or chloride of calcium) is one of those compounds which attract water from the atmosphere with great avidity, and retain it when absorbed; and, being present in the soil, it would decompose the carbonate of ammonia existing in rain-water, with the formation of sal-ammoniac and carbonate of lime. A solution of bones in muriatic acid placed on land in autumn or in winter, would therefore not only restore a necessary constituent of the soil, but would also give to it the power of retaining all the ammonia falling upon it in the rain for a period of six months.

The ashes of brown coal and of peat contain frequently silicate of potash, so that these might furnish to the straw of the cereals one of its principal constituents; these ashes contain also phosphates.

* Very favourable results have been obtained by treating seeds in the following manner:—The seeds about to be sown were steeped in the water from a dunghill, and, while still wet, were strewed with a mixture of 20 parts of fine bone-dust and 1 part of burnt gypsum, in such a manner that each grain was covered with a thin layer of the powder; by sprinkling them with water and repeating this treatment with the mixture, the coating can be increased. The seeds were allowed to dry in the air, and were then sown in the usual way. On the large scale this mode of dunging, owing to its being rather troublesome, might not answer the purpose so well as a heavy manuring with bones and gypsum.

It is of much importance to the agriculturist, that he should not deceive himself respecting the causes which give the peculiar action to the substances just mentioned. It is known that they possess a favourable action on vegetation; and it is likewise certain, that the cause of this is their containing a body, which, independently of the influence exerted by its physical properties of porosity and capability of attracting and retaining moisture, assists also in maintaining the vital processes of plants. But if the subject be treated as an unfathomable mystery, the nature of their influence will never be known.

In medicine, for many centuries, the mode of action of all remedies was supposed to be concealed by the mystic veil of Isis; but now these secrets have been explained in a very simple manner. An unpoetical hand has pointed out the cause of the wonderful and apparently inexplicable healing virtues of the springs in Savoy, by which the inhabitants cured their goître: the water was found to contain small quantities of iodine. In burnt sponges used for the same purpose, the same element was also detected. The extraordinary efficacy of Peruvian bark was found to depend on a small quantity of a crystalline body existing in it, viz. quinine; and the causes of the various effects of opium were detected in as many different ingredients of that drug.

Now all such actions depend on a definite cause, by ascertaining which, we place the actions themselves at our command.

It must be admitted as a principle of agriculture, that those substances which have been removed from a soil must be completely restored to it; but whether this restoration be effected by means of excrements, ashes, or bones, is in a great measure a matter of indifference. A time will come, when plants growing upon a field will be supplied with their appropriate manures prepared in chemical manufactories—when a plant will receive only such substances as actually serve it for food, just as at present a few grains of quinine are given to a patient afflicted with fever, instead of the ounce of wood which he was formerly compelled to swallow in addition.

There are some plants which require humus (as a source of carbonic acid), without reproducing it in any appreciable quantity; whilst others can do without it altogether, and actually enrich a soil deficient in it. Hence a rational system of agriculture would employ all the humus at command for the supply of the former, and not expend any of it for the latter; but would in fact make use of them for supplying the others with humus.

We may furnish a plant with carbonic acid, and with all the materials which it may require; we may supply it with humus in the most abundant quantity; but it will not attain complete development, unless nitrogen is also afforded to it; an herb will be formed, but no grain; even sugar and starch may be produced, but no gluten.

But, on the other hand, the supply of nitrogen, in the form of ammonia, will not suffice for the purposes of agriculture. Although ammonia is of the utmost importance for the vigorous growth of plants, it is not in itself sufficient for the production of vegetable CASEIN, FIBRIN, or ALBUMEN. These substances are not known in a free state; for they are always accompanied by alkalies, sulphates, and phosphates. We must therefore assume, that without their co-operation, ammonia could not exercise the slightest influence on the growth and formation of the seeds; that, in such a case, it is a matter of perfect indifference whether ammonia is conveyed to them or not; for it will not assist in the formation of the constituents of the blood, unless the other conditions necessary for their production be present at the same time.

All these conditions are united in liquid and solid excrements; none of them are absent. In these are present, not only ammonia, but also alkalies, phosphates, and sulphates, in the relative proportion in which they exist in our cultivated plants.

The powerful action of urine depends, therefore, not only on its compounds of nitrogen; for the phosphates and sulphates accompanying these take a decided part in the action.

Urine, in the state in which it is used as manure, does

not contain urea, as this substance has been converted into carbonate of ammonia during putrefaction. In dung reservoirs, well constructed and protected from evaporation, the carbonate of ammonia is retained in solution. When the putrefied urine is spread over the land, part of its carbonate of ammonia evaporates along with the water, while another portion is absorbed by the soil, particularly if it be clayey and ferruginous land; but, in general, only the phosphate and muriate of ammonia remain in the ground. The amount of the latter alone enables the soil to exercise a direct influence on the plants during the progress of their growth; and as they are not volatile, not a particle of them escapes being absorbed by the roots.

The existence of carbonate of ammonia in putrefied urine long since suggested the manufacture of sal-ammoniac from this material. When the latter salt possessed a high price, this manufacture was carried on by the farmer himself. For this purpose the liquid obtained from dunghills was placed in vessels of iron and subjected to distillation; the product of this distillation was then converted into muriate of ammonia by the ordinary methods (Demachy).

The carbonate of ammonia formed by the putrefaction of urine can be fixed, or be deprived of its volatility, in many ways. When a field is strewed with gypsum, and then with putrefied urine, or with the drainings of dunghills, all the carbonate of ammonia is converted into the sulphate, which remains in the soil.

But there are still simpler means of effecting this purpose; —gypsum, chloride of calcium, sulphuric or muriatic acid, and superphosphate of lime, are substances of a very low price; and if they were added to urine until the latter lost its alkalinity, the ammonia would be converted into salts, which would have no further tendency to volatilise.

When a basin, filled with concentrated muriatic acid, is placed in a common necessary, so that its surface is in free communication with the vapours issuing from below, it becomes filled after a few days with crystals of muriate of ammonia. The ammonia, the presence of which the organs of smell amply testify, combines with the muriatic acid and loses entirely its volatility, and thick clouds or fumes of the salt newly-formed hang over the basin. In stables, the same may be seen. The ammonia escaping in this manner is not only lost, as far as our vegetation is concerned, but it works also a slow, though not less certain, destruction of the walls of the building. For, when in contact with the lime of the mortar, it is converted into nitric acid, which dissolves gradually the lime. The injury thus done to a building by the formation of soluble nitrates, has received (in Germany) a special name—salpeterfrass (production of soluble nitrate of lime).

The ammonia emitted from stables and necessaries is always in combination with carbonic acid. Carbonate of ammonia and sulphate of lime (gypsum) cannot be brought together at common temperatures, without mutual decomposition. The ammonia enters into combination with the sulphuric acid, and the carbonic acid with the lime, forming compounds destitute of volatility and consequently of smell. Now, if we strew the floors of our stables, from time to time, with common gypsum, they will lose all their offensive smell, and none of the ammonia can be lost, but will be retained in a condition serviceable as manure (Mohr).

With the exception of urea, uric acid contains more nitrogen than any other substance generated by the living organism; it is soluble in water, and can be thus absorbed by the roots of plants, and its nitrogen will be assimilated in the form of ammonia from the oxalate, hydrocyanate, or carbonate of ammonia. It would be extremely interesting to study the transformations which uric acid suffers in a living plant. For the purpose of experiment, the plant should be made to grow in charcoal powder previously heated to redness, and then mixed with pure uric acid. The examination of the juice of the plant, or of the component parts of the seed or fruit, would be an easy means of detecting the differences.

In respect to the quantity of nitrogen contained in excrements, 100 parts of the urine of a healthy man are equal to 1300 parts of the fresh dung of a horse, according to the analysis of Macaire and Marcet, and to 600 parts of the fresh dung of a cow. The powerful effects of urine as a manure are well known in Flanders, and they are considered invaluable by the Chinese, who are the oldest agricultural people we know. Indeed, so much value is attached to the influence of human excrements by these people, that laws of the state forbid that any of these excrements should be thrown away, and reservoirs are placed in every house, in which they are collected with the greatest care. No other kind of manure is used for their corn-fields.

On the assumption, that the liquid and solid excrements of man amount on an average, to only $1\frac{1}{2}$ lb. daily ($\frac{5}{4}$ lb. of urine and $\frac{1}{4}$ lb. fæces), and that both taken together contain 3 per cent. of nitrogen, then, in one year, they will amount to 547 lbs., containing 16·41 lbs. of nitrogen, a quantity sufficient to yield the nitrogen of 800 lbs. of wheat, rye, oats, or of 900 lbs. of barley. (Boussingault.)

This is much more than it is necessary to add to an acre of land, in order to obtain, with the assistance of the nitrogen absorbed from the atmosphere, the richest crops every year. By adopting a system of rotation of crops, every town and farm might thus supply itself with the manure, which, besides containing the most nitrogen, contains also the most phosphates. By using, at the same time, bones and the lixiviated ashes of wood, animal excrements might be completely dispensed with on many kinds of soil.

When human excrements are treated in a proper manner, so as to remove this moisture, without permitting the escape of ammonia, they may be put into such a form as will allow them to be transported even to great distances.

This is already attempted in many towns, and the preparation of night-soil for transportation constitutes not an unimportant branch of industry.

In Paris, for example, the excrements are preserved in the houses in open casks, from which they are collected and placed in deep pits at Montfaucon, but they are not sold until they have attained a certain degree of dryness, by evaporation in the air. But whilst lying in the receptacles appropriated for them in the houses, all their urea is converted for the most part into carbonate of ammonia; the vegetable matters contained in them putrefies, all the sulphates are decomposed, and the sulphur forms sulphuretted hydrogen (volatile hydrosulphate of ammonia). The mass, when dried by exposure to the air, has lost the greatest part of its nitrogen along with its water, and the residue, besides phosphate of ammonia, consists for the most part of phosphate of lime and magnesia, together with fatty matters. This manure is sold in France under the name of *Poudrette,* and is very highly estimated on account of its powerful action. This action cannot depend on the ammonia originally contained in it, because the greatest part has escaped during the dessication. According to the analyses of Jaquemart, the Parisian poudrette does not contain more than 1·8 per cent. of ammonia.

In other manufactories of manure, the night-soil, whilst still soft, is mixed with the ashes of wood, or with earth, &c. containing a large quantity of caustic lime, and this causes a complete expulsion of all the ammonia of the excrements, depriving them in consequence of all smell. The efficacy of this manure cannot, therefore, depend upon its nitrogen.

It is evident that, if we place the solid or liquid excrements of man, or the liquid excrements of animals on our land, in equal proportion to the quantity of nitrogen removed from it in the form of plants, the sum of this element in the soil must increase every year; for to the quantity which we thus supply, another portion is added from the atmosphere. There is no proper loss of nitrogen to plants, for even the small quantity of this element which man carries with him to the grave is not finally lost to vegetation, for it escapes into the earth, and into the atmosphere, as ammonia, during the decay and putrefaction of the body.

A high degree of culture requires an increased supply of manure. With the abundance of the manure the produce in corn and cattle will augment, but must diminish with its deficiency.

The substances applicable as manure ought to be arranged according to the products desired. The alkalies are peculiarly necessary for the production of vegetable constituents destitute of nitrogen, such as sugar, starch, pectin, and gum; phosphates are peculiarly valuable for the formation of the constituents of the blood. A field richly treated with animal manure, and therefore with phosphates, produces a barley which is rejected by the brewer of beer, because it is too rich in the constituents of the blood, and proportionally poor in starch. Hence, the very ingredient which is of the highest value to the feeders of stock, is held in low estimation by the brewer; because the object of the first is to produce flesh, the object of the latter is the fabrication of alcohol.

Fresh bones, wool, hair, rags, hoofs, and horn, are manures containing nitrogen as well as phosphates, and are consequently fit to aid the process of vegetable life.

One hundred parts of dry bones contain from 32 to 33 per cent. of dry gelatine; now, supposing this to contain the same quantity of nitrogen as animal glue—viz., 5·28 per cent., then 100 parts of bones must be considered as equivalent to 250 parts of human urine.

Bones may be preserved for thousands of years, in dry, or even in moist soils, provided the access of air is prevented; as is exemplified by the bones of antediluvian animals found in loam or gypsum, the interior parts being protected by the exterior from the action of water. But they become warm when reduced to a fine powder, and moistened bones generate heat and enter into putrefaction; the gelatine is decomposed, and its nitrogen is converted into carbonate of ammonia and other ammoniacal salts, which are retained in a great measure by the powder itself. *

* Bones burnt till quite white, and recently heated to redness, absorb 7·5 times their volume of pure ammoniacal gas.

Charcoal, in a state of powder, must be considered as a very powerful means of promoting the growth of plants on heavy soils, and particularly on such as consist of argillaceous earth.

Ingenhouss proposed dilute sulphuric acid as a means of increasing the fertility of a soil. Now, when this acid is sprinkled on calcareous soils, gypsum (sulphate of lime) is immediately formed, which, of course, prevents the necessity of manuring the ground with this material. 100 parts of concentrated sulphuric acid diluted with from 800 to 1000 parts of water, are equivalent to 176 parts of gypsum.

Many kinds of ashes, of peat, and most varieties of coal ashes, contain an abundant quantity of gypsum, by which they exercise a very favourable influence on certain soils.

	Ashes of Peat from Fichtelgebirge. FIKENTSCHER.	Ashes of Peat from Bassy (Dep. de la Marne.) BERTHIER.
Silica	36·5	
Alumina	17·3	22·5 (Silica, Alumina, Peroxide of iron)
Peroxide of iron	33·0	
Carbonate of lime	2·0	51·5 (Carbonate of lime, Magnesia)
Magnesia	3·5	
Gypsum	4·5	26·0
Chloride of calcium	0·5	
Carbonaceous residue	2·7	

CHAPTER XIII.

RETROSPECTIVE VIEW OF THE PRECEDING THEORIES.

The knowledge of the processes of nutrition, in the case of the culture of meadow and of forest land, indicates that the atmosphere contains an inexhaustible quantity of carbonic acid.

On equal surfaces of wood or of meadow land, in which exist the constituents of the soil indispensable to vegetation, we obtain crops without the application of carbonaceous manures; and these crops contain, in the form of wood and hay, a quantity of carbon equal to, or, in many cases, greater than that produced by cultivated land in the form of straw, corn, and roots.

It is obvious that the cultivated land must have presented to it as much carbonic acid as is furnished to an equal surface of wood or of meadow land; that the carbon of this carbonic acid becomes assimilated, or is capable of assimilation, if the conditions exist for its reception and conversion into a constituent of plants.

However great may be the supply of food in a soil, it will be sterile for most plants, if water be deficient. At certain seasons of the year rain fructifies our fields; seeds neither germinate nor grow without a certain quantity of moisture.

The action of rain is much more striking and wonderful to the superficial observer than that of manure. For weeks and months, the influence which it exerts on the crops is appreciable, and yet very small quantities of carbonic acid and ammonia are introduced to the soil by means of rain.

Water plays, doubtless, a decided part in the growth of plants, by virtue of its elements; but, at the same time, it is a mediating member of all organic life. Plants receive from the soil, by the aid of water, the alkalies, alkaline earths and phosphates necessary to the formation of their organs. If these substances, which are necessary for the passage of atmospheric food into the organism of the plant, be deficient, its growth must be impeded. Its proper growth, in dry seasons, stands in exact relation to the quantity of the substances taken up from the soil during the first period of its development. But on a soil poor in mineral food, cultivated plants do not flourish, however abundantly water may be supplied to them.

The crop of a meadow, or of an equal surface of wood-land is quite independent of carbonaceous manures, as far as regards its carbon; it is dependent on the presence of certain ingredients of the soil destitute of carbon, and also on the conditions which enable these to enter into the plants. Now, we are able to increase the crop of carbon on our cultivated land, by the use of burnt lime, ashes, or marl,—by substances, therefore, which are entirely free from carbon. This well ascertained fact indicates that we furnish to the field, in these substances, certain constituents, which enable the cultivated plants to increase in mass, and consequently in carbon—a power which they possessed formerly only in a small degree.

After these considerations, it cannot be denied that the sterility of a field, or its poverty of produce in carbon, does not arise from a deficiency of carbonic acid, or of humus; for we have seen that this produce can be increased, to a certain extent, by the supply of matters destitute of carbon. But the very same source which supplies the meadow and woodland with carbon, namely the atmosphere, can yield that element to cultivated plants. It therefore becomes especially necessary in agriculture to employ the best, and most convenient means, of enabling the carbon of the atmosphere (carbonic acid) to pass over into the plants growing on our fields. The art of agriculture, in the mineral food which it supplies, furnishes to plants the means of appropriating their carbon from sources offering an inexhaustible provision. But when these constituents of the soil are wanting, the most abundant supply of carbonic acid, or of decaying vegetable matter, cannot increase the crops on the field.

The quantity of carbonic acid that can pass from the air into plants, is limited, in a given time, by the quantity of carbonic acid entering into contact with the organs destined for its absorption. Now, the passage of carbonic acid from the air into the organism of the plant is effected by means of the leaves; but the absorption of carbonic acid cannot take place without the contact of its particles with the surface of the leaf, or of a part of the plant capable of absorbing it. Hence, in a given time the quantity of carbonic acid absorbed must stand in exact proportion to the surface of the leaves, and to the amount of it existing in the air.

Two plants of the same kind, with equal surfaces of leaves (*i. e.* surfaces of absorption) will take, during the same time, and under like conditions, the same amount of carbon. And if the air contains double the quantity of carbonic acid that it does at another time, the plants, under like conditions, will absorb double the quantity of carbon.* A plant with only half the surface of the leaves of another plant will absorb quite as much carbon as the latter, if the air supplied to the former contains twice the amount of carbonic acid.

These considerations point out to us the cause of the favourable action exerted on cultivated plants by humus, and by all decaying organic substances.

Young plants, when dependent on the air alone, can only increase their amount of carbon according to their absorbing surfaces. But it is obvious, if their roots receive, by means of humus, three times the amount of carbonic acid absorbed by their leaves in the same time, their increase in weight will be fourfold, on the assumption of the existence of all the conditions for the assimilation of the carbon. Hence, four times the quantity of stems, leaves, and buds must be formed; and,

* Boussingault remarked that leaves of a vine inclosed in a globe removed completely from the air all the carbonic acid contained in it, however rapidly the stream of air was made to pass. (DUMAS : Lectures, p. 23.)

by the increased surface thus obtained, the plants will receive in the same degree an increased power of absorbing food from the air; and this power remains in activity long after the supply of carbon to the roots has ceased.

But the use of humus as a source of carbonic acid, in arable land, is not only to increase the amount of carbon in the plant; for, by the increased SIZE attained by the plant in a given time, there is also given, in fact, SPACE for the reception of the constituents of the soil necessary for the formation of new leaves and twigs.

From the surface of young plants a constant evaporation of water takes place, the amount of which is in proportion to the temperature and surface. The numerous fibres of the roots supply the water which is evaporated, just as if they were so many pumps; so that, as long as the soil continues moist, the plants receive, by means of water, the necessary constituents of the soil. A plant with double the surface of another plant must evaporate twice the quantity of water that the latter does. The water thus absorbed is expelled again in vapour, but the salts and constituents of the soil introduced to the plant by its agency, still remain there. A plant with twice the surface of leaves of another plant, but with the same quantity of water in proportion to its size, still receives from the same soil a greater quantity of ingredients, in proportion to its water, than the latter plant receives.

The growth of the latter soon reaches a termination when the further supply ceases, while the former continues to grow, because it contains a larger quantity of the substances necessary for the assimilation of atmospheric food. But in both plants the number and size of the seeds will altogether depend upon the amount of the mineral ingredients of the seed existing in the plants; the plants containing or receiving from the soil a greater amount of alkaline and earthy phosphates than other plants obtain in the same time, will also produce a greater number of seeds than the latter.

Thus it is that, in a hot summer, when the supply of the

constituents of the soil is cut off by rain, the height and strength of the plants, and the development of the seed, stand in exact proportion to the quantity of the constituents of the soil taken up during their former period of growth.

The produce of a field in corn and in straw varies very considerably in different years. In one year we may obtain the same weight of corn of similar composition to that obtained in another year, but the crop of straw may be considerably greater; or the reverse may take place, and the crops of straw (of carbon) may be equal, while the corn may amount to double the quantity. But when we obtain twice the quantity of corn from the same surface, we must have also a corresponding increase of the constituents of the soil in the corn; or, when we obtain twice the quantity of straw, there must be twice the amount of the ingredients of the soil in the straw. In one year the wheat may be 3 feet in height, and yield 1200 lbs. of seed per acre, while, in another year, it may grow one foot higher, and yet yield only 800 lbs.

An unequal crop indicates, under all circumstances, an unequal proportion of the constituents of the soil taken up for the formation of the corn and of the straw. Straw contains and requires phosphates, as well as corn, but in much smaller proportion.

In a wet spring, when the supply of these salts is not so great as that of alkalies, of silica, and of sulphates, the crop of seeds becomes diminished; because a certain quantity of the phosphates, which would otherwise be employed in the formation of the seeds, is now used for the production of the stem and leaves; the constituents of the seeds cannot be perfected without an abundant supply of phosphates. By depriving a plant of these salts, we could produce artificially the state in which they attain a height of three feet, and blossom, without the production of seeds. The crop of corn growing on a soil rich in the constituents of straw, is often less in a wet spring than upon a soil poor in these ingredients; because the supply of mineral food on the latter lasts longer and is in better proportion for the growth of all the constituents of plants than in the former case.

On the supposition that all the conditions necessary to our cultivated plants, for the assimilation of food from the atmosphere, existed in the most favourable form, yet the action of humus would be useful in effecting a more rapid growth of the plants, and by thus GAINING TIME. In all cases, the crop of carbon is increased by means of humus; and if the conditions be absent for the conversion of this element into other constituents, it assumes the form of starch, gum, and of sugar, that is, of substances destitute of mineral ingredients.

Every moment of time is of value in the practice of farming; and, in this respect, humus is of especial importance in kitchen gardening.

Our corn plants and edible roots find in our fields, in the form of the remains of a past vegetation, sufficient vegetable matter to correspond to the mineral food existing in the soil, and therefore, with sufficient carbonic acid to produce a quick growth during spring. Any further supply of carbonic acid would be wholly useless, unless it were accompanied by a corresponding increase of the mineral constituents adapted to form parts of the plant. Upon a Hessian acre of good meadow land we obtain 2500 lbs. of hay, according to the opinion of experienced farmers. Meadows yield this crop, without any supply of organic matters, or without any manures containing nitrogen or carbon. By proper irrigation, and by treatment with ashes and gypsum, the crop can be increased to double the amount. Let us assume, however, that the 2500 lbs. of hay form the maximum crop; still, it is certain that all the carbon and nitrogen of the plants constituting it must have been obtained from the air.

According to Boussingault, hay, dried at the temperature of boiling water, contains 45·8 per cent. of carbon (a result agreeing with analyses made in this laboratory), and 1·5 per cent. of nitrogen; hay dried in air still retains 14 per cent. of water, which escapes at the heat of boiling water.

2500 lbs. of hay dried in air correspond to 2150 lbs. of

hay dried at the temperature of boiling water. With the 984 lbs. of carbon contained in the crop of 2150lbs. of hay, we have also removed from the acre of meadow-land 32·2 lbs. of nitrogen. If we assume that this nitrogen has entered the plant in the form of ammonia, it is obvious that for every 3640 lbs. of carbonic acid (calculated at 27 per cent. of carbon) the air contains 39·1 lbs. ammonia (taken at 82 per cent. of nitrogen); or that, for every 1000 lbs. of carbonic acid, the air contains $10\frac{7}{10}$ lbs. ammonia—a quantity corresponding to about $\frac{1}{100000}$ of the weight of the air, or of $\frac{1}{600000}$ of its volume.

Thus, for every 100 parts of carbonic acid absorbed by the surface of the leaves of the meadow-plants, there must also be absorbed from the air above one part of ammonia. When we calculate how much nitrogen different plants obtain from equal surfaces of land, basing our calculations on known analyses, the following results are obtained:—

1000 lbs of carbon remove in nitrogen—

From meadow-land, in hay	32·7
" arable land, in wheat	21·5
" " oats	22·3
" " rye	15·2
" " potatoes	34·1
" " mangel wurzel	39·1
" " clover	44
" " peas	62

These facts lead to certain conclusions of high importance to agriculture. We observed especially that the proportion of nitrogen absorbed, relatively to that of carbon, stands in a fixed relation to the surface of the leaves.

1. PLANTS CONTAINING NEARLY ALL THEIR NITROGEN CONCENTRATED IN THEIR SEEDS, SUCH AS THE CEREALS, CONTAIN ALTOGETHER LESS NITROGEN THAN THE LEGUMINOUS PLANTS, PEAS AND CLOVER.

2. THE CROP OF NITROGEN FROM A MEADOW TO WHICH NO AZOTISED MANURE HAS BEEN GIVEN, IS MUCH GREATER THAN THAT FROM A MANURED FIELD OF WHEAT.

3. THE CROP OF NITROGEN IN CLOVER OR IN PEAS IS

MUCH GREATER THAN THAT OF A HIGHLY-MANURED FIELD OF POTATOES OR OF TURNIPS.

Boussingault obtained in five years, from his farm in Bechelbronn, Alsace, in the form of potatoes, wheat, clover, turnips, and oats, 8383 carbon, and 250·7 nitrogen; in the succeeding five years,* 8192 carbon, 284.2 nitrogen; in a third rotation of six years,† 10949 carbon, 353·6 nitrogen; or, in sixteen years, 27424 carbon, and 858·5 nitrogen; or ALTOGETHER, IN THE PROPORTION OF 1000 CARBON TO 31·3 NITROGEN.

A most remarkable and important result follows from this experiment—that when potatoes, wheat, turnips, peas, and clover (POTASH, LIME, and SILICA PLANTS), are cultivated successively on the same field, although this field had been thrice manured in the course of sixteen years, the same relation of nitrogen to a given quantity of carbon is obtained, as in the case of a meadow which had received no manure.

	Carbon.	Nitrogen.
Upon an acre of meadow land there is cropped of silica, lime, and potash plants, taken together . .	984	32·2
On an acre of arable land, on a sixteen years' average, of silica, lime, and potash plants	857	26·8

When we take into consideration the amount of carbon and nitrogen in the leaves of the beet and potato (for the leaves were not calculated in the produce of the arable land), then it follows that, notwithstanding all the supply of carbon and of nitrogen furnished in the manure, the arable land has not produced more of these elements than an equal surface of meadow land, WHICH RECEIVED ONLY MINERAL FOOD (constituents of the soil).

THEN, ON WHAT DEPENDS THE PECULIAR ACTION OF MANURES, AND OF THE LIQUID AND SOLID EXCREMENTS OF ANIMALS?

This question is susceptible of a simple solution. These manures have a very decided action on our arable land, from which for centuries we have removed, in the form of cattle

* Beet, wheat, clover, wheat, late turnips, oats, rye.

† Potatoes, wheat, clover, wheat, late turnips, peas, rye.

and of corn, a certain quantity of constituents of the soil which have not been restored.

If no manure had been applied to the land on the sixteenth year of the above experiment, the crop would have amounted to only a half or third part of the carbon and nitrogen.

The liquid and solid excrements used as manure enabled this surface of arable land to produce as much as the meadow land. But notwithstanding the amount of manure supplied, the field was no richer in the mineral food of plants on the sixth year, when it was manured anew, than it was in the first year. In the second year after manuring, it contained less mineral food than on the first year; and after the fifth year it became so much exhausted that it was necessary, in order to obtain crops as rich as the first year, to give back to the field all the constituents that had been removed during the five years' rotation; this was done by means of the manure.

Our supply of manure, therefore, effects only this result, that the soil of our arable land is not rendered poorer than that of meadow land capable of yielding on the same surface 25 cwt. of hay. From a meadow we remove annually, in the hay, as great an amount of the constituents of the soil as we do in the crops obtained from the arable land; and we know that the fertility of meadow land is as dependent on the restoration of the constituents of the soil, as that of arable land is upon the supply of manure. Two meadows of equal surfaces, but containing unequal quantities of inorganic food, are of unequal fertility under like circumstances. The meadow containing the greatest quantity of the mineral food yields more hay, in a certain number of years, than the other which is poorer in mineral ingredients.

But if we do not restore to a meadow the constituents of the soil removed from it, its fertility decreases.

The fertility of a meadow remains the same, not only by treating it with solid or with liquid excrements, but it may be retained, or may be even increased in fertility by the application of mineral substances left behind after the combustion of wood or of other plants. By means of ASHES we can restore the impaired fertility of our meadow land. But by the term ASHES, we understand the mineral food which plants received from the soil. When we furnish them to our meadows we enable the plants growing on them to condense carbon and nitrogen on their surface.

Now, does the action of liquid and solid excrements not depend on the same cause; FOR THESE ARE BUT THE ASHES OF PLANTS BURNT IN THE BODIES OF MAN AND OF OTHER ANIMALS?

Is fertility not quite independent of the ammonia conveyed to the soil? If we evaporated urine, dried and burned the solid excrements, and supplied to our land the salts of the urine, and the ashes of the solid excrements, would not the cultivated plants grown on it—the gramineæ and leguminosæ—obtain their carbon and nitrogen from the same sources whence they are obtained by the gramineæ and leguminosæ of our meadows?

There can scarcely be a doubt with regard to these questions, when we unite the information furnished by science to that supplied by the practice of agriculture.

The following rotation is adopted in Alsace, as being the most advantageous; it extends over a period of five years, during which the land is only once manured.

1st Year. Manured.	2nd Year.	3rd Year.	4th Year	5th Year.	6th Year. Manured.
Potatoes or Beet	Wheat	Clover	Wheat with Fallow turnips	Oats, or Rye, or Barley.	Potatoes.
POTASH PLANTS.	SILICA PLANT.	LIME PLANT.	SILICA POTASH. } PLANTS	SILICA LIME } PLANTS.	

Now, if we suppose that the action of the manure depends upon its ammonia, or amount of nitrogen, then it is obvious that a progressive diminution must ensue; that the nitrogen in the crops of the first and second years must amount to more than that contained in the crops of the fourth and fifth years. But this opinion is completely opposed to the following proportions, as indicated by analysis:—

	1st Year.	2nd Year.	3rd Year.	4th Year.	5th Year.
Nitrogen in the crop . .	46	35·4	84·6	56·0	28·4

Thus, in the third and fourth years the nitrogen in the crops amounted to much more than the quantity contained in the crops of the first and second years; and in the fifth year the quantity was only one-fourth less than it was in the second year. Now, is it possible or conceivable that the ammonia given in the first year, being a body of great volatility and very apt to evaporate along with water, could be present in greater quantity in the soil during the fourth year than it was in the first and second years; or that it could yield to the oats of the fifth year the necessary quantity of nitrogen for their growth?

But let us admit that the nitrogen conveyed to the soil by strong manuring was actually exhausted in the fifth year by the different plants cultivated upon it; and let us then compare the rotation employed in Alsace, with that adopted on one of the most fertile districts of the Rhine. In Bingen there is a nine-years rotation followed, the plants succeeding each other in the following order:—

1st Year. Manured.	2nd Year.	3rd, 4th, 5th, 6th Years.	7th Year.	8th Year.	9th Year. Manured.
Turnips.	Barley with Lucern.	Lucern.	Potatoes.	Wheat.	Barley.

Six years after manuring, after the supply of ammonia and manure containing nitrogen, after four succeeding crops of clover, and after a crop of barley and one of oats, the soil of Bingen yields rich crops of potatoes, wheat, and barley, and these succeed each other at a time when, according to our assumption, the manured field in Alsace was to be viewed as completely exhausted of its nitrogen. Can it be conceived that the ammonia of the manure could, after the lapse of of 8—9 years, furnish the nitrogen to the crops of wheat and barley? But even admitting this to be the case, we have then to inquire whence do the corn-fields in Hungary, in Sicily, or in the vicinity of Naples, receive their nitrogen, for these fields have never been manured? Are we actually to believe that the nutrition of plants in the fields of moderate climates are subject to different laws from those governing the warmer and tropical regions?

In Virginia the annual crop of nitrogen in wheat amounted to 22 lbs. an acre, on the smallest calculation, or in 100 years to 2200 lbs. If we were to suppose that this nitrogen was furnished by the field, each acre must have contained it in the form of hundreds of thousand pounds of animal excrements.

The whole population of Limousin subsist upon milk and sweet chesnuts, the production of which, being unattended with trouble, is ascribed by Dupin as the cause of their low state of intellect. Without being subjected to any system of farming, this district produces enormous quantities of the constituents of the blood, the nitrogen of which cannot have been produced from manure.

For centuries, in Hungary, wheat and tobacco have been cultivated on the same field, without any supply of nitrogen. Is it possible that this nitrogen can have had its origin in the soil? Our forests of beeches, chesnuts, and oaks become covered with leaves every year; the leaves, sap, the acorns, chesnuts, cocoa-nuts, the fruit of the bread-tree, are rich in nitrogen. This nitrogen is not contained in the soil, nor is it conveyed to the wild plants by the hand of man. Then, it is impossible to doubt the source whence the nitrogen is obtained. The source of the nitrogen can only be the atmosphere. It matters not in what form it is contained therein, or in what form it is taken from it; the conclusion is the same, that the nitrogen of wild-growing plants must be derived from the atmosphere.

Are the fields of Virginia, the fields of Hungary, our own cultivated plants, not able to receive it from the same sources as the wild growing vegetation? Is the supply of nitrogen in animal excrements a matter of absolute indifference; OR DO WE OBTAIN IN OUR FIELDS A QUANTITY OF THE CONSTITUENTS OF THE BLOOD, ACTUALLY CORRESPONDING TO THE SUPPLY OF AMMONIA?

These questions are completely solved by the investigations of Boussingault; which are so much the more valuable, as they were instituted with a totally distinct object in view.

From the known quantity of manure (common stable manure) which Boussingault put every five years upon his field (amounting to four Hessian acres), he estimated, by the analysis of the manure, the total quantity of nitrogen furnished for the rotation of five years. For this purpose, the moist stable manure was first dried by exposure to the air and to the sun, and afterwards was further dried in vacuo, by exposing it to a temperature of 230° F.; the manure thus treated was subjected to an ultimate analysis. The average crops of the field, treated with manure, were then determined; and the products, corn and straw, turnips, potatoes, peas, clover, &c., were analysed for the purpose of ascertaining their composition with reference to nitrogen, carbon, hydrogen, and ashes.*

In this manner the quantities of nitrogen conveyed to the field in the form of manure, and reaped from it in the crops, were ascertained, and could be compared together. If the plants depended for their nitrogen upon the manure, and did not receive any of that element from the air; the nitrogen of the crops could not correspond to more than the quantity in the manure. If the crops contained more than this quantity, the excess must have been obtained from other sources, and these could only be in the atmosphere; such were the suppositions on which Boussingault proceeded. According to his estimation, the three rotations† yielded:—

	1st Rotation.	2nd Rotation.	3rd Rotation.	In 16 years.
Nitrogen in lbs. . . .	501·4	508·4	707·2	1717·0
——— in stable manure .	406·4	406·4	487·6	1300·4
Excess of nitrogen obtained, lbs.	95	102	219·6	416·6

* The greatest number of these analyses—viz., the composition of potatoes (Boeckmann); of beet and turnips (Will); of wheat straw (Will); of the carbon and nitrogen of peas (Noll and Zytowieki); and of their carbon (Playfair), were repeated in this laboratory, and ascertained to be perfectly correct.

	1st Rotation.	2nd Rotation.	3rd Rotation.
† 1 Year	Potatoes	Beet	Potatoes
2 ,,	Wheat	Wheat	Wheat
3 ,,	Clover	Clover	Clover
4 ,,	Wheat, late Turnips	Wheat, late Turnips	Wheat, late Turnips
5 ,,	Oats	Oats	Peas
6 ,,	—	—	Rye

In the first and second rotation, the excess of nitrogen obtained was nearly equal; in the third, it was twice as much.

Now, asked Boussingault—did each of these plants possess the power of absorbing and appropriating to their organism nitrogen from the air, or was this power confined only to one of them; and was the excess of nitrogen due to all the various kinds of plants, or was it yielded by only one of them? A new experiment seemed to him to decide the question. Two successive crops of corn were taken from a fallow-field, well manured, and the produce amounted to:—

Nitrogen in the crops . .	174·8 lbs.
———— the manure . .	165·6
An excess of nitrogen .	9·2 lbs.

But this excess in the crop is too small not to be liable to errors in the experiment. Boussingault concluded from it, that cereals do not absorb nitrogen from the air, and that the amount of nitrogen yielded in the crops is only equivalent to that contained in the manures.

Now, when it is known that the quantity of nitrogen obtained in crops of potatoes and turnips scarcely corresponds to more than the quantity in crops of wheat, it follows that they could not have the power to form their azotised constituents without manure; so that nothing remains, except to ascribe to the clover the excess of nitrogen obtained. This explains, also, why the excess is so much greater in the third rotation than in any of the preceding; for it will be remarked, that in the third rotation a sixth crop was introduced, corresponding to the same family as clover. If, therefore, there had been neither peas nor clover in the third rotation, but, instead of these plants, one of another family, the nitrogen of the crop would have amounted only to the quantity supplied in the manure. Boussingault concludes that leguminous plants alone possess the power of appropriating, as food, nitrogen from the air, and that other cultivated plants do not at all possess this property. Hence the great importance which Boussingault ascribes to manures containing nitrogen, for, according to his view, the commercial value of a manure depends on its amount of nitrogen. But all these conclusions are thoroughly erroneous; for, if they were not so, it must follow that potash, lime, and silica plants, unless they belonged to the Leguminosæ, would not produce any nitrogen, unless they were supplied with manure containing that element.

The conclusions of Boussingault are not only erroneous in their applications to agriculture, but are incorrect in the methods which he employs; for the manure was not given to the fields in the form in which he analysed it.

Let us assume that the manure which he put upon his fields possessed the same state in which it was analysed (dried at 230° F. in vacuo); then the field would receive in the sixteen years 1300 lbs. of nitrogen. But the manure was not put upon the field in an anhydrous state, but, on the contrary, in its natural moist condition, soaked with water; and we know that all the nitrogen contained in the manure in the form of carbonate of ammonia is volatilised when it is dried at a high temperature.

Human excrements, dried in the air at ordinary temperatures, (*poudrette*) lose, at 230°, half of all the nitrogen contained in them, in the form of carbonate of ammonia. Common stable manure, when heated to 230° in vacuo, must lose at least three times as much nitrogen as it retains; that is, 3-4ths of all the nitrogen originally present in it. The nitrogen of the urine in the manure, which is converted by putrefaction into carbonate of ammonia, is not included in the 1300 lbs. of the above calculation. But if we estimate it at half of the quantity present in the dried excrements, then the field must have received, in sixteen years, 1950 lbs. of nitrogen.

BUT IN SIXTEEN YEARS, 1517 lbs. OF NITROGEN ONLY WERE OBTAINED, IN THE FORM OF CORN, STRAW, AND TUBERS; much less, therefore, than the quantity furnished to the field. Hence his erroneous conclusion, that the Leguminosæ alone possess the power of condensing nitrogen from the air; and that it is necessary to furnish nitrogen to the Gramineæ, and to plants such as turnips and potatoes. But in the same time, and upon the same surface of a good meadow, not receiving nitrogen, we may obtain (on 1 hectare,) 2060 lbs. of this element.

It is well known that dried excrements form the principal fuel in Egypt, where wood is scarce. For centuries the sal ammoniac used in Europe was supplied from the soot of these excrements, until Grapenhorst, in the latter part of last century, discovered how to prepare it, and instituted a manufactory at Brunswick.

The fields in the valley of the Nile receive no manure of animal origin except the fixed ingredients, (which contain no nitrogen,) of the ashes of the burnt dung; and yet these fields have been so fertile, for periods long before our history commences, that this fertility has become a proverb, and is quite as remarkable at the present day as it was in former times. These fields become renovated by the mud deposited during the inundations of the Nile; the mineral ingredients of the soil removed in the crops are thus restored to it. The mud of the Nile contains as little nitrogen as the mud from the Alps, in Switzerland, deposited on, and fertilising our own fields by the inundations of the Rhine.

In fact, if the mud of the Nile fertilises the soil, in consequence of its containing nitrogen, we must suppose immense strata of nitrogenised animal and vegetable matter to exist in the mountains of the interior of Africa, at heights above the line of perpetual congelation, where, owing to the absence of all vegetation, no animal, not even a bird, can now find nourishment.

Cheese must be formed from the plants upon which cows feed. The meadows of Holland must, of course, obtain their nitrogen from the air. The milch cows in Holland remain on the fields both day and night; all the salts contained in their fodder must remain upon the fields in the form of urine and of solid excrements, a small quantity proportionately being removed by the cheese.

The condition of fertility of these meadows can change as little as that of our fields, which, although not grazed upon, receive, in the form of manure, the greatest part of the ingredients removed from them.

In the cheese districts of Holland, these ingredients remain on the meadows; while in our system of farming, they are collected at home, and carried, from time to time, to our fields. The nitrogen of the urine, and that of the solid excrements of the cow, are obtained in Holland from the air; and from the same source must be obtained the quantity of that element contained in all the kinds of cheese prepared in Holland, Switzerland, and other countries.

The meadows in Holland, for centuries, have produced millions of cwts. of cheese: there are annually exported from this country thousands of cwts. of this substance; and yet this exportation does not in any way diminish the productiveness of their meadows, although they have never received more nitrogen than they originally contained.

Hence it is quite certain, that our fields cannot be exhausted by the exportation of products containing nitrogen, (unless these products contain at the same time a large amount of mineral ingredients,) because the nitrogen of vegetation is furnished by the atmosphere, and not by the soil. Hence also we cannot augment the fertility of our fields, or their powers of production, by supplying them with manures rich in nitrogen, or with ammoniacal salts alone. The crops on a field diminish or increase in exact proportion to the diminution or increase of the mineral substances conveyed to it in manure.

The formation of the constituents of the blood, and of the vegetable substances containing nitrogen existing in cultivated plants, depends upon the presence of certain substances contained in the soil. When these ingredients are absent, nitrogen will not be assimilated, however abundantly it may be supplied. The ammonia of animal excrements exerts a favourable influence only because it is accompanied by other substances necessary for its conversion into the constituents of blood. When these conditions are furnished with ammonia, the latter becomes assimilated. But when the ammonia is absent from the manure, the plants extract their nitrogen from the ammonia of the air; to which it is again restored by the decay and putrefaction of dead animal and vegetable remains.

Ammonia accelerates and favours the growth of plants on all kinds of soil, in which exist the conditions for its assimilation; but it is quite without action upon the production of the constituents of the blood, when these conditions do not exist.

It is possible to conceive that asparagin (the active ingredient of asparagus) and the ingredients so rich in nitrogen and sulphur, of mustard and of all Cruciferæ, could be generated without the co-operation of the ingredients of the soil. But if it were possible to form the organic constituents of blood existing in plants, without the aid of the inorganic ingredients of the blood, such as potash, soda, phosphates of soda and of lime, they would be of very little use to animals, and could not fulfil the purposes for which they were destined by the wisdom of the Creator. Blood, milk, and muscular fibre cannot be formed without the aid of alkalies and of phosphates; and bones cannot be produced without phosphate of lime.

In urine, and in the solid excrements of animals, and in guano, we furnish ammonia, and therefore nitrogen, to our plants. This nitrogen is accompanied by the mineral food of plants, and actually in the same proportion as both exist in the plants which served the animals for food; or, what is the same thing, in the same proportion in which both are capable of being applied for a new generation of plants.

The action of an artificial supply of ammonia as a source of nitrogen, is limited, like that of humus as a source of carbonic acid, to a gain in point of time; in other words, to the acceleration of the development, in a given time, of our cultivated plants.

By means of ammonia, in the form of animal excrements, we increase the quantity of the constituents of blood in our cultivated plants—an action which the carbonate or sulphate of ammonia *alone* never exerts.

In order to obviate any misunderstanding, we must again draw attention to the fact that this explanation is not in any way contradicted by the effects produced on the application of artificial ammonia, or of its salts. Ammonia was, and is still considered, as the source of all the nitrogen of plants: its supply is never injurious; on the contrary, it is always useful, and, for certain purposes, indispensable. But, at the same time, it is of great importance for agriculture, to know with certainty that the supply of ammonia is unnecessary for most of our cultivated plants, and that it may be even superfluous, if only the soil contain a sufficient supply of the mineral food of plants, when the ammonia required for their development will be furnished by the atmosphere. It is also of importance to know, that the rule usually adopted in France and in Germany of estimating the value of a manure according to the amount of its nitrogen, is quite fallacious, and that its value does not stand in proportion to its nitrogen.

By an exact estimation of the quantity of ashes in cultivated plants, growing on various kinds of soils, and by their analysis, we will learn those constituents of the plants which are variable, and those which remain constant.* Thus also

* The following analyses of ashes may be added to those formerly given :—

	Ashes of Clover (*Trifolium prutense*).	Ashes of Sainfoin.
Silica	5·438	2·79
Sulphate of potash .	3·080	3·87
Chloride of sodium .	1·670	2·37
Carbonate of potash .	12·728	9·93
Carbonate of soda .	13·528	17·16
Carbonate of lime . .	38·216	32·55
Magnesia . . .	4·160	9·11
Phosphate of iron . .	1·240	0·64
Phosphate of lime .	11·970	15·37
Phosphate of magnesia .	6·790	3·98
Carbonaceous matter .	0·160	0·36
	98·980	98·13

we will attain a knowledge of the quantities of all the constituents removed from the soil by different crops.

The farmer will thus be enabled, like a systematic manufacturer, to have a book attached to each field, in which he will note the amount of the various ingredients removed from the land in the form of crops, and therefore how much he must restore to bring it to its original state of fertility. He will also be able to express in pounds weight, how much of one or of another ingredient of soils he must add to his own land, in order to increase its fertility for certain kinds of plants.

These investigations are a necessity of the times in which we live; but in a few years, by the united diligence of chemists of all countries, we may expect to see the realisation of these views; and by the aid of intelligent farmers, we may confidently expect to see established, on an immovable foundation, a rational system of farming for all countries and for all soils.

SUPPLEMENTARY CHAPTERS.

I.—The Sources of Ammonia.

When animals appeared on the surface of the earth, it cannot be doubted that means must have been provided for their sustenance and increase, or in other words, that plants must have existed to furnish them with food. But it is quite as obvious that, at the period of the formation of the vegetable world itself, the conditions must have existed in the soil and in the atmosphere, necessary for the exercise of vegetable life. With the same certainty with which we presuppose the existence of a compound of carbon to furnish that element to vegetation, we must also assume the contemporaneous presence of a compound of nitrogen, such as at the present day yields that element to plants.

If we disregard the fundamental principle on which all inquiries into nature ought to proceed, then we may assume, *à priori*, according to our will and pleasure, that other compounds of carbon, differing from carbonic acid, formerly took part in the vital processes of plants; but if we still retain the foundation of all scientific inquiry, namely, induction from facts, then we cannot admit the existence of these hypothetical compounds of carbon, either because they are totally unknown to us, or that their existence is doubtful.

The same reasoning must be adopted in the case of nitrogen. Science is at present ignorant of any compound of nitrogen besides ammonia, capable of yielding nitrogen to wild plants on all parts of the earth's surface. No other such compound of nitrogen has been indicated, or even hypothetically supposed to exist, and designated by a name, in the case of cultivated plants; and therefore, until a second source of nitrogen is discovered, we must, in science, view ammonia as the *only* source.

Now, it may be asked, Is there no means of increasing the amount of ammonia which exists in the atmosphere, as well as in the form of plants and animals, and which we shall assume to be a limited quantity? This question may be repeated in another form, viz., Whether there exist undoubted facts for the opinion that the nitrogen of the air possesses, under any condition, the power of assuming the form of ammonia, or of any other compound of nitrogen? Besides nitric acid and ammonia, we do not know any other compounds of nitrogen, except those existing in plants and animals, or which may be prepared from them. With the exception of these compounds, nitrogen exists only in the form of a gas, which has been recognised as one of the principal constituents of air.

An ignorance of the proper sources whence vegetation received its nitrogen, led philosophers long since to the opinion, that plants must possess the power, in some way or another, of appropriating the nitrogen of the air in their vital processes. In fact, until it was known that ammonia formed a constituent of the air, there was scarcely any reason to doubt this power of plants; for how otherwise were wild plants to obtain the nitrogen of their azotised constituents?

Ammonia is known and considered only as a product of the destruction and decomposition of the organism. The production and formation of ammonia presupposes the existence of plants or animals. Hence there have arisen two views respecting the origin of ammonia, the correctness of which we have as little means of establishing by decisive evidence, as we have of answering the questions—Whether the hen existed before the egg, or the egg before the hen; or whether water was first created as water, or as hydrogen and oxygen? We have sufficient reason to believe that the vegetable preceded the animal kingdom; and we assume, that before plants were formed, the conditions essential for their life and increase must have existed; and that then, as well as now, ammonia must have been a constituent of the air; and that the destruction of plants did not precede the formation of ammonia. Now, it is obvious that if the same causes now continue in action, as those which effected the formation of ammonia at the commencement of vegetation,—if their action resulted in the conversion of the gaseous nitrogen into ammonia,—then at the present day, during every moment, ammonia must be forming, and the amount of that previously existing will be increased. It is natural to the mind of man to endeavour to solve questions of this kind, however small may be the expectation of success. It is known that the crust of the earth consists of compounds of oxygen with metals or with other radicals; and the view appears quite admissible, that silica has been formed from silicon and oxygen; peroxide of iron from iron and oxygen; and, to follow up the idea, magnesia and potash have been produced from oxygen, magnesium, and potassium. And yet it is utterly impossible to assign a cause which prevented the union of the oxygen with potassium or magnesium before the time that this combination actually took place. Was there, it may be asked, a time when the individual elements floated together in Chaos, without possessing any kind of affinity? In what condition then was the chlorine of common salt or the carbon of carbonic acid? It is obvious that no answers can be given to questions respecting the origin of matter. If, then, we are unable to afford any more satisfactory explanation of the origin of ammonia, than we are able to do of the other compounds occurring on the earth, we must rest satisfied that these questions will either never be solved, or that they will not be so until a future period.

The ferruginous earths in the primitive rocks of South America (Boussingault), and of Sweden (Berzelius),—in fact all ferruginous earths hitherto examined,—yield, on being heated, a certain amount of water containing appreciable quantities of ammonia. Whence has this ammonia had its origin? According to the logic of Aristotle, the occurrence of ammonia in the ferruginous earths is susceptible of a satisfactory explanation.

We may assume that water is the only original compound of hydrogen occurring in nature; other bodies containing that element are products of the decomposition of water, from which they have procured this hyrdogen.

Ammonia has been formed like other compounds of hydrogen; the ferruginous earth was formerly iron, which we may suppose to have become oxidised at the expense of water, in which case peroxide of iron would be formed, and hydrogen become liberated. Now, if we assume that hydrogen, at the moment of its liberation, is able to unite with nitrogen gas in contact with it, and dissolved in water, then ammonia would be produced, and would remain in union with the peroxide of iron. It is certain that this explanation of the origin of ammonia in peroxide of iron would be perfectly satisfactory, if it were ascertained with some degree of probability that peroxide of iron has had its origin by oxidation at the expense of water, and that the nitrogen of air is capable of uniting with hydrogen at the instant of its liberation. On this view we might suppose, that although there was a limit to the formation of ammonia, under former conditions, when ferruginous earth was produced, that by the simultaneous occurrence of the same or of analogous conditions at the present day, ammonia might still be produced.

But the decomposition of water, by means of iron, is effected under such circumstances as appear to exclude the simultaneous production of ammonia.

Iron does not decompose water at the ordinary temperatures, and at higher temperatures—at the boiling point of water, for example—nitrogen does not remain any longer in solution. When a stream of nitrogen is made to pass along with water over iron filings heated to redness, the nitrogen is again obtained unaltered, although it be mixed with hydrogen. It is easily explained why ammonia cannot be formed in this case, for ammoniacal gas in contact with iron at high temperatures, is decomposed into its constituents.

When finely divided hydrate of peroxide of iron is placed

in contact with metallic iron, a decomposition of water takes place at a slightly elevated temperature, and hydrogen gas is evolved, while magnetic oxide of iron is produced. As hydrated peroxide of iron acts as an acid in this case, we should here, as indeed universally, when metals are dissolved in acids with the evolution of hydrogen, obtain in the solution a salt of ammonia, if ammonia had been formed.

But hitherto the presence of ammonia under the circumstances has not been detected; and it has further been shown satisfactorily, that when water holding air in solution is decomposed by a stream of electricity, the hydrogen evolved is accompanied by a certain quantity of nitrogen gas, which could not be the case if nascent hydrogen was able to form ammonia.

It has been considered as a certain proof of the formation of ammonia from the nitrogen of the air, that peroxide of iron, formed by the rusting of iron in the air, contains a certain quantity of ammonia; but air itself contains ammonia, which possesses a considerable affinity for peroxide of iron. Marshall Hall has shown the inaccuracy of the view that water is decomposed in this case; and further experiments, instituted in this laboratory for the especial purpose of deciding this question, have shown, that when air is freed from its ammonia, by being conducted through concentrated sulphuric acid, before being brought in contact with the rusting iron, the oxide then formed does not contain a trace of ammonia.

Braconnot* has shown that most basalts, trap, granite from Rochepon, and from Bresse; syenite, amphibolite, wakit (a lava); basalt, from Baden; quarz, from Gerordines; pegmatite, and many other minerals, yield, by dry distillation, water containing a sensible quantity of ammonia.

These facts cannot be explained by the interpretation above given to the occurrence of ammonia in ferruginous earth, namely, the oxidation of iron at the expense of water; but there cannot be any doubt that the ammonia has had a similar origin in all these cases, although that origin cannot be ascribed to an oxidation of iron.

---

* *Annales de Chimie et de Physique*, tome lxvii. p. 104, &c.

The question—Whether the nitrogen of the air possesses the power of uniting with hydrogen at the moment of its liberation from water? has been lately made the subject of exact experiments, although with very different objects in view. Will and Varrentrapp applied to the quantitive estimation of nitrogen in organic bodies, the known fact that the nitrogen of bodies containing that element is evolved in the form of ammonia, when they are heated to redness, mixed with potash. By combining the ammonia with an acid, and converting it into the salt termed chloride of platinum and ammonia, the ammonia generated may be weighed with ease, and the quantity of nitrogen may be calculated from the known composition of this salt. A great number of analyses of compounds, in which the quantity of nitrogen was known, showed that this mode of procedure answered completely the object in view; until certain experiments by Reiset were published, in which he obtained ammonia by this process from substances such as sugar, &c., which were quite destitute of nitrogen. Reiset, therefore, assumed that the nitrogen of air contained in the pores of the mixture was the cause of the formation of ammonia, and that, unless this air were excluded, the method of analysis was incorrect and objectionable.

New experiments, repeated with the utmost care by Will, have shown that in circumstances similar to those formerly observed by Faraday, ammonia is actually obtained from matters destitute of nitrogen, when they are heated to redness with potash; but that, by excluding ammonia itself, nitrogen cannot be made to unite with hydrogen in a nascent state, and that ammonia cannot be produced from these elements.

The admirable experiments of Faraday (Quarterly Journal of Science, xix. p. 16) prove that, in all the cases in which ammonia was obtained by heating to redness a substance destitute of nitrogen with hydrate of potash, the ammonia existed ready formed in the substance, or in the hydrate of potash. There are no observations more convincing of the extraordinary diffusion of ammonia, which exists in all places where atmospheric air is to be found. That the reader may judge properly of Faraday's experiments, I consider it important to describe them here in detail.

After Faraday had observed that woody fibre, linen, oxalate of potash, and a number of other substances free from nitrogen, evolved ammonia on being heated with soda, potash, hydrate of lime, &c., he endeavoured to ascertain the conditions under which the formation of ammonia ensued; and in the first place, he examined the alkalies. Hydrate of potash, whether made from potashes, cream of tartar, or potassium, behaved exactly in the same manner. The organic substances, when heated alone, had no reaction on turmeric; but when heated with the alkalies, a disengagement of ammonia ensued.

It was then to be supposed that the nitrogen of the air surrounding the substances might take a part in the formation of ammonia; but this was very improbable: for it is known that the air contains oxygen, which was never observed to unite with the liberated hydrogen under the same circumstances, although its affinity for that element is infinitely greater than for nitrogen.

According to this supposition, the nitrogen of the air must have formed ammonia by uniting with the hydrogen of the decomposed water, although at the same time there was present oxygen, for which hydrogen has a much greater affinity.

The experiments were repeated in an atmosphere of pure hydrogen, prepared from water which was previously freed from all air by long-continued boiling. But in this case also, where all nitrogen was excluded, the presence of ammonia was observed. Hence Faraday concluded that there must be an unknown cause of the formation of ammonia.

Now, when it is known that ammonia is a constituent of the air; that it is present wherever air is found; that it is itself a coercible gas, which is condensed on the surface of solid bodies in much larger proportion than air; and further, when it is known that it exists in distilled water, these, and other still more incomprehensible experiments of Faraday, are explained in a very simple manner.

Fine and bright iron wire, introduced into fused potash,

causes the evolution of ammonia, which soon ceases; but a new evolution takes place when a second portion of polished iron-wire is introduced (Faraday).

Zinc introduced into potash in a state of fusion, occasions an immediate evolution of ammonia and hydrogen gas; but although the conditions for the possible formation of ammonia continue (zinc, air, and nascent hydrogen), the quantity of ammonia generated does not increase; but, by the addition of fresh zinc, or of hydrate of potash, a new quantity of ammonia may be detected.

Some potash and zinc were heated together; a part of the mixture was then placed in a flask, which was immediately closed, while another part was dissolved in water, the clear solution dried, and laid aside for 24 hours. After this time had elapsed, the first portion gave scarcely perceptible traces of ammonia. The other gave very appreciable indications of its presence, apparently as if the substances which were the source of ammonia were derived from the air, during the operation (Faraday).

White clay from Cornwall, after being heated to redness and exposed for a week to the air, yielded ammonia abundantly, when heated in a tube. But when the clay was preserved in a good stoppered bottle, after being heated to redness, this effect was not produced.

The observations which prove most undoubtedly that in all these cases the ammonia was obtained from the air and condensed on the surface of these materials, are the following (Faraday):—

Sea-sand was heated to redness in a crucible, and allowed to cool on a plate of copper; 12 grains of the sand were then placed in a clean glass tube; and an equal quantity, shaken upon the hand, was allowed to remain there for a few moments, being stirred about with the fingers, after which, it was introduced into a second tube by means of platinum foil, taking care that the grains of sand were not brought in contact with any other animal substance (Faraday).

When the first tube was heated, it gave no sign of ammonia to turmeric paper; but the second tube did so in very appreciable quantity. For the sake of precaution the tubes used in these experiments were not cleansed by tow or cloth, but unused tubes were taken, and before being employed they were heated to redness in a stream of air (Faraday).

Some asbestos heated to redness, and introduced into a tube with metallic tongs, gave, when heated, no indication of ammonia; while, on the contrary, another portion, which had been simply pressed with the finger, yielded immediate indications of ammonia when heated in a tube (Faraday).

Now it is known that ammonia evaporates by the skin, that sweat contains salts of ammonia; and nothing can be more certain than that, in the experiments last described, and also in those of the burnt sand exposed to air, ammonia must have condensed on the surface of the sand or of the asbestos.

These experiments explain in a natural manner the existence of ammonia in earth from which plants and animals are entirely absent, and also of the formation of nitre in mixtures of earths containing vegetable matter.

All observations in our times lead to the conclusion that the nitrogen of the air does not possess the property of being converted into ammonia; and, whatever reasons there may exist for the probability of this conversion, we are by no means entitled to elevate to the rank of a principle the mere opinion that a part of the nitrogen of plants arises from this source, as it is an hypothesis standing in complete contradiction to all the knowledge which we have yet attained.

All experiments which appear to prove that the nitrogen of the air becomes fixed in the organism of certain plants,—that peas and beans, for example, vegetating in a soil perfectly destitute of animal matters, must possess the power of appropriating the nitrogen of the atmosphere,—cannot now have the smallest value, when it is known that the air contains ammonia as a constant ingredient. It must be recollected that these experiments were instituted in districts in which the atmosphere is much richer in ammonia than in the free fields, and that the distilled water, with which the plants were treated, was obtained from spring-water, and contained a much larger quantity of carbonate of ammonia than rain-water. Hence, there is no reason to ascribe the increase of nitrogen in the seeds, leaves, and stems to a source which was only imagined to exist, because the quantity of ammonia in the water and air was not considered, and the foundation, therefore, of the true explanation was altogether wanting.

Chemical experiments have shown that ammonia is not only the product of the decay and putrefaction of animal bodies, but that it is also capable of being generated in many chemical processes, when nitrogen, at the moment of its liberation from compounds containing it, is offered to hydrogen; in such a case, they unite together and form ammonia.

Compound gases containing nitrogen as a constituent (cyanogen, nitric oxides, nitrous oxides), are converted into ammonia when they are mixed and conducted over spongy platinum heated to redness (Kuhlmann), or over peroxide of iron (Reiset).

When steam is conducted over red-hot wood charcoal containing nitrogen, there is obtained, among other products, hydrocyanic acid, which is converted into ammonia and formic acid when treated with alkalies.

The nitrogen of nitric acid, when placed in contact with hydrogen at the moment of its liberation, as in the solution of tin, or by fusing nitrates with potash and organic substances, is converted into the compound of hydrogen. In all cases in which we expose to a high temperature a body containing nitrogen and caustic potash, its nitrogen assumes the form of ammonia.

The nitrogen of an organic body, of vegetable or animal matter, or of the charcoal produced from them, arises from the ammonia which the plant contained and abstracted from the atmosphere: it enters, in the processes of decomposition alluded to, into its original form, and assumes the condition of ammonia.

But these instances cannot be cited as proper examples of the formation of ammonia, nor can they be considered with reference to the question which we have now been discussing.

IS NITRIC ACID FOOD FOR PLANTS?

BEFORE we can examine the opinion whether nitric acid be a means by which nitrogen is furnished to plants in nature, it is most important to consider the origin of nitric acid.

At the request of the French government, the Academy of Sciences of Paris, in the year 1770, offered a prize for the best treatise on the formation of nitric acid and its production in artificial nitre-beds. The judges appointed by the Academy, including Lavoisier, subjected to trial 70 treatises, the results of which, after the experience of 50 years, were stated in a small work published by Gay Lussac, in the year 1825,* in the following sentences :—

1. "All the nitrogen necessary for the formation of nitric acid is yielded to it by animal matter."

2. "Nitre is never generated from the air in substances adapted for its formation, without the co-operation of animal matter."

This result of very numerous and correct experiments contradicts completely the view that nitre may be generated in mixtures of earth destitute of animal matter, and therefore, at the expense of the constituents of the air. The advocates of this view cite in defence of it the following experiments :— When earth forming nitre is freed from all its soluble salts by lixiviation, and is then exposed for several years to the action of the air, it yields a second crop of nitre, and these crops may be obtained three or four times in succession, although in different proportions. The advocates of this theory, considering that all the substances containing nitrogen are removed, argue that the nitrogen of the nitre formed after-

* *Instruction sur la fabrication du salpetre publié par la Commission des poudres et salpetres*, 1825.

wards, must have been derived from the air. But this conclusion is opposed to all rules of inductive science. When a known cause produces the same action in all cases submitted to examination, we must revert to the same cause in considering the same action in cases not examined; for we have no right to assign to it a new cause, in order to save us the trouble of a closer investigation.

The advocates of the opinion that the nitrogen of the air is converted into nitric acid in the nitre-beds, have never estimated the amount of substances containing nitrogen existing in those beds; and they have never compared with this amount the quantity of nitric acid actually generated. Those who, like Gay Lussac, have taken this trouble, found that the quantity of nitric acid formed corresponded to the quantity of animal matters present in the mixture; less nitre being formed, when the amount of the latter was decreased, and by its increase, a greater quantity of nitre was produced.

Another reason for the opinion was, that nitrates were formed in certain limestone caverns in Ceylon, where, according to Dr. Davy, nitrates of potash and lime occur in a limestone containing felspar, but quite destitute of animal matter. But the latter assertion is very questionable, as there is scarcely a limestone in existence that does not yield ammoniacal liquid on being subjected to distillation. An experiment with materials expressly prepared for this purpose (carbonate of lime, felspar, and water free from ammonia), and conducted in this cavern, in order to see whether nitric acid would be formed, would have completely decided the question, if nitric acid had been found in the mixture after a certain time; but this experiment was not made, neither was the water which filtered through the roof of the cavern subjected to examination. The conclusion that nitric acid is formed in these cases, at the expense of the nitrogen of the air, is not in any way confirmed,—it is only certain that the cause of the formation of nitre in these caves remains unknown to those who have examined them.

It often happens that the well-water of towns contains

a considerable quantity of nitre, which does not exist in the wells and springs outside the towns. Berzelius detected nitrates in the well-water of the city of Stockholm. Margraf also mentions its existence; and I, myself, have shown the presence of nitrates in the waters of twelve wells in the town of Giessen,* although they could not be detected in the waters of six wells separated 2300 paces from the town. Animal matter, in a state of decay and putrefaction, existed abundantly in the soil in all the places where nitrates were found, and its nitrogen was converted into nitric acid wherever the conditions for this conversion were found united.

A large proportion of the nitre used in France, for the manufacture of gunpowder and for other purposes, is obtained at Paris. The manufacturers of nitre use in its preparation the lower parts of old broken-up houses, which have been in constant contact with the liquids of the street. Nitre exists in large quantity in the lower parts of houses, while the upper parts do not contain a trace of it.

It cannot be denied that plants grow more powerfully and luxuriantly in a soil capable of forming nitre, than they do in a soil unfit for its formation.

The favourable influence of such a soil on vegetation is justly ascribed to the animal matters contained in it, to the alkalies, and to the phosphates existing in the animal matter. Out of the animal matter also, is formed the ammonia so necessary for the support of vegetation, and without the presence of which nitric acid could not be formed.

The presence of alkaline nitrates in a soil indicates with the greatest certainty, that the most important conditions for the growth of plants are united in it; but these salts are not the primary causes of the growth, because both the formation of nitre, and the luxuriant growth of plants, are effects of similar causes acting on the earth. It is certain that the vicinity of the salt mines of Quarta Jaga and Santa Rosa, described by Darwin, although saturated with nitrates, forms a complete waste, in which a small cactus is scarcely able to grow. The

* *Annales de Chimie et de Physique*, vol. xxxv. 332.

cause of its sterility may be the want of rain; but if it were moist, and obtained abundant supplies of rain, the nitrates would have disappeared long since; and, even without their presence, vegetation would flourish luxuriantly in this climate.

The common error is to confound a soil, in which nitrates EXIST, with one in which they are in the act of FORMING. If the first soil be wanting in the conditions (animal matter) necessary for a further formation of nitric acid, it will prove sterile, but will, on the contrary, be fertile if these conditions exist. The latter, and not the nitrates, are therefore the causes of the better growth of vegetation.

It follows from the preceding observations, that, as far as our experiments extend, the formation of nitric acid on the surface of the earth is dependent on the presence of animal matter.

But as animal substances receive their nitrogen from the atmosphere in the form of ammonia, the primary origin of the nitric acid of nitrates must be the ammonia of the atmosphere. But it may be affirmed, in addition to this, that ammonia is not only the ultimate source of, but that it is actually the immediate source of nitric acid. We have reason to believe that the nitrogen of decaying animal substances assumes the form of ammonia, before being converted into nitric acid; and that it must first be in the state of ammonia, before it is able to form nitric acid with the oxygen of the air.* Hence we must view ammonia as the principal source of the formation of nitric acid on the surface of the earth; and we may expect the production of the latter wherever ammonia, and the conditions for its oxidation, are found united.

The occurrence of large beds of nitrates in America cannot afford the most distant reason for the assumption that they are formed in an unusual way; it is unnecessary to call in the assistance of the nitrogen of the air, in order to explain their great extent. We find in nature whole mountains consisting of shell-fish, and of remains of

* See the Chapter on Eremacausis in second part of this book.

microscopical animals, which must have contained a certain quantity of nitrogen when alive. We find also large layers of animal excrements (Coprolites), which place beyond all doubt the former existence of innumerable individuals of species now extinct. In the processes of decay and putrefaction, to which they have been subjected, the nitrogen of their bodies could have escaped only in two forms; in cold climates, it would assume the form of ammonia, and in warmer countries, the form of nitric acid, which must accumulate wherever the salts formed by means of it are not carried off by water.

Ammonia, however, is not the only source of the formation of nitric acid. In the action exerted by the electric spark on the constituents of air, (which are also the constituents of nitric acid) we recognise a second source, which, to all appearance, is very extended.

Cavendish was the first to observe, that by a continued passage of electric sparks through moist air, its volume diminished, and an acid, soluble in water, was formed at the same time. This great philosopher proved, by a series of decisive experiments, that the constituents of the air, the nitrogen and oxygen, united to form nitric acid when exposed to the influence of electricity.

Now, it is probable that lightning (the most powerful electric spark known), in its passage through moist air, may effect a combination of the constituents of air, in consequence of which nitric acid would be formed.

In an examination of rain-water, which the author of the present work undertook in the years 1826—1827 (Annales de Chimie et de Physique, xxxv. 329), it was actually found that out of seventy-seven analyses made of the residue of rain-water, seventeen of them, obtained by the evaporation of the rain of thunder-storms, contained more or less nitric acid, partly in combination with lime, and partly with ammonia. In the sixty others, only two contained traces of nitric acid.

The occurrence of nitric acid in rain-water as nitrate of ammonia, renders it uncertain whether the nitrogen of the former was obtained from the atmospheric air itself, or from the ammonia existing in it, in the state of a gas. Henry observed that ammoniacal gas, mixed with oxygen, and exposed to electric sparks, is likewise converted into nitric acid. It is obvious, that, if the rain contains carbonate of lime mechanically mixed with it in the form of dust, the nitrate of ammonia also present will be converted during evaporation into carbonate of ammonia, which will escape, and into nitrate of lime, which remains in the residue. The quantity of nitric acid contained in the rain of a thunder-storm cannot be estimated. Two or three hundred pounds of filtered rain-water yield only a few grains of a coloured residue, and the nitrates contained in the latter form only a fractional part of its weight.

The analysis of the water of springs and of rivers is much more adapted to give us a clear conception of the quantity of nitric acid formed by the influence of electricity in the atmosphere. If we suppose the nitric acid to exist in water in a free state, as it is a volatile acid, it must escape during the evaporation of the water in porcelain vessels, so that the residue will not contain a trace of it, if the bases necessary for its fixation be deficient. The water of our springs, streams, and rivers, is rain-water, which, if nitric acid were originally present in it, must now contain nitrates, by filtering through the earth, which invariably contains lime and alkaline bases.

It follows, from the interesting observations made by Göbel, in his journey to Southern Russia, that, by the evaporation of the river Charysacha, which falls into the Lake Elton, the latter must receive annually 47,777 millions of pounds of salts. The water of the Charysacha contains scarcely 5 per cent. of salts; so some conception may be formed of the quantity of water which must evaporate, in order to furnish the above quantity. The river has its source about forty wersts from Lake Elton, and obtains its water from the rain and snow falling on the mountains.

If nitric acid be a constant and generally appreciable constituent of rain-water, it is obvious that we ought to find sensible traces of it in the mother liquor remaining behind after the crystallisation of the salt. But Göbel did not observe the presence of nitrates either in the water of the river or in the deposited salt.

In the water of the Artesian Well* of Grenelle; in the water of the Nile;† in that of the Seine, which contains carbonate of ammonia in dry seasons; in the waters of the Thames, or of the Rhine, no one has yet proved the presence of nitrates.

We may assume, from these facts, that the nitric acid furnished to the earth in Europe, by means of rain, is extremely small in amount; so that, even if the nitric acid formed by lightning exercise a favourable action on vegetation, still this influence cannot be considered as a source of the nitrogen of plants. When it is considered that the number of thunder-storms in a year does not amount in most districts to above twelve on an average, and in many to only eight, it must be obvious from this, that it would be quite impossible to prove the presence of nitric acid in the waters of rivers or of springs.

Under the tropics, where thunder-storms are much more frequent than with us, we might suppose that the quantity of nitric acid in rain-water would be appreciably greater. But the known examinations of the spring and river waters of those regions; for example, of the waters of Paipa, near Tunga,

* Payen found in 10,000 parts of this water:—

Carbonate of lime	6·80
,, magnesia	1·42
,, potash	2·96
Sulphate of potash	1·20
Chloride of potassium	1·09

† Regnault found in 22 lbs of water of the Nile:—

Carbonate of lime	5·30
,, magnesia	7·43
Peroxide of iron	0·53
Chloride of sodium	4·77
Sulphate of magnesia	0·53
Silica	1·06
Alumina	1·59
Extractive matter	0·53
Carbonic acid	12·19
	33·93 Grammes.

of the water of the Rio Vinagre, and of the hot mineral springs of the Cordilleras, the analyses of which were instituted by Boussingault, in South America, without the presence of nitrates being detected, show that there is no foundation for the opinion that a sensibly greater quantity of nitric acid is generated in those regions, by the action of lightning, than in the temperate zones.

It follows, from the preceding observations, that nitric acid, or its salts, are not destined by nature to yield nitrogen to plants. If it were actually the case that nitric acid did yield to plants their nitrogen, we must assume that this source was accessible to all plants without distinction. But it is completely excluded from marine plants; and even in the case of the terrestrial plants of the temperate and cold zones, the rare occurrence of thunder-storms would prevent us from considering that any appreciable quantity of their nitrogen could arise from nitric acid generated by the action of lightning on the constituents of air.

But, even on the assumption that nitric acid does take a decided part in vegetable life, ammonia still remains as the ultimate source of the nitrogen of plants; for, as far as our knowledge at present extends, all the nitric acid on the surface of the earth is formed by the eremacausis of ammonia, and it is not improbable that the nitric acid, which occurs in the rain of thunder-storms, may be dependent on the presence of the same body.

Although we thus trace back the action of all animal and other substances containing nitrogen, to the only compound which furnishes this element to all plants, in a state of nature, we do not of course mean to exclude the application of these other matters to the purposes of agriculture. When we know that woollen rags, horn, and hair, in the progress of decay, offer a slow but continued supply of ammonia, it follows, that we may use them wherever their price, in comparison with the advantage anticipated, does not exclude their application.

The same reasoning holds good in the case of nitrates. In

these, nitrogen exists in another form than that of ammonia. Nitric acid, or rather nitrous acid, is, in its chemical relations, exactly opposed to ammonia; but we see, that in the organism of plants, carbonic acid and water suffer decomposition, although their constituents are united by a much greater power. We have considered sulphuric acid as a source of sulphur. Why, then, should not nitric acid suffer a similar decomposition by the same causes; why should not its nitrogen, like the carbon or sulphur, become a component part of a plant?

By strewing nitrate of soda over fields, a greater crop has been obtained, particularly on grass land. Upon corn-fields and on roots, it has had less influence.

It is not yet decided to what constituent of the salt its favourable influence is due.

When the crops of hay and straw obtained with this manure by Mr. Gray, of Dilston, and Mr. Hyett (Journal of the Royal Agricultural Society), are expressed with regard to their quantity of nitrogen, the singular result is obtained, that the amount of nitrogen in these crops amounts to double the quantity of that contained in the nitrate used as manure!

Now, when it is remembered that the crop of many meadows is rendered a half, twice, or even three times greater, by manuring with burnt bones or with wood ashes—with matters, therefore, containing no nitrogen, it still remains doubtful, whether the action of nitrate of soda should be ascribed to its nitric acid.

A number of plants, such as *Borago officinalis*, *Mesembryanthemum crystallinum*, *Apium graveolens*, the sun-flower, and tobacco, contain dissolved in their juices considerable quantities of nitre, which does not exist in other plants growing on the same soil. The presence of a nitrate in plants permits only one conclusion—that the nitrogen of nitric acid is not employed in their organism for the formation of compounds containing that element, because, if it were, at a certain period of the life of the plant, it would disappear on account of this conversion.

Whatever be the case in this respect, nitrates are manures, which do not replace those constituents of the soil which are removed in the crops. Hence, although either by means of their acid, or of their alkalies, the growth of plants may be increased for one or two years, this very increase must cause an earlier period of exhaustion and poverty to the soil. A proper and lasting advantage cannot be expected from the use of nitrates.

DOES THE NITROGEN OF THE AIR TAKE PART IN VEGETATION?

PRIESTLEY and Ingenhouss assumed that plants possess the power of assimilating the nitrogen of the air. The former states that a specimen of *Epilobium hirsutum* kept under a glass globe of ten inches in height, and of one inch in width, absorbed within a month $\frac{7}{8}$ of the air contained in it.

These experiments have been repeated by Saussure with every care (RECHERCHES, p. 189), both in pure nitrogen and in atmospheric air, exactly according to the method described by Priestley, but the results were quite the reverse. Saussure observes, "I have continued the experiments for a long time, but I never could detect a diminution of the nitrogen. The same was the case with all kinds of plants which I submitted to the same experiment. Plants, therefore, do not sensibly diminish the bulk of the air; and these experiments are confirmed by those of Woodhouse and Sennebier."

Hence, we have not any direct proof for the opinion, that the nitrogen of the air is converted into a component part of a plant by its vital processes. In the present state of our knowledge, indirect proofs are equally wanting.

Many writers on agriculture cite, as decisive proofs of the assimilation of the nitrogen of the air by plants, the experiments of Boussingault, but their interpretation in favour of this view is not supported by facts. This distinguished philosopher instituted a number of experiments in order to decide the question regarding the origin of nitrogen in plants, and we give the results of these experiments in his own words, (Ann. de Chimie et de Physique, LXIX.) :—

"I believe that I have proved by numerous experiments, that the nitrogen of a rotation of plants is greater and often much greater than the quantity contained in the manure. This excess arises doubtless from the air, and it is more than probable that, in this case, a part of the excess of nitrogen is taken up in the form of nitrate of ammonia, which M. Liebig has shown to exist as a frequent constituent of the rain of thunder-storms. But before this can be assumed, it will be necessary to examine the action of this salt on vegetation."

In a later treatise on this subject, Boussingault says, (Annales de Chimie et de Physique, 3 Série, t. I. p. 240) :—

"When these tables are examined, it follows that the nitrogen in the plants obtained amount to more than that present in the manure. I assume, as a general proposition, that this excess arises from the air. BUT IN WHAT WAY AND MANNER THIS ELEMENT IS TAKEN UP BY PLANTS, I AM UNABLE TO STATE. The nitrogen may be taken up directly as a gas, or dissolved in water, or, what is possible, and as some philosophers (Saussure for example) believe, the air may contain an infinitely small quantity of ammonia."

The experiments of Boussingault are, therefore, proofs that the nitrogen of cultivated plants is not obtained from manure alone, but that, besides this, they contain an excess which can only be derived from the atmosphere. That the nitrogen of wild plants must be derived from the air is so obvious, that it requires neither proof nor experiments.

Boussingault had not the slightest intention of making his experiments the foundation for the opinion that the nitrogen of air might be converted into parts of the plant, but only employed them as proofs that the nitrogen of cultivated plants is derived from the atmosphere.

GIANT SEA-WEED.

(From Darwin's Journal of the Voyage of the Beagle, p. 303—304.)

"There is one marine production, which from its importance is worthy of a particular history. It is the kelp or *Fucus giganteus* of Solander. This plant grows on every rock from low-water mark to a great depth, both on the outer coast and within the channels. I believe, during the voyage of the Adventure and the Beagle, not one rock near the surface was discovered, which was not buoyed by this floating weed. The good service it thus affords to vessels navigating near the stormy land is evident, and it certainly has saved many a one from being wrecked. I know few things more surprising than to see this plant growing and flourishing amidst those great breakers of the Western Ocean, which no mass of rock, let it be ever so hard, can long resist. The stem is round, slimy, and smooth, and seldom has a diameter of so much as an inch. A few taken together are sufficiently strong to support the weight of the large loose stones to which, in the inland channels, they grow attached; and some of these stones are so heavy, that, when drawn to the surface, they can scarcely be lifted into a boat by one person.

"Captain Cook, in his second voyage, says, that at Kerguelen Land 'some of this weed is of a most enormous length, though the stem is not much thicker than a man's thumb. I have mentioned, that upon some of the shoals on which it grows, we did not strike ground with a line of twenty-four fathoms. The depth of water, therefore, must have been greater. And as this weed does not grow in a perpendicular direction, but makes a very acute angle with the bottom, and much of it afterwards spreads many fathoms on the surface of the sea, I am well warranted to say that some of it grows to the length of sixty fathoms and upwards.' Certainly, at the Falkland Islands, and about Terra del Fuego, extensive beds frequently spring up from ten and fifteen fathom water. I do not suppose the stem of any other plant attains so great a length as 360 feet, as stated by

Captain Cook. The geographical range is very considerable; it is found from the extreme southern islets near Cape Horn, as far north, on the eastern coast (according to information given me by Mr. Stokes) as lat. 43°—and on the western it was tolerably abundant, but far from luxuriant, at Chiloe, in lat. 42°. It may possibly extend a little further northward, but is soon succeeded by a different species. We thus have a range of 15° in latitude; and as Cook, who must have been well acquainted with the species, found it at Kerguelen Land, no less than 140°. in longitude.

"The number of living creatures, of all orders, whose existence intimately depends on that of the kelp, is wonderful. A great volume might be written, describing the inhabitants of one of these beds of sea-weeds. Almost every leaf, excepting those that float on the surface, is so thickly incrusted with corallines as to be of a white colour. We find exquisitely delicate structures, some inhabited by simple hydra-like polypi, others by more organised kinds, and beautiful compound Ascidiæ. On the flat surfaces of the leaves, various patelliform shells, Trochi, uncovered molluscs, and some bivalves are attached. Innumerable crustacea frequent every part of the plant. On shaking the great entangled roots, a pile of small fish, shells, cuttle-fish, crabs of all orders, sea-eggs, star-fish, beautiful Holuthuriæ (some taking the external form of the nudibranch molluscs), Planariæ, and crawling nereidous animals, of a multitude of forms, all fall out together.

"I can only compare these great aquatic forests of the southern atmosphere with the terrestrial ones in the intertropical regions. Yet, if the latter should be destroyed in any country, I do not believe nearly so many species of animals would perish, as, under similar circumstances, would happen with the kelp. Amidst the leaves of this plant, numerous species of fish live, which nowhere else would find food or shelter; with their destruction, the many cormorants, divers, and other fishing-birds, the otters, seals, and porpoises, would soon perish also; and lastly, the Fuegian savage, the miserable lord of this miserable land, would redouble his cannibal feast, decrease in numbers, and perhaps cease to exist."

APPENDIX.

EXPERIMENTS OF WIEGMANN AND POLSTORF.

THE composition of the artificial soil used in the experiments of Wiegmann and Polstorf, on the organic ingredients of Plants, was as follows (Preischrift, p. 9):—

Quarzy sand	861·26
Sulphate of potash	0·34
Chloride of sodium	0·13
Gypsum (anhydrous)	1·25
Chalk (elutriated)	10·00
Carbonate of magnesia	5·00
Peroxide of Manganese	2·50
Peroxide of iron	10·00
Hydrated alumina	15·00
Phosphate of lime	15·60
Acid of peat with potash*	3·41
,, ,, soda	2·22
,, ,, ammonia	10·29
,, ,, lime	3·07
,, ,, magnesia	1·97
,, ,, peroxide of iron	3·32
,, ,, alumina	4·64
Insoluble acid of peat	50·00

* This salt was made by boiling common peat with weak potash ley, and precipitating by means of sulphuric acid the dark-coloured solution. This precipitate is that termed Torfsaeure (acid of peat), in the above analysis. The salts of this acid referred to in the analysis were obtained by dissolving this acid in potash, soda, or ammonia, and by evaporating the solutions ; the salts of magnesia, lime, peroxide of iron, and alumina, were obtained by saturating this solution with their respective bases, by which means double decomposition was effected. HUMUS is the substance remaining by the decay of animal and of vegetable matters, which are seldom absent from a soil. This was replaced by the acid of peat in the experiments of Wiegmann and Polstorf. When the acid of peat is boiled for some time with water, it passes into an insoluble modification denoted above as insoluble acid of peat.

The following experiments were instituted in pure sand, and in the artificial soil:—

VICIA SATIVA.

A.

In Pure Sand.

The vetches attained by the 4th of July a height of ten inches, and seemed disposed to put out blossoms. On the 6th of the same month, the blossoms unfolded; and on the 11th they formed small pods, which, however, did not contain seeds, and withered away by the 15th. Similar plants, which had already begun to have yellow leaves below, were drawn with their roots out of the sand, the roots washed with distilled water, and then dried and incinerated.

B.

In Artificial Soil.

The plants reached the height of eighteen inches by the middle of June, so that it became necessary to support them with sticks; they blossomed luxuriantly on the 16th of June; and about the 26th put out many healthy pods, which contained on the 8th of August ripe seeds, capable of germinating. Similar plants to the above were taken with their roots from the soil; they were then washed and incinerated.

HORDEUM VULGARE.

A.

In Pure Sand.

The barley reached on the 25th of June, when it blossomed imperfectly, a height of 1¼ foot, but it did not produce seed; and in the month of July, the points of the leaves became yellow; on which account, on the 1st of August, we removed the plants from the soil, and treated them as before.

B.

In Artificial Soil.

The Barley, by the 25th of June, had reached a height of 2¼ feet, by which time it had blossomed perfectly; and yielded, on the 10th of August, ripe and perfect seeds; upon which the plants, together with their roots, were taken from the soil, and treated as formerly.

AVENA SATIVA.

A.

In Pure Sand.

The oats, on the 30th of June, were 1½ foot in height, but had blossomed very imperfectly; they did not produce fruit; and, in the course of July, the points of their leaves became yellow, as in the case of the barley; on which account the stalks were removed from the soil on the 1st of August, and treated as formerly.

B.

In Artificial Soil.

The oats reached 2½ feet on the 28th of June, having blossomed perfectly. By the 16th of August, they had produced ripe and perfect seeds; the stalks and roots were, therefore, removed from the soil, and treated as above.

POLYGONUM FAGOPYRUM.

A.

In Pure Sand.

The buck-wheat, on the 8th of May, seemed to flourish the best of all the plants grown on pure sand. By the end of June, it had reached a height of 1½ foot, and branched out considerably. On the 28th of June, it began to blossom, and continued to blossom till September, without producing seeds. It would certainly have continued to blossom still longer, had we not removed it from the soil on the 4th of September, as it lost too many leaves: it was treated as before.

B.

In Artificial Soil.

The buck-wheat grew very quickly in this soil, and reached a height of 2½ feet. It branched out so strongly, that it was necessary to support it with a stick; it began to blossom on the 15th of June, and produced perfect seeds, the greater number of which were ripe on the 12th of August. On the 4th of September, it was taken from the soil along with the roots, and treated as before, on account of losing too many leaves from below; although it was partly still in blossom, and with unripe fruit.

NICOTIANA TABACUM.

A

In Pure Sand.

The tobacco-plant sown on the 10th of May did not appear till the 2d of June, although it then grew in the normal manner; when the plants had obtained their second pair of leaves, I removed the superfluous plants, leaving only the five strongest specimens. These continued to grow very slowly till the occurrence of frost in October, and obtained only a height of 5 inches, without forming a stem. They were removed along with their roots from the sand on the 21st October, and treated as the above.

B.

In Artificial Soil.

The tobacco sown on the 10th of May came up on the 22d of the same month and grew luxuriantly. When the plants obtained the second pair of leaves, I withdrew the superfluous plants, and allowed only the three strongest to remain. These obtained stems of above 3 feet in height, with many leaves; on the 25th of July they began to blossom; on the 10th of August, they put forth seeds; and, on the 8th of September, ripe seed capsules, with completely ripe seeds, were obtained. On the 27th of October, the plants were removed from the soil, and treated as above.

TRIFOLIUM PRATENSE.

A

In Pure Sand.

The clover, which appeared on the 5th of May, grew at first pretty luxuriantly, but reached a height of only $3\frac{1}{2}$ inches by the 15th of October, when its leaves became suddenly brown, in consequence of which I removed it from the soil, and treated it as above.

B

In Artificial Soil.

The clover reached a height of 10 inches by the 15th of October; it was bushy and its colour was dark green. It was taken from the soil, in order to compare it with the former experiments, and was treated in the same way.

CONSTITUENTS OF THE ASHES OF THE SEEDS.

100 parts of dry seeds yield—

	Soluble in water.	Soluble in muriatic acid.	Silica.		Ashes in 100 parts.
Vicia faba . .	1·562	0·563	0·442	=	2·567
Hordeum vulgare . .	0·746	0·563	1·123	=	2·432
Avena sativa . .	0·465	0·277	2·122	=	2·864
Polygonum fagopyrum	0·823	0·547	0·152	=	1·522
Trifolium pratense .	1·218	3·187	0·282	=	4·687

CONSTITUENTS OF THE PLANTS GROWN IN PURE SAND AND IN THE ARTIFICIAL SOIL.

		Soluble in water.	Soluble in muriatic acid.	Insoluble in water and muriatic acid. (Silica).		Ashes:
Vicia sativa, 15 grains plants dried in air . .	In sand . .	0·516	0·375	0·135	=	1·026
	In artificial soil	0·693	0·821	0·320	=	1·834
Hordeum vulgare, 12·5 grains plants . .	Sand . . .	0·123	0·195	0·355	=	0·673
	Soil . .	0·167	0·226	0·487		0·880
Avena sativa, 13 grains plants	Sand . . .	0·216	0·024	0·354	=	0·594
	Soil . .	0·255	0·030	0·461	=	0·746
Polygonum fagopyrum . .	Sand (12 grains plants)	0·086	0·094	0·045	=	0·225
	Soil (12·7 gr. plants) .	0·148	0·226	0·133	=	0·507
Nicotiana tabacum . .	Sand (4 grains plants) . .	0·223	0·252	0·031	=	0·506
	Soil (12·5 plants . . .	1·146	2·228	0·549	=	3·923
Trifolium pratense, 14·5 grains plants .	Sand . .	0·522	0·350	0·091		0·963
	Soil . . .	0·659	0·943	0·082	=	1·684

The above numbers express the unequal weight of mineral nutritive substances taken up from the sand and artificial soil by equal weights of the different plants mentioned. The relative and not the absolute weight of the component parts of the ashes is given. For example, the five tobacco plants grown in sand gave 0·506 gr. in ashes, whilst the three which grew in the artificial soil gave 3·923; five would, therefore give 6·525 gr. The proportion of the mineral ingredients taken up by five tobacco plants from the sand, and that taken up from the artificial soil is as 10 : 120. In an equal space of time, those which grew in the artificial soil absorbed nearly thirteen times more of inorganic ingredients than those in the sand, and the whole development of the plant was exactly in proportion to the supply of food. Wiegmann and Polstorf subtracted the ashes of the seed used from the numbers in the last line, which show the amount of ashes in a given weight of the grown plant, but this has caused a small error in the numbers, as all the plants grown in the sand were reduced to ashes, and a corresponding part only of those grown in the artificial soil. The weight of the seed of every plant grown was 3 grammes, if we except the tobacco, which was not weighed.

TABLE

Showing the Amount of Moisture in Vegetable Substances according to the Analyses of Boussingault.

	Subst. dried at 110° C.	Water.		Subst. dried at 110° C.	Water.
Wheat	0·855	0·145	Beet	0·122	0·878
Rye	0·834	0·166	Turnips	0·075	0·925
Oats	0·792	0·208	Helianthus Tub.	0·208	0·792
Wheat straw	0·740	0·260	Peas	0·914	0·086
Rye straw	0·813	0·187	Pea Straw	0·882	0·118
Oat straw	0·713	0·287	Clover Stalk	0·790	0·210
Potatoes	0·241	0·759	Stalk of Hel. Tub.	0·871	0·129

COMPOSITION OF MANURE DRIED IN VACUO AT 110° C.

	Carbon.	Hydrogen.	Oxygen.	Nitrogen.	Salts & Earth.
I.	32·4	3·8	25·8	1·7	36·3
II.	32·5	4.1	26·0	1·7	35·7
III.	38·7	4·5	28·7	1·7	26·4
IV.	36·4	4·0	19·1	2·4	38·1
V.	40·0	4·3	27·6	2·4	25·7
VI.	34·5	4·3	27·7	2·0	31·5
Mean.	35·8	4·2	25·8	2·0	32·2

These results show that the quantity of manure necessary for one hectare of land during five years ought to contain:—

	Kilogrammes.
Carbon	3637.6
Hydrogen	426.8
Oxygen	2521.5
Nitrogen	203.2
Salts and earths	3271·9
	10061·0

COMPOSITION OF THE PRODUCE OF THE LAND, DRIED IN VACUO AT 110°. C.

	With the Ashes.					Without the Ashes.			
	Carbon.	Hydrogen.	Oxygen.	Nitrogen.	Ashes.	Carbon.	Hydrogen.	Oxygen.	Nitrogen.
Wheat	46·1	5·8	43·4	2·3	2·4	47·2	6·0	44·4	2·4
Rye	46·2	5·6	44·2	1·7	2·3	47·3	5·7	45·3	1·7
Oats	50·7	6·4	36·7	2·2	4·0	52·9	6·6	38·2	2·3
Wheat straw	48·4	5·3	38·9	0·4	7·0	52·1	5·7	41·8	0·4
Rye straw	49·9	5·6	40·6	0·3	3·6	51·8	5·8	42·1	0·3
Oat straw	50·1	5·4	39·0	0·4	5·1	52·8	5·7	41·1	0·4
Potatoes	44·0	5·8	44·7	1·5	4·0	45·9	6·1	46·4	1·6
Beet	42·8	5·8	43·4	1·7	6·3	45·7	6·2	46·3	1·8
Turnips	42·9	5·5	42·3	1·7	7·6	46·3	6·0	45·9	1·8
Helianthus tub.	43·3	5·8	43·3	1·6	6·0	46·0	6·2	46·1	1.7
Yellow peas	46·5	6·2	40·0	4·2	3·1	48·0	6·4	41·3	4·3
Pea straw	45·8	5·0	35·6	2·3	11·3	51·5	5·6	40·3	2·6
Red clover stalk	47·4	5·0	37·8	2·1	7·7	51·3	5·4	41·1	2·2
Stalk of Hel. tub.	45·7	5·4	45·7	0·4	2·8	47·0	5·6	47·0	0·4

1.—ROTATION.

Year.	Substances.	Produce of a Hectare.	Dry Produce.	Carbon.	Hydrog.	Oxyg.	Nitrog.	Salts and Earths.
		Kilogr.	Kilogr.	Kilogr.	Kilogr.	Kilogr.	Kilogr.	Kilogr.
1	Potatoes	12800	3085	1357·4	178·9	1379·0	46·3	123·4
2	Wheat	1343	1148	529·3	66·6	498·2	26·4	27·5
	Wheat straw	3052	2258	1093·0	119·7	878·2	9·0	158·1
3	Clover	5100	4029	1909·7	201·5	1523·0	84·6	310·2
4	Wheat	1659	1418	653·8	82·2	615·4	32·6	34·0
	Wheat straw	3770	2790	1350·4	147·8	1085·3	11·2	195·3
	Turnips	9550	716	307·2	39·3	302·9	12·2	54·4
5	Oats	1344	1064	539·5	68·0	390·5	23·3	42·6
	Oat straw	1800	1283	642·8	69·3	500·4	5·1	65·4
	Total	40418	17791	8383·1	973·3	7172·9	250·7	1010·9
Manure used		49086	10161	3637·6	426·8	2621·5	203·2	3271·9
Difference			+7630	+4745·5	+546·5	+4551·4	+47·5	—2261·0

2.—ROTATION

Year.	Substances.	Produce of a Hectare.	Dry Produce.	Carbon.	Hydr.	Oxygen.	Nitro.	Salts and Earths.
		Kilogr.	Kilogr.	Kilogr.	Kilogr.	Kilogr.	Kilogr.	Kilogr.
1	Turnips . .	26000	3172	1357·7	184·0	1376·7	53·9	199·8
2	Wheat . .	1185	1013	467.0	58·8	439·6	23·3	24·3
	Wheat straw	2693	1993	964·0	105·6	775·3	8·0	139·5
3	Clover . .	5100	4029	1909·7	201·5	1523·0	84·6	310·2
4	Wheat . .	1659	1418	653·8	82·2	615·4	32·6	34·0
	Wheat straw	3770	2790	1350·4	147·8	1085·3	11·2	195·3
	Beet . . .	9550	716	307·2	39·3	302·9	12·2	54·4
5	Oats . . .	1344	1064	539·5	68·0	390·5	23·3	42·6
	Oat straw .	1800	1283	642·8	69·3	500·4	5·1	65·4
	Total .	53101	17478	8192·1	956·5	7009·1	254·2	1065·5
Manure used . .		49086	10161	3637·6	426·8	2621·5	203·2	3271·9
Difference . . .			+7317	+4554·5	+529·7	+4387·6	+51·0	—2206·4

3.—ROTATION.

Year.	Substances.	Prod. of a Hectare.	Dry Produce.	Carbon.	Hydr.	Oxygen.	Nitro.	Salts and Earths.
		Kilogr.	Kilogr.	Kilogr.	Kilogr.	Kilogr.	Kilogr.	Kilogr.
1	Potatoes. .	12800	3085	1357·4	178·9	1379·0	46·3	123·4
2	Wheat . .	1343	1148	529·3	66·6	498·2	26·4	27·5
	Wheat straw	3052	2258	1093·0	119·7	878·2	9·0	158·1
3	Clover stalk	5100	4029	1909·7	201·5	1523·0	84·6	310·2
4	Wheat . .	1659	1418	653·8	82·2	615·4	32·6	34·0
	Wheat straw	3770	2790	1350·4	147·8	1085·3	11·2	195·3
	Turnips . .	9550	716	307·2	39·3	302·9	12·2	54·4
5	Peas . . .	1092	998	464·1	61·9	399·2	41·9	30·9
	Pea straw .	2790	2461	1127·3	123·0	876·1	56·6	278·1
6	Rye . . .	1679	1394	644·0	78·1	616·1	23·7	32·1
	Rye straw .	3731	3033	1513·5	169·8	1231·4	9·1	109·2
	Total .	46566	23330	10949·7	1268·8	9404·8	353·6	1353·2
	Manure used	58900	12192	4364·2	512·2	3145·5	243·8	3925·8
	Difference .		+11138	+ 6585·5	+756·6	+6259·3	+109·8	—2572·6

4.—ROTATION.

Year.	Substances.	Produce of a Hectare	Dry Produce.	Carbon.	Hydro.	Oxygen.	Nitro.	Salts and Earths.
		Kilogr.	Kilogr.	Kilogr.	Kilogr.	Kilogr.	Kilogr	Kilogr.
1	Manured fallow	—	—	—	—	—	—	—
2 & 3	Wheat . .	3318	2836	1037·4	164·5	1230·8	65·2	68·1
	Wheat straw . .	7500	5550	2686·2	294·2	2159·0	22·2	388·5
	Total . .	10818	8386	3723·6	458·7	3389·8	87·4	456·6
	Manure used . .	20000	4140	1482·1	173·9	1068·1	82·8	1333·1
	Difference . .		+ 4246	+2241·5	+284·8	+2321·7	+4·6	—876·5

5.—CULTIVATION OF HELIANTHUS TUBEROSUS.

Year.	Substances.	Produce of a Hectare.	Dry Produce.	Carbon.	Hydro.	Oxygen.	Nitro.	Salts and Earths.
		Kilogr.	Kilogr.	Kilogr.	Kilogr.	Kilogr.	Kilogr.	Kilogr.
1 & 2	Helianthus Tub.	52880	11000	4763·0	638·0	4763·0	176·0	660·0
	Its woody stalk	28200	24562	11224·7	1326·3	11224·7	98·2	687·2
	Total . .	81080	35562	15987·7	1964·3	15987·7	274·2	1347·2
	Manure used .	45450	9408	3368·1	395·1	2427·3	188·2	3029·3
	Difference .		+26154	+12619·6	+1569·2	+13560·4	+86·0	+1682·1

REVIEW OF THE EXPERIMENTS.

Rotation.	Dry manure on 1 Hectare in 1 year.	Nitrogen of the Manure.	Dry Produce in one year.	Nitrogen of the Produce.	Gain of Organic Matter.	Gain of Nitrogen.
	Kilogr.	Kilogr.	Kilogr.	Kilogr.	Kilogr.	Kilogr.
1.	2032	40·6	3558	50·1	1526	9·5
2.	2032	40·6	3495	50·8	1463	10·2
3.	2032	40·6	3888	58·9	1856	18·4
4	1360	25·8	2795	29·1	1435	3·3
5.	4704	94·1	17781	137·1	13087	43·0

ANALYSIS

Of the Ashes of some Plants made in the Giessen Laboratory, by Hertwig (Annalen de Pharmacie), 46.

100 Parts contain	Beechwood Ashes.	Beech Bark.	Fir Wood.	Fir Bark.	Fir Leaves.	Pine Leaves.	Havana Tobacco.	Hanover Tobacco.	Straw of Field Beans.	Pea Straw. 1.	Pea Straw. 2.	Potatoe Straw.	Clover, Lucerne.
Carb. potash	11·72		11·30				6·18		13·32	4·16	4·34		23·47
Carb. soda	12·37	3·02	7·42	2·95		10·72	19·4	1·61	16·06	8·27		4·69	8·16
Sulph. potash	3·49					1·95		11·11	32·4	10·75	11·99		2·23
Common salt							8·64	9·24	0·28	4·63	3·72	2·28	2·27
Sulphuric soda							7·39	1·09					
Silicate soda						3·90							
Carb. lime	49·54	64·76	50·94	64·98		63·32	51·38	40·00	39·50	47·81	49·73	43·68	41·61
Magnesia	7·74	16·90	5·60	0·93		18·6	7·09	4·27	1·92	4·05	1·38	3·76	6·41
Phosphate of lime . . .	3·32	2·71	3·43	5·03		6·35	9·04	17·95	6·43	5·15	1·15	5·73	11·80
Phosphate of magnesia . .	2·92	0·66	2·90	4·18					6·66	4·37	7·82		0·91
Phosphate of iron . . .	0·76	0·46	1·04	1·04		0·88			3·49	0.90	3·64	1·30	0·81
Phosphate of alum	1·51	0·84	1·75	2.42		0·71				1·20		2·75	
Phosphate of manganese .	1·59												
Silica	2·46	9·04	13·37	17·28		10·31	8·26	15·25	7·97	7·81	15·54	29·81	2·26

ANALYSIS OF THE ASHES OF THE STRAW OF RYE, BY DR. FRESENIUS.

A.—*Ingredients soluble in water and muriatic acid.*

Potash united to silicic acid	6·88	
Sulphate of potash	1·75	
Chloride of potassium	0·25	
Chloride of sodium	0·56	
Lime united to silicic acid	4·19	
Magnesia	0·76	
Phosphate of lime	2·50	
Phosphate of magnesia	1·28	
Phosphate of oxide of iron	1·57	
Small quantity of phosphate of protox. of manganese.		19·74

B.—*Residue insoluble in water and muriatic acid.*

Potash united to silicic acid	9·21	
Lime united to silicic acid	3·43	
Magnesia united to silicic acid	1·16	
Phosphate of iron	1·63	
Phosphate of protoxide of manganese	Traces.	
Silicic acid	63·89	
Carbonaceous matter	0·94	
		80·26
		100·00

Soluble and insoluble together.

Potash united to silicic acid	16·09
Sulphate of potash	1·75
Chloride of potash	0·25
Chloride of sodium	0·56
Lime united to $Si\ O^3$	7·62
Magnesia	1·92
Phosphate of lime	2·50
Phosphate of magnesia	1·28
Phosphate of oxide of iron	3·20
Small quantity of phosphate of protoxide of manganese.	
Silicic acid	63·89
Carbonaceous matter	0·94
	100·00

ASHES ANALYSED BY BERTHIER.

—	Wood of White Beech.	Charcoal of White Beech.	Charcoal of Red Beech.	Oak Wood.	Oak Bark.	Linden Wood.	Mohaleb Cherry Wood.	Wood of Red-Berried Elder.	Wood of Cervis Siliquastrum.	Alder Wood.	Fir Wood.		Pine Charcoal.	Wheat Straw.	Rhine Fern.	Potato Straw.	Tobacco Root.
Per cent of ashes		2·65	3·00	2·50	6·00	5·00	1·60	1·64	1·70			0·83	1·24	4·40		15·00	
100 *ashes contain*—																	
Soluble in water	18·9	1·72	1·60	12·0	5·0	10·8	1·60	31·5	19·0	18·8	25·7	50·0	13·6	9·0	29·0	4·2	12·3
Insoluble in water	81·1	78·9	82·0	88·0	95·0	89·2	84·0	68·5	81·0	81·2	74·3	50·0	86·4	81·0	71·0	95·8	87·7
100 *soluble parts contain*—																	
Carbonic acid		24·60	22·4	24·0	23·2	27·42	20·0	24·0	24·9	6·6	30·2	13·5	20·75	Trace.	15·0	6·2	10·0
Sulphuric acid		7·25	7·3	8·1	6·0	7·53	6·0	6·4	3·1	0.3	3·1	6·9	12·00	2·0	5·6	23·0	10·3
Muriatic acid		4·61	5·2	0·1	0·7	1·80	10·0	0·4	0·5		0·3	0	6·63	13·0	11·0	12·0	18·26
Silica		1·00	1·0	0·2	0·8	1·61	1·0	0·2	1·0		1·0	2·0	1.33	35·0			
Potash		50·65										28·2	31·66				
Soda		12·10	64·1	67·6	69·3	60·64	63·0	67·0	70·5		65·4	41·5	15·33	50·0	68·4	58·8	61·44
Water												7·9					
100 *insoluble parts contain*—																	
Carbonic acid	33·2	29·20	32·9	39·6	38·5	39·8	34·0	31·4	34·0	31·0	23·0	21·5	36·0		25·3		
Phosphoric acid	10·0	8·80	5·7	0·8		2·8	6·3	8·3	7·5	7·7	4·2	1·8	1·0	1·2	8·0		
Silica	5·0	3·90	5·8	3·8	1·1	2·0	1·8	3·2	2·4	5·0	8·0	13·0	4·6	75·0	21·8	38·0	
Lime	38·6	42·70	42·6	54·8	50·1	51·8	48·8	49·2	46·0	50·2	39·8	27·2	42·3	5·8	42·8		
Magnesia	7·8	6·95	7·0	0·6	0·8	2·2	70·	2·5	7·2	2·5	4·4	8·7	10·5		0·7		
Oxide of iron	1·4	0·10	1·5			0·1	0·5	1·1	1·3	3.6	14·1	22·3	0·1	2·5	0·7		
Oxide of manganese	3·4	6·95	4·5		7·4	0·6	0·8	1·8	0·7		6·0	5·5	0·4		0·7		
Carbon		1·40			2·1								4·8	7·5			
Potash														8·0			

ANALYSES OF ASHES BY BERTHIER.

—	Wood of Walnut Tree.	Wood of the White Mulberry-tree.		Orange-tree Wood.	Wood of Quercus Alba.	Wood of Holm Oak.	Birch Wood.	Bean-tree Wood.	Chesnut-tree Wood.
Per cent of ashes	1·57	1·60					1·00	1·25	
100 *ashes contain*—									
Soluble in water	15·4	15·0	25·0	9·6	7·5		16,0	31·5	14·6
insoluble in water	84·6	85·0	75·0	90·4	92·5		84·0	68·5	85·4
100 *soluble parts contain*—									
Carbonic acid	20·2		23·0	37·0			17·0		18·75
Sulphuric acid	5·1		8·3				2·3	8·2	8·75
Muriatic acid	0·5		4·0	4·0			0·2	2·0	0·50
Silica	0·5						1·0	1·7	2·75
Potash } Soda and water }	73·2		5·20	5·9			79·5		69·25
Soda and water			11·5						
100 *insoluble parts contain*—									
Carbonic acid	3·70	27·1	42·0	33·5	41·4	35·81	31·0	19·0	30·60
Phosphoric acid	4·8	11·6	1·8	1·9	3·0	2·81	4·3	18·4	1·81
Silica	4·2	7·7	2·9	6·0	3·3	5.05	5·5	8·0	8·45
Lime	42·4	46·7	46·1	45·0	50·3	48·34	52·2	45·6	51·10
Magnesia	4·4	5·2	4·6	7·0	1·0	2·40	3·0	9·0	3·80
Oxide of iron } Oxide of manganese }	4·0	0·3	0·5	1·0	1.0	4·78	0·5		3·50
Oxide of manganese		0·5	1·3				3·5		
Carbon				5·6					

ANALYSES OF ASHES, BY BERTHIER.

	Fern.	Wheat Straw.	Share Grass.	Heath.	Rhine Fern.
Sulphate of potash	0·70	0·4	12·0	5·0	3·3
Chloride of potassium	Trace.	3·2	11·4	1·2	9·0
Carbonate of potash		Trace.		6·8	16·7
Potash with silica		13·0			
Silica	73·0	71·5	50·8	37·5	16·5
Carbonate of lime	24·8	9·6	6·2	28·0	43·4
Sulphate of lime			14·4		
Phosphate of lime	1·0	2·3	2·2	13·0	10·0
Magnesia	0·5		3·0	1·0	0·2
Oxide of iron				1·4	0·7
Oxide of manganese				6·1	0·2
	100·0	100·0	100·00	100·0	100·0

ANALYSES OF THE ASHES OF SOME PLANTS COLLECTED IN SOLY, IN THE ELECTORATE OF HESSE—1842. BY THON, IN THE GIESSEN LABORATORY.

	Wild Peas.	Yellow Kitchen Peas.	Small Field Beans.	Wheat.
Carbon	—	—	—	0·522
Silica	0·56	0·29	1·48	1·914
Phosphate of iron	0·68	1·94	0·34	0·525
Lime	2·96	2·70	5·38	3·011
Magnesia	7·75	6·91	7·35	13·575
Phosphoric acid	38·34	34·01	35·33	45·531
Sulphuric acid	2·63	4·28	2·28	—
Potash	27·12	35·20	21·71	24·170
Soda	17·43	10·32	21·07	10·340
Chloride of sodium	1·88	2·56	3·32	—
	99·35	98·21	98·26	99·588

The wheat ashes were from the reddish yellow wheat without a beard.

The wild peas and the small field beans are not eatable, but make remarkably good fodder for cattle.

The soil on which the plants grew consisted of clay and loam mixed with a little sand and carbonate of potash.

DE SAUSSURE'S INQUIRIES INTO THE ORIGIN OF THE MINERAL INGREDIENTS OF PLANTS.

	A. Granite from Mount Breven.	B. Stone of the La Salle Mount.	C. Limestone of Reculey de Thoiry.
Lime	1.74	24·36—(Carbonate.)	9·8
Alumina	13·25	4·	0·625
Silica	73·25	30·	
Oxide of iron and manganese	9·	13·	0·625
Carbonic acid		27·	
Petroleum			0·25
Loss	2·76	1·64	0·5

According to De Saussure the proportion of water, carbon, and ashes in the plants of these three mountains is as follows:—

100 parts fresh branches with leaves lost by drying in the air—

	Water.		Carbon.		Ashes.	
	A.	B.	A.	B.	A.	B.
Pine	51·17	48·24	10·62	11·11	1·187	1·128
Larch	58·07	57·13	10·16	10·39	0·961	0·926
Oleander	59·73	52·78	9·05	9·62	0·654	0·339
Bilberry	50·11	47·50	11.69	12·32	1·096	1·048
Juniper	55·19	49·45	10·63	11·46	1·081	1·082

100 Ashes contain	Bilberry.		Pine.		
	a	c	a	b	c
Carb. pot.			3·60	7·36	
Sulphate of potash	16·38	23·50	4·24	12·63	15
Chloride of potassium					
Carb. lime	40·35	53·70	46·34	51·19	63
Carbonate of magnesia	5·85		6·77		
Alumina	17·54	14·25	14·86	11·95	16
Silica	13·45	17 5	13·49	6·87	
Oxide of iron and manganese	6·43	6·80	10·52	10·00	3
	100·00	100·00	99·82	100·00	97

	Oleander.			Juniper.	
	a	b	c	a	c
Carb. sulph. and chloride of pot.	10·82	12·25	17·76	15·25	14·0
Carbonate of lime	30·02	57·00	71·54	64·25	66·6
Carbonate of magnesia	5·00				
Alumina	28·80	13·31	5·93		
Silica	14·86	5·44		0·25	
Oxide of iron and manganese	8·40	11·00	4·86		
	97·90	99·00	100·09		

100 *Parts of the Ashes of the Humus on which those Plants grew, gave—*

	Earth of Pine.		Earth of Oleander.	Earth of Juniper.
	a	c	c	c
Salts of potash		4·57	1·85	
Carbonate of lime	1·16	23·20	16·65	
Carbonate of magnesia	0·37			
Alumina	14·00	37·10	43·70	
Silica	60·50	13·71	14·27	13·0
Oxide of iron and manganese	16·00	16·10	23·83	
	92·03	94·68	100·30	

ANALYSES OF THE ASHES OF SOME PLANTS,

BY DE SAUSSURE.

CHEMICAL INQUIRIES INTO VEGETATION. LEIPZIG, 1805.

The method of analysis employed by De Saussure consisted of the following :—

A. The ashes were treated with water, and the parts soluble in it were introduced into the calculations, in the second and following columns.

B. The residue remaining undissolved in the last operation was dissolved in nitric acid, and evaporated to dryness; the portion now insoluble in water was silica.

C. By precipitating the solution obtained in B, with prussiate of potash, the iron and manganese was obtained, the amount of iron supplied by the re-agent being subtracted in the calculation.

D. By a further precipitation of the solution with ammonia, the earthy phosphates were obtained (lime and magnesia.)

E. By treating this precipitate with caustic potash, neutralising it with an acid, and precipitating it with ammonia, the earthy phosphates mixed with alumina (phosphate) were procured.

F. By a further precipitation of the liquid D with carbonate of soda, and by continued boiling, the earthy carbonates were obtained.

G. The difference of the products of these different operations, when compared with the total weight of the ashes analysed, expressed the few per cent. loss; and the quantity of salts with alkaline bases which were not dissolved by the first treatment with water.

According to the second mode of procedure, which Saussure considers to be the most exact, the ashes containing alkaline phosphates were analysed.

The ashes were dissolved in nitric acid, the lime and magnesia separated as phosphates, the liquid evaporated to dryness, and heated to redness with the addition of charcoal.

The residual salts were now saturated with acetic acid, dried and treated with alcohol; the phosphates and sulphates of potash, and chloride of potassium were left behind.

b. The residue was taken up by water and mixed with acetate of lime; the residue being dried and heated to redness, was treated with acetic acid (c), and the portion not dissolved was estimated as pure phosphate of lime, from which it was assumed that 100 parts (phosphate of lime), corresponded to 129 parts phosphate of potash; for 8 Ca O 3 P_2 O_5 gives 3 (P_2 O_5, 3 K O).

The solutions a and c, and also that remaining after the precipitation with acetate of lime, were evaporated and heated to redness; the residue was weighed and the chlorine and sulphuric acid estimated, and calculated as chloride of potassium and sulphate of potash. By subtracting the two latter salts, and also the potash calculated from the phosphate of lime, from the weight of the whole residue, the quantity of potash not existing as phosphate of potash was obtained.

Neither of these two methods can be considered accurate in the present day. But as all the analyses were executed according to similar methods, the results are always of value, and in so far are they comparable with each other.

ANALYSES OF ASHES OF PLANTS, BY SAUSSURE.

100 parts Ashes contain—	Ripe Horse Chesnuts.	Seeds of *Vicia Faba* (Beans).	Wheat Straw.	Grains of Wheat.	Bran of Wheat.	Stalk of Maize.	Seeds of Maize.	Barley Straw.	Barley Grains.	*Vicia Faba* in Blossom.	*Vicia Faba*, Straw without Seeds.
Potash	51·	22·45	12·5	15·	14·	59·	14·	16·	18·	57·25 Carbonate.	31
Phosphate of lime	28·	43·93	5.	32·	30·	9·7	47·5		9·2		
Chloride of potassium	} 3·	0·9	3·	0·16	0·16	2·5	0.25	0·5	0·25	} 12·	14
Sulphate of potash		2·	2·			1·25	0·25	3·5	1·5		2
Earthy phosphates	12·	27·92	6·2	44·5	46·5	5·	36·	7·75	32·5	18·	6
Earthy carbonates			1·			1·		12·5		5·	37·5
Silica	0·5		61·5	0·5	0·5	18·	1·	57·	35·5	2·	2·75
Metallic oxides	0·25	0·5	1·	0·25	0·25	0·5	0·12	0·5	0·25	0·5	0·57
Loss	5·25	2·3	7·8	7·59	8·59	3·05	0·88	2·25	2·8	8·25	6·

100 PARTS ASHES CONTAIN

Name of the Plant.	Ashes from 1000 parts of the dry plant.	Alkalies and Salts with alkaline bases.	Phosph. of Lime and Mag.	Metallic Oxide, Iron and Mang.	Carbonate of Earths.	Silica.	100 parts of Ashes give Soluble Salts.
Oak leaves from 10th May.	53·	72·24	24·	0.64	0·12	3·	47·
,, ——27th Sept.	55·	42·5	18·25	1·75	23·	14·5	17·
,, pealed branches .	4·	58·58	28·5	1·	12·25	0·12	26·
,, the bark of above . .	60·	29·75	4·5	1·75	63·25	0·25	7·
,, wood of same . . .	2·	59·25	4·5	2.25	32·	2·	38·6
,, sap of same . . .	4·	55·3	24·	2·	11·	7·5	32·
,, bark of same . . .	60·	28·5	3·	2·	66·	1·5	7·
,, mast of same . .	73·	29·75	3·75	1·	65·	0·5	7·
,, extract of the wood .	61·	51·					
Oakwood, mould . .	41·	32·5	10·5	14·	10·	32·	
Aqueous extract of the mould	111·	66·					
Leaves of the Populus Nigra from 26th May	66·	51·5	13·	125.	29·	5·	36·

100 PARTS ASHES CONTAIN

Name of Plant.	Ashes of 1000 parts of dried Plant.	Alkalies and Salts wth alkaline bases.	Phosphates of Lime and Magn.	Metallic Oxide, Iron Magn.	Carbonate of Earths.	Silica.	Soluble of Salts from 100 parts of ashes.
Leaves of the Poplar (*Populus Nigra*) 12 Sept.	93·	44·	7·	1·5	36·	11·5	26·
Stem of the Poplar . . .	8·	50·5	16·75	1·5	27·	3·3	26·
Bark of the same . . .	72·	29·2	5·3	1·5	60·	4·	6·
Leaves of the Hazelnut (*Coryllus Avellana*) 1st May	61·	50·7	23·3	1·5	22·	2·5	26·
Ditto, 22nd June . . .	62·	30·	19·5	2·	44·1	4·	22·7
,, 20th Sept.	70·	44·	14·	1·5	29·	11·3	11·
Peeled branches . . .	5·	28·	12·	2·	36·	22·	24·5
Bark of same	62·	56.7	35·	0·12	8·	0·25	12·5

100 P. ASHES CONTAIN

Name of plant.	Ashes of 1000 p. of dried plant.	Alkalies, and their salts.	Phosph. of lime and magn.	Phosph. of metal oxides.	Carbonate of Earths.	Silica.	Soluble salts, from 100 p. of ashes.
Wood of Morus nigra	7·	41.38	2·25	0·25	56·	0·12	21·
Sap of the same	13·	47·5	27·25	0·25	24·	1·	26·
Bark of the same	89·	30·13	8·5	1·12	45·	15·25	7·
Inner part of the bark	88·	34·38	16·5	1·	48·	0·12	10·
Wood of the white beech, Carpinus betulus	6·	48·63	23·	2·25	26·	0·12	22·
Sap of same	7·	47·	36·	1·	15·	1·	18·
Bark of same	134·	34·88	4·5	0·12	59·	1·5	4·5
Horse chesnut	35·	9·5	Only those salts which are soluble in water are certain.				
Leaves of same, 10th May	72·	50·					
From 23d May to 23d July	84·	24·					
From 27th September	86.	13·5					

100 P. ASHES CONTAIN

Name of plant.	Ashes in 100 p. of the dry plant.	Alkalies and their salts.	Phosph. of lime and magnesia.	Phosph. of metal oxides.	Carbonate of earths.	Silica.	Soluble salts from 100 ashes.
Chesnut blossoms	71·	50·					
Sunflower, before blossoming, from 25th June	147·	79·67	6·7	0·12	11·56	1·5	68·
Ditto, from 23d July	137·	79·78	6·	0·12	12·	1·5	61·
Ditto with seed	69·25	22.5	0·5	4·	3·75		51·5
Pine leaves from sand from 20th June	29·	40·13	12·27	1·6	43·5	2·5	16·
Ditto, from siliceous land	29·	34·5	12·	5·5	29·	19·	15·
Bilberries (chalk soil) from 20th August	26·	36·38	18·	3·12	42·	0·5	17·
Bilberries (siliceous soils)	22·	41·5	22·	9·5	22·	5·	24·

PART II.

THE CHEMICAL PROCESSES OF FERMENTATION, DECAY, AND PUTREFACTION.

CHAPTER I.

CHEMICAL TRANSFORMATIONS.

WOODY fibre, sugar, gum, and all such organic compounds, suffer certain changes when in contact with other bodies—that is, they suffer DECOMPOSITION.

There are two distinct modes in which these decompositions take place in organic chemistry.

When a substance composed of two compound bodies, crystallised oxalic acid for example, is brought in contact with concentrated sulphuric acid, a complete decomposition is effected upon the application of a gentle heat. Now, crystallised oxalic acid is a combination of water with the anhydrous acid; but concentrated sulphuric acid possesses a much greater affinity for water than oxalic acid, so that it attracts all the water of crystallisation from that substance. In consequence of this abstraction of the water, anhydrous oxalic acid is set free; but, as this acid cannot exist in a free state; a division of its constituents necessarily ensues, by which carbonic acid and carbonic oxide are produced, and evolved in the gaseous form in equal volumes. In this example, the decomposition is the consequence of the removal of two constituents (the elements of water), which unite with the sulphuric acid, and its cause is the superior affinity of the acting body (the sulphuric acid) for water. In consequence of the removal of the component parts of water, the remaining elements enter into a new form; in place of oxalic acid, we have its elements in the form of carbonic acid and carbonic oxide.

This form of decomposition, in which the change is effected by the agency of a body which unites with one or more of the constituents of a compound, is quite analogous to the decomposition of inorganic substances. When we bring sulphuric acid and nitrate of potash together, nitric acid is separated in consequence of the affinity of sulphuric acid for potash; in consequence, therefore, of the formation of a new compound (sulphate of potash).

In the second form of these decompositions, the chemical affinity of the acting body causes the component parts of the decomposing body to combine, so as to form new compounds, of which either both, or only one, combine with the acting body. Let us take dry wood, for example, and moisten it with sulphuric acid; after a short time the wood is carbonised, while the sulphuric acid remains unchanged, with the exception of its being united with more water than it possessed before. Now, this water did not exist as such in the wood, although its elements, oxygen and hydrogen, were present; but by the chemical attraction of sulphuric acid for water, they were in a certain measure compelled to unite in this form; and, in consequence of this, the carbon of the wood was separated as charcoal.

HYDROCYANIC ACID and WATER, in contact with hydrochloric acid, are mutually decomposed. The nitrogen of the hydrocyanic acid, and the hydrogen of a certain quantity of the water, unite together and form AMMONIA; whilst the carbon and hydrogen of the hydrocyanic acid combine with the oxygen of the water and produce FORMIC ACID. The ammonia combines with the muriatic acid. Here the contact of muriatic acid with water and hydrocyanic acid causes a disturbance in the attraction of the elements of both compounds, in consequence of which they arrange themselves into new combinations, one of which—ammonia—possesses the power of uniting with the acting body.

Inorganic chemistry can present instances analogous to this class of decomposition also: but there are forms of organic chemical decomposition of a very different kind, in which none of the component parts of the decomposing matter enter into combination with the body which determines the decomposition. In cases of this kind a disturbance is produced in the mutual attraction of the elements of a compound, and they, in consequence, arrange themselves into one or into several new combinations, which are incapable of suffering further change under the same conditions.

When, by means of the chemical affinity of a second body, by the influence of heat, or through any other causes, the composition of an organic compound is made to undergo such a change, that its elements form two or more new compounds, this manner of decomposition is called a chemical TRANSFORMATION or METAMORPHOSIS. It is an essential character of chemical transformations, that none of the elements of the body decomposed are singly set at liberty.

The changes designated by the terms FERMENTATION, DECAY, and PUTREFACTION, are chemical transformations effected by an agency which has hitherto escaped attention, but the existence of which will be proved in the following pages.

CHAPTER II.

ON THE CAUSES WHICH EFFECT FERMENTATION, DECAY,* AND PUTREFACTION.

ATTENTION has been only recently directed to the fact, that a body in the act of combination or decomposition exercises an influence upon any other body with which it may be in contact. Platinum, for example, does not decompose nitric acid; it may be boiled with this acid without being oxidised by it, even when in a state of such fine division that it no longer reflects light (black spongy platinum). But an alloy of silver and platinum dissolves with great ease in nitric acid; the oxidation which the silver suffers, causes the platinum to undergo the same change; or, in other words, the latter body, from its contact with the oxidising silver, acquires the property of decomposing nitric acid.

Copper does not decompose water, even when boiled in dilute sulphuric acid; but an alloy of copper, zinc, and nickel, dissolves easily in this acid with evolution of hydrogen gas.

Tin decomposes nitric acid with great facility, but water with difficulty; and yet, when tin is dissolved in nitric acid, hydrogen is evolved at the same time, from a decomposition of the water contained in the acid, and ammonia is formed in addition to oxide of tin.

In the examples here given, the only combination or decom-

* An essential distinction is drawn in the following part of the work, between *decay* and *putrefaction* (*Verwesung und Fäulniss*), and they are shown to depend on different causes; but as the word *decay* is not generally applied to a distinct species of decomposition, and does not indicate its true nature, I shall in future, at the suggestion of the author, employ the term *eremacausis* (from ηρέμα by degrees, and καῦσις burning).—ED.

position which can be explained by chemical affinity is the last. In the other cases, electrical action ought to have retarded or prevented the oxidation of the platinum or copper while they were in contact with silver or zinc, but, as experience shows, the influence of the opposite electrical conditions is more than counterbalanced by chemical actions.

The same phenomena are seen in a less dubious form in compounds, the elements of which are held together by a feeble affinity. It is well known that there are chemical compounds of so unstable a nature, that changes in temperature and electrical condition, or even simple mechanical friction, or contact with bodies apparently totally indifferent, cause such a disturbance in the attraction of their constituents, that the latter enter into new forms, without any of them combining with the acting body. These compounds appear to stand but just within the limits of chemical combination, and agents exercise a powerful influence on them, which are completely devoid of action on compounds of a stronger affinity. Thus, by a slight increase of temperature, the elements of hypochlorous acid separate from one another with evolution of heat and light; chloride of nitrogen explodes by contact with many bodies, which combine neither with chlorine nor nitrogen at common temperatures; and the contact of any solid substance is sufficient to cause the explosion of iodide of nitrogen, or of fulminating silver.

It has never been supposed that the causes of the decomposition of these bodies should be ascribed to a peculiar power, different from that which regulates chemical affinity,—a power which mere contact with the down of a feather is sufficient to set in activity, and which, once in action, gives rise to the decomposition. These substances have always been viewed as chemical compounds of a very unstable nature, in which the component parts are in a state of such tension, that the least disturbance overcomes their chemical affinity. They exist only by the *vis inertiæ,* and any shock or movement is sufficient to destroy the attraction of their component parts, and consequently their existence as definite compounds.

Peroxide of hydrogen belongs to this class of bodies; it is decomposed by all substances capable of attracting oxygen from it, and even by contact with many bodies, such as platinum or silver, which do not enter into combination with any of its constituents. In this respect, its decomposition depends evidently upon the same causes as those which effect that of iodide of nitrogen, or of fulminating silver. Yet it is singular that the cause of the sudden separation of the component parts of peroxide of hydrogen has been viewed as different from those of common decomposition, and has been ascribed to a new power termed the CATALYTIC FORCE. Now, it has not been considered, that the presence of the platinum and silver serves here only to accelerate the decomsition; for without the contact of these metals, the peroxide of hydrogen decomposes spontaneously, although very slowly. The sudden separation of the constituents of peroxide of hydrogen differs from the decomposition of gaseous hypochlorous acid, or solid iodide of nitrogen, only in so far as the decomposition takes place in a liquid.

A remarkable action of peroxide of hydrogen has attracted much attention, because it differs from ordinary chemical phenomena. This is the reduction which certain oxides suffer by contact with this substance, on the instant at which the oxygen separates from the water. The oxides thus easily reduced, are those of which the whole, or part at least, of their oxygen is retained merely by a feeble affinity, such as the oxides of silver and of gold, and peroxide of lead.

Now, other oxides very stable in composition, effect the decomposition of peroxide of hydrogen, without experiencing the smallest change; but when oxide of silver is employed to effect the decomposition, all the oxygen of the silver is carried away with that evolved from the peroxide of hydrogen, and as a result of the decomposition, water and metallic silver remain. When peroxide of lead is used for the same purpose, half its oxygen escapes as a gas. Peroxide of manganese may in the same manner be reduced to the protoxide, with the liberation of oxygen, if there be present an acid to exercise an affinity for the protoxide and convert it into a soluble salt. If, for example, we add to peroxide of hydrogen sulphuric acid, and then peroxide of manganese in the state of fine powder, much more oxygen is evolved than the compound of oxygen and hydrogen could yield; and on examining the solution, we find a salt of the protoxide of manganese, so that half of the oxygen has been evolved from the peroxide of that metal.

A similar phenomenon occurs, when carbonate of silver is treated with several organic acids. Pyruvic acid, for example, combines readily with pure oxide of silver, and forms a salt of sparing solubility in water. But when this acid is brought in contact with carbonate of silver, the oxygen of part of the oxide escapes with the carbonic acid, and metallic silver remains in the state of a black powder. (Berzelius.)

Now no other explanation of these phenomena can be given, than that a body in the act of combination or decomposition enables another body, with which it is in contact, to enter into the same state. It is evident that the active state of the atoms of one body has an influence upon the atoms of a body in contact with it; and if these atoms are capable of the same change as the former, they likewise undergo that change; and combinations and decompositions are the consequence. But when the atoms of the second body are not of themselves capable of such an action, any further disposition to change ceases from the moment at which the atoms of the first body assume the state of rest, that is, when the changes or transformations of this body are quite completed.

This influence exerted by one compound upon the other, is exactly similar to that which a body in the act of combustion exercises upon a combustible body in its vicinity; with this difference only, that the causes which determine the commencement and duration of the condition of change are different. For the cause, in the case of the combustible body, is heat, which is generated every moment anew; whilst in the phenomena of decomposition and combination, which we are considering at present, the cause is a body in the state of chemical action, which exerts the decomposing influence only so long as this action continues.

Numerous facts show that motion alone exercises a considerable influence on chemical forces. Thus, the power of cohesion does not act in many saline solutions, even when they are fully saturated with salts, if they are permitted to cool whilst at rest. In such a case, the salt dissolved in a liquid does not crystallise; but when a grain of sand is thrown into the solution, or when it receives the slightest movement, the whole liquid becomes suddenly solid with the evolution of heat. The same phenomenon happens with water, for this liquid may be cooled much under 32° F. (0° C.), if kept completely undisturbed, but solidifies in a moment when put in motion.

The atoms of a body must in fact be set in motion before they can overcome the *vis inertiæ* so as to arrange themselves into certain forms. A dilute solution of a salt of potash, mixed with tartaric acid, yields no precipitate whilst at rest; but if motion is communicated to the solution by agitating it briskly, crystals of cream of tartar are instantly deposited. A solution of a salt of magnesia also, though not rendered turbid by the addition of phosphate of ammonia, deposits the phosphate of magnesia and ammonia on those parts of the vessel touched with the rod employed in stirring.

In the processes of combination and decomposition under consideration, motion, by overcoming the *vis inertiæ*, gives rise immediately to another arrangement of the atoms of a body, that is, to the production of a compound which did not before exist in it. Of course these atoms must previously possess the power of arranging themselves in a certain order, otherwise both friction and motion would be without the smallest influence.

The simple permanence in position of the atoms of a body, is the reason that so many compounds appear to present themselves, in conditions, and with properties, different from those which they possess when they obey the natural attractions of their atoms. Thus sugar and glass, when melted and cooled rapidly, are transparent, of a conchoidal fracture, and elastic and flexible to a certain degree. But the former becomes dull and opaque on keeping, and exhibits, by cleavage, crystalline faces which belong to crystallised sugar. Glass assumes also the same condition, when kept soft by heat for a long period; it becomes white, opaque, and so hard as to strike fire with steel. Now, in both these bodies, the atoms evidently have different positions in the two forms. In the first form their attraction did not act in the direction in which their power of cohesion was strongest. It is known also, that when sulphur is melted and cooled rapidly by throwing it into cold water, it remains transparent, elastic, and so soft that it may be drawn out into long threads; but that, after a few hours or days, it becomes again hard and crystalline.

The remarkable fact here is, that the amorphous sugar or sulphur returns again into the crystalline condition, without any assistance from an exterior cause; a fact which shows that their molecules have assumed another position, and that they possess, therefore, a certain degree of mobility, even in the condition of a solid. A very rapid transposition or transformation of this kind is seen in arragonite, a mineral which possesses exactly the same composition as calcareous spar, but of which the hardness and crystalline form prove that its molecules are arranged in a different manner. When a crystal of arragonite is heated, an interior motion of its molecules is caused by the expansion; the permanence of their arrangement is destroyed; and the crystal splinters with much violence, and falls into a heap of small crystals of calcareous spar.

It is impossible for us to be deceived regarding the causes of these changes. They are owing to a disturbance of the state of the equilibrium, in consequence of which the particles of the body put in motion obey either other affinities, or their own natural attractions.

But if it be true, as we have just shown it to be, that mechanical motion is sufficient to cause a change of condition in many bodies, it cannot be doubted that a body in the act of combination or decomposition is capable of imparting the same condition of motion or activity, in which its atoms are, to those of certain other bodies: or, in other words, of enabling other bodies with which it is in contact to enter into combinations, or suffer decompositions.

The reality of this influence has been already sufficiently proved by the facts derived from inorganic chemistry; but it is of much more frequent occurrence in the relations of organic matter, and causes very striking and wonderful phenomena.

By the terms FERMENTATION, PUTREFACTION, and EREMACAUSIS, are meant those changes in form and properties which compound organic substances undergo when separated from the organism, and exposed to the influence of water and a certain temperature. Fermentation and putrefaction are examples of the kind of decomposition which we have named transformations: the elements of the bodies capable of undergoing these changes arrange themselves into new combinations, in which the constituents of water generally take a part.

EREMACAUSIS (or decay) differs from fermentation and putrefaction, inasmuch as it cannot take place without the access of air, the oxygen of which is absorbed by the decaying bodies. Hence it is a process of slow combustion, in which heat is uniformly evolved, and occasionally even light. In the processes of decomposition termed fermentation and putrefaction, gaseous products are very frequently formed, which are either inodorous, or possess a very offensive smell.

The transformations of those matters which evolve gaseous products without odour, are now, by pretty general consent, designated by the term FERMENTATION; whilst to the spontaneous decomposition of bodies which emit gases of a disagreeable smell, the term PUTREFACTION is applied. But the smell is, of course, no distinctive character of the nature of the decomposition, for both fermentation and putrefaction are processes of decomposition of a similar kind, the one of substances destitute of nitrogen, the other of substances containing that element.

It has also been customary to distinguish from fermentation and putrefaction a particular class of transformations, viz., those in which conversions and transpositions are effected without the evolution of gaseous products. But the conditions under which the products of the decomposition present themselves are purely accidental; there is therefore no reason for the distinction just mentioned.

CHAPTER III.

FERMENTATION AND PUTREFACTION.

Several bodies appear to enter spontaneously into the states of fermentation and putrefaction, particularly such as contain nitrogen. Now, it is very remarkable, that very small quantities of these substances, in a state of fermentation or putrefaction, possess the power of causing unlimited quantities of similar matters to pass into the same state. Thus, a small quantity of the juice of grapes in the act of fermentation, added to a large quantity of the same fluid, which is not fermenting, induces the state of fermentation in the whole mass. So likewise the most minute portion of milk, paste, juice of the beet-root, flesh, or blood, in the state of putrefaction, causes fresh milk, paste, juice of the beet-root, flesh or blood, to pass into the same condition when in contact with them.

These changes evidently differ from the class of common decompositions effected by chemical affinity; they are chemical actions, conversions, or decompositions, excited by contact with bodies already in the same condition. In order to form a clear idea of these processes, analogous but less complicated phenomena must previously be studied.

The compound nature of the molecules of an organic body, and the phenomena presented by them when in relation with other matters, point out the true cause of these transformations. Evidence is afforded even by simple bodies, that in the formation of combinations, the force with which the combining elements adhere to one another is inversely proportional to the number of simple atoms in the compound molecule. Thus, protoxide of manganese by absorption of oxygen is converted into the sesquioxide, the peroxide, manganic and hypermanganic acids, the number of atoms of oxygen being augmented by ½, by 1, by 2, and by 5. But all the oxygen contained in these compounds, beyond that which belongs to the protoxide, is bound to the manganese by a much more feeble affinity; a red heat causes an evolution of oxygen from the peroxide, and the manganic and hypermanganic acids cannot be separated from their bases without undergoing immediate decomposition.

There are many facts which prove, that the most simple inorganic compounds are also the most stable, and undergo decomposition with the greatest difficulty, whilst those of a complex composition yield easily to changes and decompositions. The cause of this evidently is, that in proportion to the number of atoms which enter into a compound, the directions in which their attractions act will be more numerous.

Whatever ideas we may entertain regarding the infinite divisibility of matter in general, the existence of chemical proportions removes every doubt respecting the presence of certain limited groups or masses of matter which we have not the power of dividing. The particles of matter called equivalents in chemistry are not infinitely small, for they possess a weight, and are capable of arranging themselves in the most various ways, and of thus forming innumerable compound atoms. The properties of these compound atoms differ in organic nature, not only according to the form, but also in many instances according to the direction and place, which the simple atoms take in the compound molecules.

When we compare the composition of organic compounds with inorganic, we are quite amazed at the existence of combinations, in one single molecule of which, ninety or several hundred atoms or equivalents are united. Thus, the compound atom of an organic acid of very simple composition, acetic acid for example, contains twelve equivalents of simple elements; one atom of kinovic acid contains thirty-three; one of sugar thirty-six; one of amygdalin ninety; and one of stearic acid 138 equivalents. The component parts of animal bodies are infinitely more complex even than these.

Inorganic compounds differ from organic in as great a degree in their other characters as in their simplicity of constitution. Thus, the decomposition of a compound atom of sulphate of potash is aided by numerous causes, such as the power of cohesion, or the capability of its constituents to form solid, insoluble, or at certain temperatures volatile compounds with the body brought into contact with it, and nevertheless a vast number of other substances produce in it not the slightest change. Now, in the decomposition of a complex organic atom, there is nothing similar to this.

The empirical formula of sulphate of potash is SKO_4. It contains only 1 eq. of sulphur, and 1 eq. of potassium. We may suppose the oxygen to be differently distributed in the compound, and by a decomposition we may remove a part or all of it, or replace one of the constituents of the compound by another substance. But we cannot produce a different arrangement of the atoms, because they are already disposed in the simplest form in which it is possible for them to combine. Now, let us compare the composition of sugar of grapes with the above: here 12 eq. of carbon, 12 eq. of hydrogen, and 12 eq. of oxygen, are united together, and we know that they are capable of combining with each other in the most various ways. From the formula of sugar, we might consider it either as a hydrate of carbon, wood, starch, or sugar of milk, or further, as a compound of either with alcohol, or of formic acid with sachulmin.* Indeed we may calculate almost all the known organic compounds destitute of nitrogen from sugar, by simply adding the elements of water, or by replacing any one of its elementary constituents by a different substance. The elements necessary to form these compounds are therefore contained in the sugar, and they must also possess the power of forming numerous combinations amongst themselves by their mutual attractions.

* The black precipitate obtained by the action of hydrochloric acid on sugar.

Now, when we examine what changes sugar undergoes when brought into contact with other bodies which exercise a marked influence upon it, we find, that these changes are not confined to any narrow limits, like those of inorganic bodies, but are in fact unlimited.

The elements of sugar yield to every attraction, and to each in a peculiar manner. In inorganic compounds, an acid acts upon a particular constituent of the body, which it decomposes, by virtue of its affinity for that constituent, and never resigns its proper chemical character, in whatever form it may be applied. But when it acts upon sugar, and induces great changes in that compound, it does this not by any superior affinity for a base existing in the sugar, but by disturbing the equilibrium in the mutual attraction of the elements of the sugar amongst themselves. Muriatic and sulphuric acids, which differ so much from one another, both in characters and composition, act in the same manner upon sugar. But the action of both varies according to the state in which they are; thus, they act in one way when dilute, in another when concentrated, and even differences in their temperature cause a change in their action. Thus, sulphuric acid of a moderate degree of concentration converts sugar into a black carbonaceous matter, forming at the same time acetic and formic acids. But when the acid is more diluted, the sugar is converted into two brown substances, both of them containing carbon and the elements of water. Again, when sugar is subjected to the action of alkalies, a whole series of different new products are obtained; while oxidising agents, such as nitric acid, produce from it carbonic acid, acetic acid, oxalic acid, formic acid, saccharic acid, and many other products which have not yet been examined.

If from the facts here stated we estimate the power with which the elements of sugar are united together, and judge of the force of their attraction by the resistance which they offer to the action of bodies brought into contact with them, we must regard the atom of sugar as belonging to that class of compound atoms, which exist only by the *vis inertiæ* of their elements. Its elements seem merely to retain passively the position and condition in which they had been placed, for we do not observe that they resist a change of this condition by their own mutual attraction, as is the case with sulphate of potash.

Now it is only such compounds as sugar, compounds therefore possessing a very complex molecule, which are capable of undergoing the decompositions named fermentation and putrefaction.

We have seen that certain metals acquire a power which they do not of themselves possess, namely, that of decomposing water and nitric acid, by simple contact with other metals in the act of chemical combination. We have also seen, that peroxide of hydrogen and the persulphuret of the same element, in the act of decomposition, cause other compounds of a similar kind, but of which the elements are much more strongly combined, to undergo the same decomposition, although they exert no chemical affinity or attraction for them or their constituents. The cause producing these phenomena will be also recognised, by attentive observation, in those matters which excite fermentation or putrefaction. All bodies in the act of combination or of decomposition have the property of inducing those processes; or, in other words, of causing a disturbance of the statical equilibrium in the attractions of the elements of complex organic molecules, in consequence of which those elements group themselves anew, according to their special affinities.

The proofs of the existence of this cause of action can be easily produced; they are found in the characters of the bodies which effect fermentation and putrefaction, and in the regularity with which the distribution of the elements takes place in the subsequent transformations. This regularity depends exclusively on the unequal affinity which they possess for each other in an isolated condition. The action of water on wood, charcoal, and cyanogen, the simplest of the compounds of nitrogen, suffices to illustrate the whole of the transformations of organic bodies; of those in which nitrogen is a constituent, and of those in which it is absent.

CHAPTER IV.

ON THE TRANSFORMATION OF BODIES WHICH DO NOT CONTAIN NITROGEN AS A CONSTITUENT, AND OF THOSE IN WHICH IT IS PRESENT.

WHEN oxygen and hydrogen combined in equal equivalents, as in steam, are conducted over charcoal, heated to the temperature at which it possesses the power to enter into combination with one of these elements, a decomposition of the steam ensues. An oxide of carbon (either carbonic oxide or carbonic acid) is under all circumstances formed, while the hydrogen of the water is liberated. This proves that the attraction between carbon and oxygen is more powerful, at a high temperature, than that between oxygen and hydrogen. The carbon here is not shared between the elements of the water; for no carburetted hydrogen is formed.

Acetic and meconic* acids suffer a true transformation under the influence of heat, that is, their component elements are disunited, and form new compounds without any of them being singly disengaged. Acetic acid is converted into acetone and carbonic acid ($C_4 H_3 O_3 = C_3 H_3 O + CO^2$), and meconic acid into carbonic acid and komenic acid; whilst, by the influence of a higher temperature, the latter is further decomposed into pyro-meconic acid and carbonic acid.

Now, in these cases, the carbon of the bodies decomposed is shared between the oxygen and hydrogen; part of it unites with the oxygen and forms carbonic acid, whilst the other portion enters into combination with the hydrogen, and an oxide of a hydro-carbon is formed, in which all the hydrogen is contained.

* An acid existing in opium, and named from the Greek for poppy.

In a similar manner, when alcohol is exposed to a gentle red heat, its carbon is shared between the elements of the water—an oxide of a hydro-carbon which contains all the oxygen, aldehyde, and some gaseous compounds of carbon and hydrogen being produced.

It is evident that during transformations caused by heat, no foreign affinities can be in play, so that the new compounds must result merely from the elements arranging themselves, according to the degree of their mutual affinities, into new combinations which are constant and unchangeable in the conditions under which they were originally formed, but undergo changes when these conditions become different. If we compare the products of two bodies, similar in composition but different in properties, subjected to transformations under the influence of two different causes, we find that the manner in which the atoms are transposed is absolutely the same in both.

In the transformation of wood in marshy soils, by what we call putrefaction, its carbon is shared between the oxygen and hydrogen of its own substance, and of the water—carburetted hydrogen is consequently evolved, as well as carbonic acid, both of which compounds have an analogous composition (CH_2, CO_2).

Thus also, in the transformation of sugar called fermentation, its elements are divided into two portions; the one, carbonic acid, contains $\frac{2}{3}$ of the oxygen of sugar; and the other, alcohol, contains all its hydrogen.

In the transformation of acetic acid, produced by a red heat, carbonic acid, containing $\frac{2}{3}$ of the oxygen of the acetic acid, is formed, and acetone, containing all its hydrogen.

It is evident, from these facts, that the elements of a complex compound are left to their special attractions whenever their equilibrium is disturbed, from whatever cause this disturbance may proceed. It appears also, that the subsequent distribution of the elements, so as to form new combinations, always takes place in the same way, with this difference only, that the nature of the products formed is dependent upon the number of atoms of the elements entering into action; or, in other words, that the products differ *ad infinitum*, according to the composition of the original substance.

ON THE TRANSFORMATION OF BODIES CONTAINING NITROGEN.

By the examination of the substances most prone to fermentation and putrefaction, it is found that they are all, without exception, bodies containing nitrogen. In many of these compounds, a transposition of their elements occurs spontaneously as soon as they cease to form part of a living organism; that is, when they are drawn out of the sphere of attraction in which alone they are able to exist.

There are, indeed, bodies destitute of nitrogen which possess a certain degree of stability only when in combination, but which are unknown in an isolated condition, because their elements, freed from the power by which they were held together, arrange themselves according to their own natural attractions. Hypermanganic, manganic, and hyposulphurous acids, belong to this class of substances, which however are rare.

The case is very different with azotised bodies. It would appear that there is, in the nature of nitrogen, some peculiarity which gives its compounds the power to decompose spontaneously with so much facility. Now, nitrogen is known to be the most indifferent of all the elements: it evinces no particular attraction to any one of the simple bodies; and this character it preserves in all its compounds, a character which explains the cause of its easy separation from the matters with which it is united.

It is only when the quantity of nitrogen exceeds a certain limit, that azotised compounds have some degree of permanence, as is the case with melamin, ammelin, &c. Their liability to change is also diminished, when the quantity of nitrogen is very small in proportion to that of the other elements with which it is united, so that their mutual attractions preponderate.

This easy transposition of atoms is best seen in the fulminating silvers, in fulminating mercury, in the iodide or chloride of nitrogen, and in all fulminating compounds.

All other azotised substances acquire the same power of decomposition, when the elements of water are brought into play; and indeed the greater part of them are not capable of transformation, while this necessary condition to the transposition of their atoms is absent. Even the compounds of nitrogen most liable to change, such as those found in animal bodies, do not enter into a state of putrefaction when dry.

The result of the known transformations of azotised substances proves, that water does not merely act as a medium in which motion is permitted to the elements in the act of transposition, but that its influence depends on chemical affinity. When the decomposition of such substances is effected with the assistance of water, their nitrogen is invariably liberated in the form of ammonia. This is a fixed rule without any exceptions, whatever may be the cause which produces the decompositions. All organic compounds containing nitrogen, evolve the whole of that element in the form of ammonia when acted on by alkalies. Acids, and increase of temperature, produce the same effect. It is only when there is a deficiency of water or of its elements, that cyanogen or other azotised compounds are produced.

From these facts it may be concluded, that ammonia is the most stable compound of nitrogen; and that hydrogen and nitrogen possess a degree of affinity for each other surpassing the attraction of the latter body for any other element.

Already in considering the transformations of substances destitute of nitrogen, we have recognised the great affinity of carbon for oxygen as a powerful cause for effecting the disunion of the elements of a complex organic atom in a definite manner. But carbon is also invariably contained in azotised organic compounds, while the great affinity of nitrogen for hydrogen furnishes a new and powerful cause of change, and thus facilitates the transposition of their component parts. Thus in the bodies destitute of nitrogen we have one element, and in those containing that substance, two elements which mutually share the elements of water. Hence there are two opposite affinities at play, which mutually strengthen each other's action.

Now we know, that the most powerful attractions may be overcome by the influence of two affinities. Thus, a decomposition of alumina may be effected with the greatest facility, when the affinity of charcoal for oxygen, and of chlorine for aluminium, are both put in action, although neither of these alone has any influence upon it. There is in the nature and constitution of the compounds of nitrogen a kind of tension of their component parts, and a strong disposition to yield to transformations, which effect spontaneously the transposition of their atoms from the instant that water or its elements are brought in contact with them.

The characters of the hydrated cyanic acid, one of the simplest of all the compounds of nitrogen, are perhaps the best adapted to convey a distinct idea of the manner in which the atoms are disposed of in transformations. This acid contains carbon, nitrogen, hydrogen, and oxygen, in such proportions, that the addition of a certain quantity of the elements of water is exactly sufficient to cause the oxygen contained in the water and acid to unite with the carbon and form carbonic acid, and the hydrogen of the water and acid, to combine with the nitrogen and form ammonia. The most favourable conditions for a complete transformation are, therefore, associated in these bodies, and it is well known, that the disunion takes place on the instant in which the cyanic acid and water are brought into contact, the mixture being converted into carbonic acid and ammonia, with brisk effervescence.

This decomposition may be considered as the type of the transformations of all azotised compounds; it is putrefaction in its simplest and most perfect form, because the new products, the carbonic acid and ammonia, are incapable of further transformations.

Putrefaction assumes a totally different and much more complicated form, when the products, at first formed, undergo a further change. In these cases the process consists of several stages, of which it is impossible to determine when one ceases and the other begins.

The transformations of cyanogen, a body composed of carbon and nitrogen, and the simplest of all the compounds of nitrogen, will convey a clear idea of the great variety of products which are produced in such a case: it is the only example of the putrefaction of an azotised body which has been at all accurately studied.

A solution of cyanogen in water becomes turbid after a short time, and deposits a black, or brownish-black matter, which is a combination of ammonia with another body, produced by the simple union of cyanogen with water. This substance is insoluble in water, and is thus enabled to resist further change.

A second transformation is effected by the cyanogen being shared between the elements of the water, in consequence of which CYANIC ACID is formed by a certain quantity of the cyanogen combining with the oxygen of the water; while HYDROCYANIC ACID is also formed, by another portion of the cyanogen uniting with the hydrogen thus liberated.

Cyanogen experiences a third transformation, by which a complete disunion of its elements takes place, these being divided between the constituents of the water. OXALIC ACID is the one product of this disunion, and AMMONIA the other.

Cyanic acid, the formation of which has been mentioned above, cannot exist in contact with water, being decomposed immediately into carbonic acid and ammonia. The cyanic acid, however, newly formed in the decomposition of cyanogen, escapes this decomposition by entering into combination with the free ammonia, by which means UREA is produced.

The hydrocyanic acid is also decomposed into a brown matter containing hydrogen and cyanogen, the latter in greater proportion than it does in the gaseous state. Oxalic acid, urea, and carbonic acid, are also formed by its decomposition, and FORMIC ACID and AMMONIA are produced by the decomposition of its radical.

Thus, a substance consisting of only two elements (carbon and nitrogen) yields, in contact with water, eight totally different products. Several of these products are formed by the transformation of the original body, its elements being shared between the constituents of water; others are produced in consequence of a further change in those first formed. The urea and carbonate of ammonia are generated by the combination of two of the products, and in their formation the whole elements have assisted.

These examples show, that the results of decomposition by fermentation and putrefaction comprehend very different phenomena. The first kind of transformation is, the transposition of the elements of one complex compound, by which new compounds are produced with or without the assistance of the elements of water. In the products newly formed in this manner, either the same proportions of those component parts which were contained in the matter before transformation, are found, or with them, an excess, consisting of the constituents of water which had assisted in promoting the disunion of the elements.

The second kind of transformations consists of the transpositions of the atoms of two or more complex compounds, by which the elements of both arrange themselves mutually into new products, with or without the co-operation of the elements of water. In this kind of transformations, the new products contain the sum of the constituents of all the compounds which had taken a part in the decomposition.

CHAPTER V.

FERMENTATION OF SUGAR.

THE peculiar decomposition of sugar may be viewed as a type of all the transformations designated fermentation.*

The analysis of sugar from the cane, proves that it contains the elements of carbonic acid and alcohol, *minus* 1 atom of water. The alcohol and carbonic acid produced by the fermentation of a certain quantity of sugar, contain together one equivalent of oxygen, and one equivalent of hydrogen, the elements, therefore, of one equivalent of water, more than the sugar contained. The excess of weight in the products is thus explained most satisfactorily; it is owing, namely, to the elements of water having taken part in the metamorphosis of the sugar.

It is known that 1 atom of sugar contains 12 equivalents of carbon, both from the proportions in which it unites with bases, and from the composition of sacharric acid, the product of its oxidation. Now none of these atoms of carbon are

* When yeast is made into a thin paste with water, and 1 cubic centimètre of this mixture introduced into a graduated glass receiver filled with mercury, in which are already 10 grammes of a solution of cane-sugar, containing 1 gramme of pure solid sugar; it is found, after the mixture has been exposed for 24 hours to a temperature of from 20 to 25 C. (68—77 F.), that a volume of carbonic acid has been formed, which, at 0° C. (32° F.) and an atmospheric pressure indicated by 0·76 mètre Bar. would be from 245 to 250 cubic centimètres. But to this quantity we must add 11 cubic centimètres of carbonic acid, with which the 11 grammes of liquid would be saturated; so that in all, 255—259 cubic centimètres of carbonic acid are obtained. This volume of carbonic acid corresponds to from 0·503 to 0·5127 grammes by weight. Thénard also obtained from 1 gramme of sugar 0·5262 grammes of absolute alcohol. 100 parts of cane-sugar yield, therefore, of alcohol and carbonic acid together 103·89 parts. Now in these two products are contained 42 parts of carbon, or exactly the quantity originally present in the sugar.

contained in the sugar as carbonic acid, because the whole quantity is obtained as oxalic acid, when sugar is treated with hypermanganate of potash (Gregory); and as oxalic acid is a lower degree of the oxidation of carbon than carbonic acid, it is impossible to conceive that the lower degree should be produced from the higher, by means of one of the most powerful agents of oxidation which we possess.

It can be also proved, that the hydrogen of the sugar does not exist in it in the form of alcohol, for it is converted into water and a kind of carbonaceous matter, when treated with acids, particularly with such as contain no oxygen; and this manner of decomposition is never suffered by a compound of alcohol.

Sugar contains, therefore, neither alcohol nor carbonic acid, so that these bodies must be produced by a different arrangement of its atoms, and by their union with the elements of water.

In this metamorphosis of sugar, the elements of the yeast, by contact with which its fermentation was effected, take no appreciable part in the transposition of the elements of the sugar; for in the products resulting from the action, we find no component part of this substance. The same sugar, which in contact with yeast, yields alcohol and carbonic acid, gives rise, when in contact with putrefying white cheese, to butyric acid, hydrogen being at the same time liberated.

We may now study the fermentation of a vegetable juice, containing not only saccharine matter, but also such substances as albumen and gluten. The juices of parsneps, beet-roots, and onions, are well adapted for this purpose. When such a juice is mixed with yeast at common temperatures, it ferments like a solution of sugar. Carbonic acid gas escapes from it with effervescence, and in the liquid, alcohol is found in quantity exactly corresponding to that of the sugar originally contained in the juice. But such a juice undergoes spontaneous decomposition at a temperature of from 95° to 104° (35°—40° C.). Gases possessing an offensive smell are evolved in considerable quantity, and when the liquor is examined after the decomposition is completed, no alcohol can be detected. The sugar has also disappeared, and with it all the azotised compounds which existed in the juice previously to its fermentation. Both were decomposed at the same time; the nitrogen of the azotised compounds remains in the liquid as ammonia, and, in addition to it, there are three new products, formed from the component parts of the juice. One of these is lactic acid, the slightly volatile compound found in putrid animal mixtures; the other is the crystalline body which forms the principal constituent of manna; and the third is a mass resembling gum-arabic, which forms a thick viscous solution with water. These three products weigh more than the sugar contained in the juice, even without calculating the weight of the gaseous products. Hence they are not produced from the elements of the sugar alone. None of these three substances could be detected in the juice before fermentation. They must, therefore, have been formed by the interchange of the elements of the sugar with those of the foreign substances also present. It is this mixed transformation of two or more compounds which receives the special name of PUTREFACTION.

YEAST OR FERMENT.

When attention is directed to the condition of those substances which possess the power of inducing fermentation and putrefaction in other bodies, evidence is found in their general characters, and in the manner in which they combine, that they all are bodies, the atoms of which are in the act of transposition.

The characters of the remarkable matter deposited in an insoluble state during the fermentation of beer, wine, and vegetable juices, may first be studied.

This substance, called YEAST OR FERMENT, from the power which it possesses of causing fermentation in sugar, or saccharine vegetable juices, possesses all the characters of a COMPOUND OF NITROGEN IN THE STATE OF PUTREFACTION AND EREMACAUSIS.

Like wood in the state of eremacausis, yeast converts the oxygen of the surrounding air into carbonic acid, but it also evolves this gas from its own mass, like bodies in the state of putrefaction. (Colin.) When kept under water, it emits carbonic acid, accompanied by gases of an offensive smell, (Thénard,) and is at last converted into a substance resembling old cheese. (Proust.) But when its own putrefaction is completed, it has no longer the power of inducing fermentation in other bodies. The presence of water is quite necessary for sustaining the properties of ferment, for by simple pressure its power to excite fermentation is much diminished, and is completely destroyed by drying. Its action is arrested also by the temperature of boiling water, by alcohol, common salt, an excess of sugar, oxide of mercury, corrosive sublimate, pyroligneous acid, sulphurous acid, nitrate of silver, volatile oils, and in short substances, all of which possess antiseptic properties.

THE INSOLUBLE PART OF THE SUBSTANCE CALLED FERMENT DOES NOT CAUSE FERMENTATION. For when the yeast from wine or beer is carefully washed with water, care being taken that it is always covered with this fluid, the residue does not produce fermentation.

THE SOLUBLE PART OF FERMENT LIKEWISE DOES NOT EXCITE FERMENTATION. An aqueous infusion of yeast may be mixed with a solution of sugar, and preserved in vessels from which the air is excluded, without either experiencing the slightest change. What then, we may ask, is the matter in ferment which excites fermentation, if neither the soluble nor insoluble parts possess the power? This question has been answered by Colin in the most satisfactory manner. He has shown that in reality IT IS THE SOLUBLE PART. But before it obtains this power, the decanted infusion must be allowed to cool in contact with the air, and to remain some time exposed to its action. When introduced into a solution of sugar in this state, it produces a brisk fermentation; but without previous exposure to the air, it manifests no such property.

The infusion absorbs oxygen during its exposure to the air, and carbonic acid may be found in it after a short time.

Yeast produces fermentation in consequence of the progressive decomposition which it suffers from the action of air and water.

Now when yeast is made to act on sugar, it is found that after the completion of the transformation of the latter substance into carbonic acid and alcohol, part of the yeast itself has disappeared.

From 20 parts of fresh yeast from beer, and 100 parts of sugar, Thénard obtained, after the fermentation was completed, 13·7 parts of an insoluble residue, which diminished to 10 parts when employed in the same way with a fresh portion of sugar. These ten parts were white, possessed of the properties of woody fibre, and had no further action on sugar.

It is evident, therefore, that, during the fermentation of sugar by yeast, both of these substances suffer decomposition at the same time, and disappear in consequence. But if yeast be a body which excites fermentation by being itself in a state of decomposition, all other matters in the same condition should have a similar action upon sugar; and this is in reality the case. Muscle, urine, isinglass, osmazone, albumen, cheese, gliadine, gluten, legumin, and blood, when in a state of putrefaction, all have the power of producing the putrefaction, or fermentation of a solution of sugar. Yeast, which by continued washing has entirely lost the property of inducing fermentation, regains it when its putrefaction has recommenced, in consequence of its being kept in a warm situation for some time.

Yeast and putrefying animal and vegetable matters act as peroxide of hydrogen does on oxide of silver, when they induce bodies with which they are in contact to enter into the same state of decomposition. The disturbance in the attraction of the constituents of the peroxide of hydrogen effects a disturbance in the attraction of the elements of the oxide of silver, the one being decomposed on account of the decomposition of the other.

Peroxide of hydrogen is rapidly decomposed in contact with moist fibrin of blood, an animal substance in a continuous state of decomposition. The oxygen which it contained, in addition to that necessary to form water, escapes with violent effervescence.

Now if we consider the process of the fermentation of pure sugar, in a practical point of view, we meet with two facts of constant occurrence. When the quantity of ferment is too small in proportion to that of the sugar, its putrefaction will be completed before the transformation of all the sugar is effected. Some sugar here remains undecomposed, because the cause of its transformation is absent, viz. contact with a body in a state of decomposition.

But when the quantity of ferment predominates, a certain quantity of it remains after all the sugar has fermented, its decomposition proceeding very slowly, on account of its insolubility in water. This residue of ferment is still able to induce fermentation, when introduced into a fresh solution of sugar, and retains the same power until it has passed through all the stages of its own transformation. Hence a certain quantity of yeast is necessary in order to effect the transformation of a certain portion of sugar, not because it acts by its quantity in increasing any affinity, but because its influence depends solely on its presence, and its presence is necessary, until the last atom of sugar is decomposed.

These facts and observations point out the existence of a new cause, which effects combinations and decompositions. This cause is the action which bodies in a state of combination or decomposition exercise upon substances, the component parts of which are united together by a feeble affinity. This action resembles a peculiar power, attached to a body in the state of combination or decomposition, but exerting its influence beyond the sphere of its own attractions. We are now able to account satisfactorily for many known phenomena.

A large quantity of hippuric acid may be obtained from the fresh urine of a horse, by the addition of muriatic acid; but when the urine has undergone putrefaction, no trace of it can be discovered. The urine of man contains a considerable quantity of urea; but when the urine putrefies, the urea entirely disappears. When urea is added to a solution of sugar in the state of fermentation, it is decomposed into carbonic acid and ammonia. No asparagin can be detected in a putrefied infusion of asparagus, licorice-root, or the root of marshmallow (*Althæa officinalis*).

It has also been mentioned, that the strong affinity of nitrogen for hydrogen, and that of carbon for oxygen, are the cause of the facility with which the elements of azotised compounds are disunited; those affinities aiding each other, inasmuch as by virtue of them different elements of the compounds strive to take possession of the different elements of water. Now since it is found that no body destitute of nitrogen possesses, when pure, the property of decomposing spontaneously whilst in contact with water, we must ascribe this property which azotised bodies possess in so eminent a degree, to something peculiar in the nature of the compounds of nitrogen, and to their constituting, in a certain measure, more highly organised atoms.

Every azotised constituent of the animal or vegetable organism runs spontaneously into putrefaction, when exposed to moisture and a high temperature.

Azotised matters are, accordingly, the only causes of fermentation and putrefaction in vegetable substances.

Putrefaction, on account of its effects, as a mixed transformation of many different substances, may be classed with the most powerful processes of deoxidation, by which the strongest affinities are overcome.

When a solution of gypsum in water is mixed with a decoction of sawdust, or any other organic matter capable of putrefaction, and preserved in well-closed vessels, it is found after some time, that the solution no longer contains sulphuric acid, but in its place carbonic and free hydrosulphuric acids, between which the lime of the gypsum is shared. In stagnant water containing sulphates in solution, crystallised pyrites is observed to form on the decaying roots.

Now we know that in the putrefaction of wood under water, when air therefore is excluded, a part of its carbon combines with the oxygen of the water, as well as with the oxygen which the wood itself contains; whilst its hydrogen and that of the decomposed water are liberated either in a pure state, or as carburetted hydrogen.

It is evident, that if with the water a substance containing a large quantity of oxygen, such as sulphuric acid, be also present, the matters in the state of putrefaction will make use of the oxygen of that substance as well as that of the water, in order to form carbonic acid; and the sulphur and hydrogen being set free will combine whilst in the nascent state, producing hydrosulphuric acid, which will be again decomposed if metallic oxides be present; and the results of this second decomposition will be water and metallic sulphurets.

The putrefied leaves of woad (*Isatis tinctoria*), in contact with indigo-blue, water, and alkalies, suffer further decomposition, and the indigo is deoxidised and dissolved.

The mannite formed by the putrefaction of the juice of beet-root and other plants containing sugar, contains the same number of equivalents of carbon and hydrogen as the sugar of grapes, but two atoms less of oxygen; and it is highly probable that it is produced from sugar of grapes, contained in those plants, in precisely the same manner as indigo-blue is converted into deoxidised white indigo.

During the putrefaction of gluten, carbonic acid and pure hydrogen gases are evolved; phosphate, acetate, caseate, and lactate of ammonia being at the same time produced in such quantity, that the further decomposition of the gluten ceases. But when the supply of water is renewed, the decomposition begins again, and in addition to the salts just mentioned, carbonate of ammonia and a white crystalline micaceous matter (caseous oxide) are formed, together with hydrosulphate of ammonia, and a mucilaginous substance coagulable by chlorine. *Lactic acid* is almost always produced by the putrefaction of organic bodies.

We may now compare fermentation and putrefaction with the decomposition which organic compounds suffer under the influence of a high temperature. Dry distillation would appear to be a process of combustion or oxidation going on in the interior of a substance, in which a part of the carbon unites with all or part of the oxygen of the compound, while other new compounds containing a large proportion of hydrogen are necessarily produced. Fermentation may be considered as a process of combustion or oxidation of a similar kind, taking place in a liquid between the elements of *the same matter*, at a very slightly elevated temperature; and putrefaction as a process of oxidation, in which the oxygen of *all* the substances present comes into play.

CHAPTER VI.

EREMACAUSIS, OR DECAY.

In organic nature, besides the processes of decomposition named fermentation and putrefaction, another and not less striking class of changes occurs, which bodies suffer from the influence of the air. This is the act of gradual combination of the combustible elements of a body with the oxygen of the air; a slow combustion or oxidation, to which we shall apply the term of *eremacausis*.

The conversion of wood into humus, the formation of acetic acid out of alcohol, nitrification, and numerous other processes, are of this nature. Vegetable juices of every kind, parts of animal and vegetable substances, moist sawdust, blood, &c., cannot be exposed to the air, without suffering immediately a progressive change of colour and properties, during which oxygen is absorbed. These changes do not take place when water is excluded, or when the substances are exposed to the temperature of 32°, and it has been observed that different bodies require different degrees of heat, in order to effect the absorption of oxygen, and, consequently, their eremacausis. The tendency to undergo this change is possessed in the highest degree by substances containing nitrogen.

When vegetable juices are evaporated by a gentle heat in the air, a brown or brownish-black substance is precipitated as a product of the action of oxygen upon them. This substance, which appears to possess similar properties from whatever juice it is obtained, has received the name of *extractive matter;* it is insoluble or very sparingly soluble in water, but is dissolved with facility by alkalies. By the action of air on solid animal or vegetable matters, a similar pulverulent brown substance is formed, and is known by the name of *humus.*

The conditions which determine the commencement of eremacausis are of various kinds. Many organic substances, particularly such as are mixtures of several more simple matters, oxidise in the air when simply moistened with water; others not until they are subjected to the action of alkalies; but the greatest part of them undergo this state of slow combustion or oxidation, when brought in contact with other matters already in a state of decay.

The eremacausis of an organic matter is retarded or completely arrested by all those substances which prevent fermentation or putrefaction. Mineral acids, salts of mercury, aromatic substances, empyreumatic oils, and oil of turpentine, possess a similar action in this respect. The latter substances have the same effect on decaying bodies as on phosphuretted hydrogen, the spontaneous inflammability of which they destroy.

Many bodies which do not decay when moistened with water, enter into eremacausis when in contact with an alkali. Gallic acid, hæmatin, and many other compounds, may be dissolved in water and yet remain unaltered; but if the smallest quantity of a free alkali is present, they acquire the property of attracting oxygen, and are converted into a brown substance like humus, evolving very frequently at the same time carbonic acid. (Chevreul.)

A very remarkable kind of eremacausis takes place in many vegetable substances, when they are exposed to the influence of air, water, and ammonia. They absorb oxygen very rapidly, and form splendid violet or red-coloured liquids, as in the case of orcin and erythrin. They now contain an azotised substance, not in the form of ammonia.

All these facts show that the action of oxygen seldom affects the carbon of decaying substances, and this corresponds exactly to what happens in combustion at high temperatures. It is well known, for example, that when no more oxygen is admitted to a compound of carbon and hydrogen than is sufficient to combine with its hydrogen, the carbon is not burned, but is separated as lamp-black; while, if the quantity of oxygen is not sufficient even to consume all the hydrogen, new compounds are formed, such as naphthalin and similar matters, which contain a smaller proportion of hydrogen than those compounds of carbon and hydrogen which previously existed in the combustible substance.

There is no example of carbon combining directly with oxygen at common temperatures, but numerous facts show that hydrogen, in certain states of condensation, possesses that property. Lamp-black which has been heated to redness may be kept in contact with oxygen gas, without forming carbonic acid; but lamp-black, impregnated with oils containing a large proportion of hydrogen, gradually becomes warm, and inflames spontaneously. The spontaneous inflammability of the charcoal used in the fabrication of gunpowder has been correctly ascribed to the hydrogen contained in it in considerable quantity; for during its reduction to powder, no trace of carbonic acid can be detected in the air surrounding it; it is not formed until the temperature of the mass has reached a red heat. The heat which produces the inflammation is therefore not caused by the oxidation of the carbon.

The matters subject to eremacausis may be divided into two classes. The first class comprehends those substances which unite with the oxygen of the air, without evolving carbonic acid; and the second, such as emit carbonic acid while they absorb oxygen.

When the oil of bitter almonds is exposed to the air, it absorbs two equivalents of oxygen, and is converted into benzoic acid; but half of the oxygen absorbed combines with the hydrogen of the oil, and forms water, which remains in union with the anhydrous benzoic acid.

According to the experiments of Döbereiner, 100 parts of

pyrogallic acid absorbs 38·09 parts of oxygen when in contact with ammonia and water; the acid being changed in consequence of this absorption into a mouldy substance, which contains less oxygen than the acid itself. It is evident that the substance formed is not a higher oxide; and it is found, on comparing the quantity of the oxygen absorbed with that of the hydrogen contained in the acid, that they are exactly in the proportions for forming water.

When colourless orcin is exposed together with ammonia to the contact of oxygen gas, the beautiful red-coloured orcein is produced. Now, the only changes which take place here are, that the absorption of oxygen by the elements of orcin and ammonia causes the formation of water; 1 equivalent of orcin $C_{16}\ H_{11}\ O_7$, and 1 equivalent of ammonia NH_3, absorb 5 equivalents of oxygen, and 5 equivalents of water are produced, the composition of orcein being $C_{16}\ H_9\ O_7\ N$. (Dumas.) In this case it is evident, that the oxygen absorbed has united merely with the hydrogen.

But, although it appears very probable that the oxygen acts primarily and principally upon hydrogen, the most combustible constituent of organic matter in the state of decay; still it cannot thence be concluded that the carbon is quite devoid of the power to unite with oxygen, when every particle of it is surrounded with hydrogen, an element with which the oxygen combines with greater facility.

We know, on the contrary, that although nitrogen cannot be made to combine with oxygen directly, yet it is oxidised and forms nitric acid, when mixed with a large quantity of hydrogen, and burned in oxygen gas. In this case its affinity is evidently increased by the combustion of the hydrogen, which is in fact communicated to it. It is conceivable that, in a similar manner, the carbon may be directly oxidised in several cases, obtaining from its contact with hydrogen in eremacausis a property which it does not itself possess at common temperatures. But the formation of carbonic acid during the eremacausis of bodies containing hydrogen, must in most cases be ascribed to another cause. It appears to be formed in a manner similar to the formation of acetic acid, by the eremacausis of saliculite of potash. This salt, when exposed to a moist atmosphere, absorbs 3 atoms of oxygen; *melanic acid* is produced, a body resembling humus, in consequence of the formation of which, the elements of 1 atom of acetic acid are separated from the saliculous acid.

An alkaline solution of hæmatin being exposed to an atmosphere of oxygen, 0·2 grm. absorb 28·6 cubic centimeters of oxygen gas in twenty-four hours, the alkali acquiring at the same time 6 cubic centimeters of carbonic acid. (Chevreul.) But these 6 cubic centimeters of carbonic acid contain only an equal volume of oxygen, so that it is certain from this experiment that $\frac{3}{4}$ of the oxygen absorbed have not united with the carbon. It is highly probable, that during the oxidation of the hydrogen, a portion of the carbon had united with the oxygen contained in the hæmatin, and had separated from the other elements as carbonic acid.

The experiments of De Saussure upon the decay of woody fibre show that such a separation is highly probable. Moist woody fibre evolved one volume of carbonic acid for every volume of oxygen which it absorbed. It has just been mentioned that carbonic acid contains its own volume of oxygen. Now, woody fibre contains carbon and the elements of water, so that the result of the action of oxygen upon it is exactly the same as if pure charcoal had combined directly with oxygen. But the characters of woody fibre show, that the elements of water are not contained in it in the form of water; for, were this the case, starch, sugar, and gum must also be considered as hydrates of carbon.

But if the hydrogen does not exist in woody fibre in the form of water, the direct oxidation of the carbon cannot be considered as at all probable, without rejecting all the facts established by experiment regarding the process of combustion at low temperatures.

If we examine the action of oxygen upon a substance containing a large quantity of hydrogen, such as alcohol, we find most distinctly, that the direct formation of carbonic acid is the last stage of its oxidation, and that it is preceded by a series of changes, the last of which is a complete combustion of the hydrogen. Aldehyde, acetic, formic, oxalic, and carbonic acids, form a connected chain of products arising from the oxidation of alcohol; and the successive changes which this fluid experiences from the action of oxygen may be readily traced in them. Aldehyde is alcohol *minus* hydrogen; acetic acid is formed by the direct union of aldehyde with oxygen. Formic acid and water are formed by the union of acetic acid with oxygen. When all the hydrogen is removed from formic acid, oxalic acid is produced; and the latter acid is converted into carbonic acid by uniting with an additional portion of oxygen. All these products appear to be formed simultaneously, by the action of oxidising agents on alcohol; but it can scarcely be doubted, that the formation of the last product, the carbonic acid, does not take place until all the hydrogen has been abstracted.

The absorption of oxygen by drying oils certainly does not depend upon the oxidation of their carbon; for in raw walnut-oil, for example, which was not free from mucilage and other substances, only twenty-one volumes of carbonic acid were formed for every 146 volumes of oxygen gas absorbed.

It must be remembered, that combustion or oxidation at low temperatures produces results quite similar to combustion at high temperatures WITH LIMITED ACCESS OF AIR. The most combustible element of a compound exposed to the action of oxygen, must become oxidised first, for its superior combustibility is caused by its being enabled to unite with oxygen at a temperature at which the other elements cannot enter into that combination; this property having the same effect as a greater affinity.

The combustibility of potassium is no measure of its affinity for oxygen; we have reason to believe that the attraction of magnesium and aluminium for oxygen is greater than that of potassium for the same element; but neither of those metals oxidises either in air or water at common temperatures, whilst potassium decomposes water with great violence, and appropriates its oxygen.

Phosphorus and hydrogen combine with oxygen at ordinary temperatures, the first in moist air, the second when in contact with finely-divided platinum; while charcoal requires a red heat before it can enter into combination with oxygen. It is evident that phosphorus and hydrogen are more combustible than charcoal, that is, that their affinity for oxygen AT COMMON TEMPERATURES is greater; and this is not the less certain, because it is found, that carbon in certain other conditions shows a much greater affinity for oxygen than either of those substances.

In putrefaction, the conditions are evidently present, under which the superior affinity of carbon for oxygen comes into play; neither expansion, cohesion, nor the gaseous state, opposes it, whilst in eremacausis all these restraints have to be overcome.

The evolution of carbonic acid, during the decay or eremacausis of animal or vegetable bodies which are rich in hydrogen, must accordingly be ascribed to a transposition of the elements or disturbance in their attractions, similar to that which gives rise to the formation of carbonic acid in the processes of fermentation and putrefaction. While the hydrogen of the substance is removed and oxidised by eremacausis, carbon and oxygen separate from the remaining elements in the form of carbonic acid.

The eremacausis of such substances is, therefore, a decomposition analogous to the putrefaction of azotised bodies. For in these there are two affinities at play; the affinity of nitrogen for hydrogen, and that of carbon for oxygen, and both facilitate the disunion of the elements. Now there are two affinities also in action in those bodies which decay with the evolution of carbonic acid. One of these affinities is the attraction of the oxygen of the air for the hydrogen of the substance, which corresponds to the attraction of nitrogen for the same element; and the other is the affinity of the carbon of the substance for its oxygen, which is constant under all circumstances.

When wood putrefies in marshes, carbon and oxygen are separated from its elements in the form of carbonic acid, and hydrogen in the form of carburetted hydrogen. But when wood decays or putrefies in the air, its hydrogen does not combine with carbon, but with oxygen, for which it has a much greater affinity at common temperatures.

Now it is evidently owing to the complete similarity of these processes, that decaying and putrefying bodies can mutually replace one another in their reciprocal actions.

All putrefying bodies pass into a state of decay, when exposed freely to the air, and all decaying matters into that of putrefaction when air is excluded. All bodies, likewise, in a state of decay are capable of inducing putrefaction in other bodies, in the same manner as putrefying bodies themselves do.

CHAPTER VII.

EREMACAUSIS OR DECAY OF BODIES DESTITUTE OF NITROGEN: FORMATION OF ACETIC ACID.

All those substances which appear to possess the property of entering spontaneously into fermentation and putrefaction, do not in reality suffer those changes without some previous disturbance in the attraction of their elements. Eremacausis always precedes fermentation and putrefaction, and it is not until after the absorption of a certain quantity of oxygen that the signs of a transformation in the interior of the substances show themselves.

It is a very general error to suppose that organic substances have the power of undergoing change spontaneously, without the aid of an external cause. When they are not already in a state of change, it is necessary, before they can assume that state, that the existing equilibrium of their elements should be disturbed; and the most common cause of this disturbance is undoubtedly the atmosphere which surrounds all bodies.

The juices of the fruit or other parts of a plant prone to decomposition, retain their properties unchanged as long as they are protected from immediate contact with the air, that is, as long as the cells or organs in which they are contained resist the influence of the air. It is not until after the juices have been exposed to the air, and have absorbed a certain quantity of oxygen, that the substances dissolved in them begin to be decomposed.

The beautiful experiments of Gay-Lussac upon the fermentation of the juice of grapes, as well as the important practical improvements to which they have led, are the best proofs that the atmosphere possesses an influence upon the changes of organic substances. The juice of grapes expressed under a receiver filled with mercury, so that air was completely excluded, did not ferment. But when the smallest portion of air was introduced, a certain quantity of oxygen became absorbed, and fermentation immediately began. Although the juice was expressed from the grapes in contact with air, under the conditions therefore necessary to cause its fermentation, still this change did not ensue when the juice was heated in close vessels to the temperature of boiling water. When thus treated, it could be preserved for years without losing its property of fermenting. A fresh exposure to the air at any period caused it to ferment.

Animal food of every kind, and even the most delicate vegetables, may be preserved unchanged if heated to the temperature of boiling water in vessels from which the air is completely excluded. Food thus prepared has been kept for fifteen years, and upon opening the vessels after this long time, has been found as fresh and well-flavoured as when originally placed in them.

The action of the oxygen in these processes of decomposition is very simple; it excites changes in the composition of the azotised matters dissolved in the juices;—the mode of combination of the elements of those matters undergoes a disturbance and change in consequence of their contact with oxygen. The oxygen acts here in a similar manner to the friction or motion which effects the mutual decomposition of two salts, the crystallisation of salts from their solution, or the explosion of fulminating mercury. It causes the state of rest to be converted into a state of motion.

When this condition of intestine motion is once excited, the presence of oxygen is no longer necessary. The smallest particle of an azotised body in this act of decomposition exercises an influence upon the particles in contact with it, and the state of motion is thus propagated through the substance. The air may now be completely excluded, but the fermentation or putrefaction proceeds uninterruptedly to its completion.

Aldehyde attracts oxygen from the air, and, by the process of eremacausis, becomes vinegar; if the air be now excluded, the disturbance already begun is not arrested, but the products are very different. Two substances are then formed by a change in the arrangement of the elements. Their composition is similar, but they are very unlike in character.

The contact of ammonia and of alkalies in general may be mentioned amongst the chemical conditions which determine the commencement of eremacausis; for their presence causes many substances to absorb oxygen and to decay, in which neither oxygen nor alkalies alone produce that change.

Thus alcohol does not combine with the oxygen of the air at common temperatures. But a solution of potash in alcohol absorbs oxygen with much rapidity, and acquires a brown colour. The alcohol is found after a short time to contain acetic acid, formic acid, and the products of the decomposition of aldehyde by alkalies, including the resin of aldehyde, which gives the liquid a brown colour.

The most general condition for the production of eremacausis in organic matter is contact with a body already in the state of eremacausis or putrefaction. We have here an instance of true contagion; for the communication of the state of combustion is in reality the effect of the contact.

It is decaying wood which causes fresh wood around it to assume the same condition, and it is the very finely divided woody fibre in the act of decay which in moistened gall-nuts converts the tannic acid with such rapidity into gallic acid.

A most remarkable and decided example of this induction of combustion has been observed by De Saussure. It has already been mentioned, that moist woody fibre, cotton, silk, or vegetable mould, in the act of fermentation or eremacausis, converts the oxygen gas surrounding it into carbonic acid, without change of volume. Now, De Saussure added a certain quantity of hydrogen gas to the oxygen, and observed a diminution in volume immediately after the addition. A part of the hydrogen gas had disappeared, and along with it a portion of the oxygen, but a corresponding quantity of carbonic acid gas had not been formed. The hydrogen and oxygen had disappeared in exactly the same proportion as that in which they combine to form water; a true combustion of the hydrogen, therefore, had been induced by mere contact with matter in the state of eremacausis. The action of the decaying substance here produced results exactly similar to those effected by spongy platinum; but that they proceeded from a different cause was shown by the fact that the presence of carbonic oxide, which arrests completely the action of platinum on a mixture of oxygen and hydrogen, did not retard in the slightest degree the combustion of the hydrogen in contact with the decaying bodies.

But the same bodies were found by De Saussure not to possess the property just described, before they were in a state of fermentation or decay; and he has shown that even when they are in this state, the presence of antiseptic matter destroys completely all their influence.

Let us suppose a volatile substance containing a large quantity of hydrogen to be substituted for the hydrogen gas in De Saussure's experiments. Now, the hydrogen in such compounds being contained in a state of greater condensation would suffer a more rapid oxidation, that is, its combustion would be sooner completed. This principle is in reality attended to in the manufactories in which acetic acid is prepared according to the new plan. In the process there adopted all the conditions are afforded for the eremacausis of alcohol, and for its consequent conversion into acetic acid.

The alcohol is exposed to a moderate heat, and spread over a very extended surface, but these conditions are not sufficient to effect its oxidation. The alcohol must be mixed with a substance which is with facility changed by the oxygen of the air, and either enters into eremacausis by mere contact with oxygen, or by its fermentation or putrefaction yields products possessed of this property. A small quantity of beer, acescent wine, a decoction of malt, honey, and numerous other substances of this kind, possess the action desired.

The difference in the nature of the substances possessing this property shows, that none of them can contain a peculiar matter which has the property of exciting eremacausis; they are only the bearers of an action, the influence of which extends beyond the sphere of their own attractions. Their power consists in a condition of decomposition or eremacausis, which impresses the same condition upon the atoms of alcohol in its vicinity; exactly as in the case of an alloy of platinum and silver dissolving in nitric acid, in which the platinum becomes oxidised by virtue of an inductive action exercised upon it by the silver in the act of its oxidation. In the preparation of vinegar, the hydrogen of alcohol, with the formation of water and evolution of heat, is oxidised at the expense of the oxygen in contact with it; the residue is aldehyde, a substance possessing as great an affinity for oxygen as sulphurous acid; and by uniting directly with the latter, it produces acetic acid.

CHAPTER VIII.

EREMACAUSIS OF SUBSTANCES CONTAINING NITROGEN: NITRIFICATION.

WHEN azotised substances are burned at high temperatures, their nitrogen does not enter into direct combination with oxygen. The knowledge of this fact is of assistance in considering the process of the eremacausis of such substances. Azotised organic matter always contains carbon and hydrogen, both of which elements have a very strong affinity for oxygen.

Now nitrogen possesses a very feeble affinity for oxygen, so that it is placed, in regard to that element, in a position similar to that of the carbon of bodies containing much hydrogen during their combustion; a separation of the carbon of the latter substances in an uncombined state takes place, and in the same way the substances containing nitrogen give out that element in its gaseous form.

When moist azotised animal matter is exposed to the action of air, ammonia is constantly liberated; nitric acid is never formed under these circumstances.

But when alkalies or alkaline bases are present, a union of oxygen with the nitrogen takes place under the same circumstances, and nitrates are formed together with the other products of oxidation.

Although we see the most simple means and direct methods employed in the great processes of decomposition occurring in nature, still we find that the final result depends on a succession of actions, which are essentially influenced by the chemical nature of the bodies submitted to decomposition.

When it is observed that the character of a substance

remains unaltered in a whole series of phenomena, there is no reason to ascribe a new character to it, for the purpose of explaining a single phenomenon, especially where the explanation of that, according to known facts, offers no difficulty.

The most distinguished philosophers suppose that the nitrogen in an animal substance, when exposed to the action of air, water, and alkaline bases, obtains the power of combining directly with oxygen, and of thus forming nitric acid; but we are not acquainted with a single fact which justifies this opinion. It is only by the interposition of a large excess of hydrogen in the state of combustion or oxidation, that nitrogen can be converted into an oxide.

When a compound of nitrogen and carbon, such as cyanogen, is burned in oxygen gas, its carbon alone is oxidised; and when it is conducted over a metallic oxide heated to redness, an oxide of nitrogen is very rarely produced, and never when the carbon is in excess. Kuhlmann found in his experiments, that it was only when cyanogen was mixed with an excess of oxygen gas, and conducted over spongy platinum, that nitric acid was generated.

Kuhlmann could not succeed in causing pure nitrogen to combine directly with oxygen, even under the most favourable circumstances; thus, with the aid of spongy platinum at different temperatures, no union took place.

The carbon in the cyanogen gas must, therefore, have given rise to the combustion of the nitrogen by induction.

On the other hand, we find that ammonia (a compound of hydrogen and nitrogen) cannot be exposed to the action of oxygen, without the formation of an oxide of nitrogen, and production of nitric acid, in consequence of this union.

It is owing to the great facility with which ammonia is converted into nitric acid, that it is so difficult to obtain a correct determination of the quantity of nitrogen in a compound subjected to analysis, in which it is either contained in the form of ammonia, or from which ammonia is formed by an elevation of temperature. For when ammonia is passed over the red-hot oxide of copper, it is converted, either completely or partially, into binoxide of nitrogen.

When ammoniacal gas is conducted over peroxide of manganese or iron heated to redness, a large quantity of nitrate of ammonia is obtained, if the ammonia be in excess; and the same decomposition happens when ammonia and oxygen are together passed over red-hot spongy platinum.

It appears, therefore, that the combination of oxygen with nitrogen occurs rarely during the combustion of compounds of the latter element with carbon, but that nitric acid is always a product when ammonia is present in the substance exposed to oxidation.

The cause wherefore the nitrogen in ammonia exhibits such a strong disposition to become converted into nitric acid is undoubtedly that the two products, which are the result of the oxidation of the constituents of ammonia, possess the power of uniting with one another. Now this is not the case in the combustion of compounds of carbon and nitrogen; here one of the products is carbonic acid, which, on account of its gaseous form, must oppose the combination of the oxygen and nitrogen, by preventing their mutual contact, while the superior affinity of its carbon for the oxygen during the act of its formation will aid this effect.

When sufficient access of air is admitted during the combustion of ammonia, water is formed as well as nitric acid, and both of these bodies combine together. The presence of water may, indeed, be considered as one of the conditions essential to nitrification, since nitric acid cannot exist without it.

Eremacausis is a kind of putrefaction, differing from the common process of putrefaction, only in the part which the oxygen of the air plays in the transformations of the body in decay. When this is remembered, and when it is considered that in the transposition of the elements of azotised bodies their nitrogen always assumes the form of ammonia, and that in this form, nitrogen possesses a much greater disposition to unite with oxygen than it has in any of its other compounds;

we can with difficulty resist the conclusion, that ammonia is the source of the formation of nitric acid on the surface of the earth.

Azotised animal matter is not, therefore, the immediate cause of nitrification; it contributes to the production of nitric acid only in so far as it is a slow and continued source of ammonia.*

Now it has been shown in the former part of this work, that ammonia is always present in the atmosphere, so that nitrates might thence be formed in substances which themselves contained no azotised matter. It is known also, that porous substances possess generally the power of condensing ammonia; there are few ores of iron which do not evolve ammoniacal products when heated to redness, and ammonia is the cause of the peculiar smell perceived upon moistening aluminous minerals. Thus, ammonia, by being a constituent of the atmosphere, is a very widely diffused cause of nitrification, which will come into play whenever the different conditions necessary for the oxidation of ammonia are combined. It is probable that other organic bodies in the state of eremacausis are the means of causing the combustion of ammonia; at all events, the cases are very rare in which nitric acid is generated from ammonia, in the absence of all matter capable of eremacausis.

From the preceding observations on the causes of fermentation, putrefaction, and decay, we may now draw several conclusions calculated to correct the views generally entertained respecting the fermentation of wine and beer, and several other important processes of decomposition occurring in nature.

* According to the observations of Collard de Martigny, ammonia is converted directly into nitric acid when in contact with hydrate of lime and with air, without the intervention of any decaying substance.

CHAPTER IX.

ON VINOUS FERMENTATION:—WINE AND BEER.

It has already been mentioned that fermentation is excited in the juice of grapes by the access of air; alcohol and carbonic acid being formed by the decomposition of the sugar contained in the fluid. But it was also stated, that the process once commenced, continues until all the sugar is completely decomposed, quite independently of any further influence of the air.

In addition to the alcohol and carbonic acid formed by the fermentation of the juice, there is also produced a yellow or grey insoluble substance, containing a large quantity of nitrogen. It is this body which possesses the power of inducing fermentation in a new solution of sugar, and which has in consequence received the name of *ferment*.

The alcohol and carbonic acid are produced from the elements of the sugar, and the ferment from those azotised constituents of the juice termed gluten, or vegetable albumen.

According to the experiments of De Saussure, fresh impure gluten evolved, in five weeks, twenty-eight times its volume of a gas which consisted of $\frac{3}{4}$ of carbonic acid, and $\frac{1}{4}$ of pure hydrogen gas; ammoniacal salts of several organic acids were formed at the same time. Water must, therefore, be decomposed during the putrefaction of gluten; the oxygen of this water must enter into combination with some of its constituents, whilst hydrogen is liberated, a circumstance which happens only in decompositions of the most energetic kind. Neither ferment, nor any substance similar to it, is formed in this case; and we have seen that hydrogen is not evolved in the fermentation of saccharine vegetable juices.

It is evident that the decomposition which gluten suffers in an isolated state, and that which it undergoes when dissolved in a vegetable juice, belong to two different kinds of transformations. There is reason to believe that its change to the insoluble state depends upon an absorption of oxygen, for its separation in this state may be effected, under certain conditions, by free exposure to the air, without the presence of fermenting sugar. It is known also that the juice of grapes, or vegetable juices in general, become turbid when in contact with air, before fermentation commences; and this turbidity is owing to the formation of an insoluble precipitate of the same nature as ferment.

From the phenomena observed during the fermentation of wort,* it is known with perfect certainty that ferment is formed from gluten at the same time that the transformation of the sugar is effected; for the wort contains the azotised matter of the corn, namely, gluten in the same condition as it exists in the juice of grapes. The wort ferments by the addition of yeast, but after its decomposition is completed, the quantity of ferment or yeast is found to be thirty times greater than it originally was.

Yeast from beer and that from wine, examined under the microscope, present the same form and general appearance. They are both acted on in the same manner by alkalies and by acids, and possess the power of inducing fermentation anew in a solution of sugar; in short, they must be considered as identical.

The fact that water is decomposed during the putrefaction of gluten, has been completely proved. The tendency of the carbon of the gluten to appropriate the oxygen of water must therefore always be in action, whether the gluten is decomposed in a soluble or insoluble state. These considerations, therefore, as well as the circumstance which all the experiments made on this subject appear to point out, that the conversion of gluten to the insoluble state is the result of

* Wort is an infusion of malt; it consists of the soluble parts of this substance dissolved in water.

oxidation, lead us to conclude that the oxygen consumed in this process is derived from the elements of water, or from the sugar which contains oxygen and hydrogen in the same proportion as water. At all events, the oxygen thus consumed in the fermentation of wine and beer is not taken from the atmosphere.

The fermentation of pure sugar in contact with yeast must evidently be a very different process from the fermentation of wort or of *must*.*

In the former case, the yeast disappears during the decomposition of sugar; but in the latter, a transformation of gluten is effected at the same time, by which ferment is generated. Thus yeast is *destroyed* in the one case, but is *formed* in the other.

Now, since no free hydrogen gas can be detected during the fermentation of beer and wine, it is evident that the oxidation of the gluten, that is, its conversion into ferment, must take place at the cost either of the oxygen of the water, or of that of the sugar; whilst the hydrogen liberated must enter into new combinations, or by the deoxidation of the sugar, new compounds containing a large proportion of hydrogen, and small quantity of oxygen, together with the carbon of the sugar, must be formed.

It is well known that wine and fermented liquors generally contain, in addition to the alcohol, other substances which could not be detected before their fermentation, and which must have been formed, therefore, during that process, in a manner similar to the production of mannite. The smell and taste distinguishing wine from all other fermented liquids are known to depend upon an ether of a volatile and highly combustible acid; the ether is of an oily nature, and has received the name ŒNANTHIC ETHER. It is also ascertained that the smell and taste of brandy from corn and potato are owing to a peculiar oil, the OIL OF POTATOES. This oil is more closely allied to alcohol in its properties, than to any other organic substance.

* The liquid expressed from grapes when fully ripe is called *must*.

These bodies are products of the deoxidation of the substances dissolved in the fermenting liquids; they contain less oxygen than sugar or gluten, but are remarkable for their large proportion of hydrogen.

Œnanthic acid contains an equal number of equivalents of carbon and hydrogen, exactly the same proportions of these elements, therefore, as sugar, but by no means the same proportion of oxygen. The oil of potatoes contains much more hydrogen.

Although it cannot be doubted that these volatile liquids are formed by a mutual interchange of the elements of gluten and of sugar, in consequence, therefore, of a true process of putrefaction, still it is certain, that other causes exercise an influence upon their production and peculiarities.

The substances in wine to which its taste and smell are owing, are generated during the fermentation of the juice of such grapes as contain a certain quantity of tartaric acid; they are not found in wines free from all acid, or which contain a different organic acid, such as acetic acid.

The wines of warm climates possess no odour; wines grown in France have it in a marked degree, but in the wines from the Rhine the perfume is most intense. The kinds of grapes on the Rhine, which ripen very late, and scarcely ever completely, such as the Riessling and Orleans, have the strongest perfume or *bouquet*, and contain, proportionally, a larger quantity of tartaric acid. The wines from the earlier grapes, such as the Rulander, and others, contain a large proportion of alcohol, and are similar to Spanish wines in their flavour, but they possess no *bouquet*.

The grapes grown at the Cape from Riesslings, transplanted from the Rhine, produce an excellent wine, which does not, however, possess the aroma peculiar to the Rhenish wine.

It is evident, from these facts, that the acid of wines, and their characteristic perfumes, have some connection, for they are always found together; and it can scarcely be doubted that the presence of the former exercises a certain influence on the formation of the latter. This influence is very plainly observed in the fermentation of liquids destitute of tartaric acid, and particularly of those which are nearly neutral or alkaline, such as the *mash** of potatoes or corn.

The brandy obtained from corn and potatoes contains an ethereal oil of a similar composition in both, to which these liquors owe their peculiar smell. This oil is generated during the fermentation of the mash; it exists ready formed in the fermented liquids, and distils over with alcohol when a gentle heat is applied.

It is observed that a greater quantity of alcohol is obtained when the mash is made quite neutral by ashes or by carbonate of lime, and that the proportion of oil in the brandy also is increased.

Now, it is known that brandy made from potato starch, which has been converted into sugar by dilute sulphuric acid, is completely free from the potato oil, so that this substance must be generated in consequence of a change suffered by the cellular tissue of the potatoes during their fermentation.

Experience has shown that the simultaneous fermentation or putrefaction of the cellular tissue, by which this oil is generated, may be completely prevented in the fabrication of brandy from corn.†

The same malt, which in the preparation of brandy yields a fluid containing the oil of which we are speaking, affords, in the formation of beer, a spirituous liquor in which no trace of that oil can be detected. The principal difference in the preparation of the two liquids is, that in the fermentation of wort, an aromatic substance (hops) is added, and it is certain that its presence modifies the transformations which take place. Now, it is known that the volatile oil of mustard, and the empyreumatic oils, arrest completely the action of yeast; and although the oil of hops does not possess this property, still it diminishes, in a great degree, the influence

* *Mash* is the mixture of malt, potatoes, and water, in the *mash tun*, a large vessel in which it is infused.

† In the manufactory of M. Dubrunfaut, so considerable a quantity of this oil is obtained under certain circumstances from brandy made from potatoes, that it might be employed for the purpose of illuminating his whole manufactory.

of decomposing azotised bodies upon the conversion of alcohol into acetic acid. There is, therefore, reason to believe that some aromatic substances, when added to fermenting mixtures, are capable of producing very various modifications in the nature of the products generated.

Whatever opinion, however, may be held regarding the origin of the volatile odoriferous substances obtained in the fermentation of wine, it is quite certain that the characteristic smell of wine is owing to an ether of an organic acid, resembling one of the fatty acids (œnanthic ether).

It is only in liquids containing other very soluble acids, that the fatty acids and œnanthic acid are capable of entering into combination with the ether of alcohol, and of thus producing compounds of a peculiar smell. This ether is found in all wines containing a free acid, but is absent from those in which no acids are present. This acid, therefore, is the means by which the smell is produced; since without its presence œnanthic ether could not be formed.

The greatest part of the oil of brandy made from corn consists of a fatty acid not converted into ether; it dissolves oxide of copper and metallic oxides in general, and combines with the alkalies.

The principal constituent of this oil is an acid identical in composition with œnanthic acid, but different in properties. (Mulder.) It is formed in fermenting liquids, which, if they be acid, contain only acetic acid, a body which has no influence in causing other acids to form ethers.

The oil of brandy made from potatoes is the hydrate of an organic base analogous to ether, and capable, therefore, of entering into combination with acids. It is formed in considerable quantity in fermenting liquids possessing an alkaline reaction; under circumstances, consequently, in which it is incapable of combining with an acid.

The products of the fermentation and putrefaction of neutral vegetable and animal matters are generally accompanied by substances of an offensive odour; but the most remarkable example of the generation of a true ethereal oil is seen in the fermentation of the *Herba centaurium minus*, a plant destitute of smell. When it is exposed in water to a slightly elevated temperature it ferments, and emits an agreeable penetrating odour. By the distillation of the liquid, an ethereal oily substance of great volatility is obtained, which excites a pricking sensation in the eyes, and a flow of tears. (Büchner.)

We know that most of the blossoms and vegetable substances possessing a smell owe this property to a volatile oil existing in them; but it is not less certain, that others emit a smell only when they undergo change or decomposition.

Arsenic and arsenious acid are both quite inodorous. It is only during their oxidation that they emit their characteristic odour of garlic. The oil of the berries of the elder-tree, many kinds of oil of turpentine, and oil of lemons, possess a smell only during their oxidation or decay. The same is the case with many blossoms; and Geiger has shown, that the smell of musk is owing to its gradual putrefaction and decay.

It is also probable, that the peculiar odorous principle of many vegetable substances is newly formed during the fermentation of the saccharine juices of the plants. At all events, it is a fact, that very small quantities of the blossoms of the violet, elder, linden, or cowslip, added to a fermenting liquid, are sufficient to communicate a very strong taste and smell, which the addition of the water distilled from a quantity a hundred times greater would not effect. The various kinds of beer manufactured in Bavaria are distinguished by different flavours, which are given by allowing small quantities of the herbs and blossoms of particular plants to ferment along with the wort. On the Rhine, also, an artificial *bouquet* is often given to wine for fraudulent purposes, by the addition of several species of the sage and rue to the fermenting liquor; but the fictitious perfume thus obtained differs from the genuine aroma, by its inferior durability, and by being gradually dissipated.

The juice of grapes grown in different climates differs not

only in its proportion of free acid, but also in respect of the quantity of sugar dissolved in it. The quantity of azotised matter in the juice seems to be the same in whatever part the grapes may grow; at least, no difference has been observed in the amount of yeast formed during fermentation in the south of France, and on the Rhine.

The grapes grown in hot climates, as well as the boiled juice obtained from them, are proportionally rich in sugar. Hence, during the fermentation of the juice the complete decomposition of its azotised matters, and their separation in the insoluble state, are effected before all the sugar has been converted into alcohol and carbonic acid. A certain quantity of the sugar consequently remains mixed with the wine in an undecomposed state, the condition necessary for its further decomposition being absent.

The azotised matters in the juice of grapes of the temperate zones, on the contrary, are not completely separated in the insoluble state, when the entire transformation of the sugar is effected. The wine of these grapes, therefore, does not contain sugar, but variable quantities of undecomposed gluten in solution.

This gluten gives the wine the property of becoming spontaneously converted into vinegar, when the access of air is not prevented. For it absorbs oxygen and becomes insoluble; and its oxidation is communicated to the alcohol, which is converted into acetic acid.

By allowing the wine to remain at rest in casks with a very limited access of air, and at the lowest possible temperature, the oxidation of this azotised matter is effected without the alcohol undergoing the same change, a higher temperature being necessary to enable alcohol to combine with oxygen. As long as the wine in the *stilling-casks* deposits yeast, it can still be caused to ferment by the addition of sugar, but old well-cleared wine has lost this property, because the condition necessary for fermentation, namely, a substance in the act of decomposition or putrefaction, is no longer present in it.

In hotels and other places where wine containing much gluten is drawn gradually from a cask, and a proportional quantity of air necessarily introduced, its eremacausis, that is, its conversion into acetic acid, is prevented by the addition of a small quantity of sulphurous acid. This acid, by entering into combination with the oxygen of the air contained in the cask, or dissolved in the wine, prevents the oxidation of the organic matter.

The various kinds of beer differ from one another in the same way as the wines.

English, French, and most of the German beers, are converted into vinegar when exposed to the action of air. But this property is not possessed by Bavarian beer, which may be kept in vessels only half-filled without acidifying or experiencing any change. This valuable quality is obtained for it by a peculiar management of the fermentation of the wort. The perfection of experimental knowledge has here led to the solution of one of the most beautiful problems of the theory of fermentation.

Wort is proportionally richer in gluten than in sugar, so that during its fermentation in the common way, a great quantity of yeast is formed as a thick scum. The carbonic acid evolved during the process attaches itself to the particles of the yeast, by which they become specifically lighter than the liquid in which they are formed, and rise to its surface. Gluten in the act of oxidation comes in contact with the particles of the decomposing sugar in the interior of the liquid. The carbonic acid from the sugar and insoluble ferment from the gluten are disengaged simultaneously, and cohere together.

A great quantity of gluten remains dissolved in the fermented liquid, even after the transformation of the sugar is completed, and this gluten causes the conversion of the alcohol into acetic acid, on account of its strong disposition to attract oxygen, and to undergo decay. Now, it is plain, that with its separation, and that of all substances capable of attracting oxygen, the beer would lose the property of becoming acid. This end is completely attained in the process of fermentation adopted in Bavaria.

The wort, after having been treated with hops in the usual manner, is thrown into very wide flat vessels, in which a large surface of the liquid is exposed to the air. The fermentation is then allowed to proceed, while the temperature of the chambers in which the vessels are placed is never allowed to rise above from 45° to 50° F. The fermentation lasts from three to six weeks, and the carbonic acid evolved during its continuance is not in large bubbles which burst upon the surface of the liquid, but in small bubbles like those which escape from an acidulous mineral water, or from a liquid saturated by high pressure. The surface of the wort is scarcely covered with a scum, and all the yeast is deposited on the bottom of the vessel, in the form of a viscous sediment.

In order to obtain a clear conception of the great difference between the two kinds of fermentation, it may perhaps be sufficient to recall to mind the fact, that the transformation of gluten or of other azotised matters is a process consisting of several stages. The first stage is the conversion of the gluten into insoluble ferment in the interior of the liquid, and as the transformation of the sugar goes on at the same time, carbonic acid and yeast are simultaneously disengaged. It is known with certainty, that this formation of yeast depends upon oxygen being appropriated by the gluten in the act of decomposition; but it has not been sufficiently shown, whether this oxygen is derived from the water, from the sugar, or from the gluten itself; whether it combines directly with the gluten, or merely with its hydrogen, so as to form water. For the purpose of obtaining a definite idea of the process, we may designate the first change as the stage of oxidation. This oxidation of the gluten then, and the transposition of the atoms of the sugar into alcohol and carbonic acid, are necessarily attendant on each other, so that if the one is arrested the other must also cease.

Now, the yeast which rises to the surface of the liquid is

not the product of a complete decomposition, but is oxidised gluten still capable of undergoing a new transformation by the transposition of its constituent elements. By virtue of this condition it has the power to excite fermentation in a solution of sugar; and if the gluten be also present, the decomposing sugar induces its conversion into fresh yeast, so that, in a certain sense, the yeast appears to reproduce itself.

Yeast of this kind is oxidised gluten in a state of *putrefaction,* and by virtue of this state it induces a similar transformation in the elements of the sugar.

The yeast formed during the fermentation of Bavarian beer is oxidised gluten in a state of *decay*. The process of decomposition which its constituents are suffering, gives rise to a very protracted putrefaction (*fermentation*) in the sugar. The intensity of the action is diminished in so great a degree, that the gluten which the fluid still holds in solution takes no part in it; the sugar in fermentation does not excite a similar state in the gluten.

But the contact of the already decaying and precipitated gluten or yeast, causes the eremacausis of the gluten dissolved in the wort; oxygen gas is absorbed from the air, and all the gluten in solution is deposited as yeast.

The ordinary frothy yeast may be removed from fermenting beer by filtration, without the fermentation being thereby arrested; but the precipitated yeast of Bavarian beer cannot be removed without the whole process of its fermentation being interrupted. The beer ceases to ferment altogether, or, if the temperature is raised, undergoes the ordinary fermentation.

The precipitated yeast does not excite ordinary fermentation, and, consequently, is quite unfitted for the purpose of baking; but the common frothy yeast can cause the kind of fermentation by which the former kind of yeast is produced.

When common yeast is added to wort at a temperature of between 40° and 50° F., a slow tranquil fermentation takes place, and a matter is deposited on the bottom of the vessel, which may be employed to excite new fermentation; and when the same operation is repeated several times in succession, the ordinary fermentation changes into that process by which only precipitated yeast is formed. The yeast now deposited has lost the property of exciting ordinary fermentation, but it produces the other process even at a temperature of 50° F.

In wort subjected to fermentation, at a low temperature, with this kind of yeast, the condition necessary for the transformation of the sugar is the presence of that yeast; but for the conversion of gluten into ferment by a process of oxidation, something more is required.

When the power of gluten to attract oxygen is increased by contact with precipitated yeast in a state of decay, the unrestrained access of air is the only other condition necessary for its own conversion into the same state of decay, that is, for its oxidation. We have already seen that the presence of free oxygen and of gluten are conditions which determine the eremacausis of alcohol and its conversion into acetic acid, but they are incapable of exerting this influence at low temperatures. A low temperature retards the slow combustion of alcohol, while the gluten combines spontaneously with the oxygen of the air, just as sulphurous acid does when dissolved in water. Alcohol undergoes no such change at low temperatures, but during the oxidation of the gluten in contact with it, is placed in the same condition as the gluten itself when sulphurous acid is added to the wine in which it is contained. The oxygen of the air unites both with the gluten and alcohol of wine not treated with sulphurous acid; but when this acid is present it combines with neither of them, being altogether absorbed by the acid. The same thing happens in the peculiar process of fermentation adopted in Bavaria. The oxygen of the air unites only with the gluten and not with the alcohol, although it would have combined with both at higher temperatures, so as to form acetic acid.

Thus, then, this remarkable process of fermentation with the precipitation of a mucous-like ferment consists of a simultaneous putrefaction and decay in the same liquid. The sugar is in the state of putrefaction, and the gluten in that of decay.

Appert's method of preserving food, and this kind of fermentation of beer, depend on the same principle.

In the fermentation of beer after this manner, all the substances capable of decay are separated from it by means of an unrestrained access of air, while the temperature is kept sufficiently low to prevent the alcohol from combining with oxygen. The removal of these substances diminishes the tendency of the beer to become acescent, or, in other words, to suffer a further transformation.

In Appert's mode of preserving food, oxygen is allowed to enter into combination with the substance of the food, at a temperature at which decay, but neither putrefaction nor fermentation, can take place. With the subsequent exclusion of the oxygen and the completion of the decay, every cause which could effect further decomposition of the food is removed. The conditions for putrefaction are rendered insufficient in both cases: in the one, by the removal of the substances susceptible of decay; in the other, by the exclusion of the oxygen which would effect it.

It has been stated to be uncertain whether gluten, during its conversion into common yeast, that is, into the insoluble state in which it separates from fermenting liquids, really combines directly with oxygen. If it does combine with oxygen, then the difference between gluten and ferment would be, that the latter would contain a larger proportion of oxygen. Now it is very difficult to ascertain this, and even the analysis of these substances cannot decide the question. Let us consider, for example, the relations of alloxan and alloxantin* to one another. Both of these bodies contain the same elements as gluten, although in different proportions. Now they are known to be convertible into each other by oxygen being absorbed in the one case, and in the other extracted. Both are composed of absolutely the same elements, in equal proportions; with the single excep-

* Compounds obtained by the action of nitric acid on uric acid.

tion, that alloxantin contains 1 equivalent of hydrogen more than alloxan.

When alloxantin is treated with chlorine or nitric acid, it is converted into alloxan; into a body, therefore, which is alloxantin *minus* 1 equivalent of hydrogen. If, on the other hand, a stream of sulphuretted hydrogen is conducted through alloxan, sulphur is precipitated, and alloxantin produced. It may be said that, in the first case, hydrogen is abstracted; in the other, added. But it would be quite as simple an explanation, if we considered them as oxides of the same radical: the alloxan being regarded as a combination of a body composed of C_8 N_2 H_2 O_8 with 2 equivalents of water, and alloxantin as a combination of 3 atoms of water with a compound consisting of $C_8 N_2 H_2 O_8$. The conversion of alloxan into alloxantin would in this case result from its eight atoms of oxygen being reduced to seven; while alloxan would be formed out of alloxantin, by its combining with an additional atom of oxygen.

Now, oxides are known which combine with water, and present the same phenomena as alloxan and alloxantin. But compounds of hydrogen are not known to form hydrates; and custom, which rejects all dissimilarity until the claim to peculiarity is quite proved, leads us to prefer an opinion for which there is no further foundation than that of analogy. The woad (*Isatis tinctoria*) and several species of the *Nerium* contain a substance similar in many respects to gluten; this is deposited as indigo blue, when an aqueous infusion of the dried leaves is exposed to the action of the air. Now it is very doubtful whether the blue insoluble indigo is an oxide of the colourless soluble indigo, or the latter a combination of hydrogen with the indigo blue. Dumas has found the same elements in both, except that the soluble compound contained 1 equivalent of hydrogen more than the blue.

In the same manner the soluble gluten may be considered a compound of hydrogen, which becomes ferment by losing a certain quantity of this element when exposed to the action of the oxygen of the air under favourable circumstances. At all events, it is certain that oxygen is the cause of the insoluble condition of gluten; for yeast is not deposited on keeping wine, or during the fermentation of Bavarian beer, unless oxygen has access to the fluid.

Now, whatever be the form in which the oxygen unites with the gluten—whether it combines directly with it, or extracts a portion of its hydrogen, forming water—the products formed in the interior of the liquid, in consequence of the conversion of the gluten into ferment, will still be the same. Let us suppose that gluten is a compound of another substance with hydrogen, then this hydrogen must be removed during the ordinary fermentation of must and wort, by combining with oxygen, exactly as in the conversion of alcohol into aldehyde by eremacausis.

In both cases the atmosphere is excluded; the oxygen cannot, then, be derived from the air, neither can it be supplied by the elements of water, for it is impossible to suppose that the oxygen will separate from the hydrogen of water, for the purpose of uniting with the hydrogen of gluten, in order again to form water. The oxygen must, therefore, be obtained from the elements of sugar, a portion of which substance must, in order to the formation of ferment, undergo a different decomposition from that which produces alcohol. Hence a certain part of the sugar will not be converted into carbonic acid and alcohol, but will yield other products containing less oxygen than sugar itself contains. These products, as has already been mentioned, are the cause of the great difference in the qualities of fermented liquids, and particularly in their quantity of alcohol.

Must and wort do not, therefore, in ordinary fermentation, yield alcohol in proportion to the quantity of sugar which they hold in solution, a part of the sugar being employed in the conversion of gluten into ferment, and not in the formation of alcohol. But in the fermentation of Bavarian beer, all the sugar is expended in the production of alcohol; and this is especially the case whenever the transformation of the sugar is not accompanied by the formation of yeast.

It is quite certain that in the distilleries of brandy from potatoes, where no yeast is formed, or only a quantity corresponding to the malt which has been added, the proportion of alcohol and carbonic acid obtained during the fermentation of the *mash* corresponds exactly to that of the carbon contained in the starch. It is also known that the volume of carbonic acid evolved during the fermentation of beet-roots gives no exact indication of the proportion of sugar contained in them, for less carbonic acid is obtained than the same quantity of pure sugar would yield.

Beer obtained by the mode of fermentation adopted in Bavaria contains more alcohol, and possesses more intoxicating properties, than that made by the ordinary method of fermentation, when the quantities of malt used are the same. The strong taste of the former beer is generally ascribed to its containing carbonic acid in larger quantity, and in a state of more intimate combination; but this opinion is erroneous. Both kinds of beer are, at the conclusion of the fermentation, completely saturated with carbonic acid, the one as much as the other. Like all other liquids, they both must retain such a portion of the carbonic acid evolved as corresponds to their temperature and power of solution, that is, to their volumes.

The temperature of the fluid during fermentation has a very important influence on the quantity of alcohol generated. It has been mentioned, that the juice of beet-roots allowed to ferment at from 86° to 95° (30° to 35° C.) does not yield alcohol; and that afterwards, in the place of the sugar, mannite, a substance incapable of fermentation, and containing less oxygen than sugar, is found, together with lactic acid and mucilage. The formation of these products diminishes in proportion as the temperature is lower. But in vegetable juices, containing nitrogen, it is impossible to fix a limit, where the transformation of the sugar is undisturbed by a different process of decomposition.

It is known that in the fermentation of Bavarian beer the action of the oxygen of the air, and the low temperature, cause *complete* transformation of the sugar into alcohol; the cause which would prevent that result, namely, the extraction of the oxygen of part of the sugar by the gluten, in its conversion into ferment, being avoided by the introduction of oxygen from without.

The quantity of matters in the act of transformation is naturally greatest at the beginning of the fermentation of must and wort; and all the phenomena which accompany the process, such as evolution of gas, and heat, are most distinct at that time. These signs of the changes proceeding in the fluid diminish when the greater part of the sugar has undergone decomposition; but they must cease entirely before the process can be regarded as completed.

The less rapid process of decomposition which succeeds the violent evolution of gas, continues in wine and beer until the sugar has completely disappeared; and hence it is observed, that the specific gravity of the liquid diminishes during many months. This slow fermentation, in most cases, resembles the fermentation of Bavarian beer, the transformation of the dissolved sugar being in part the result of a slow and continued decomposition of the precipitated yeast; but a complete separation of the azotised substances dissolved in it cannot take place when air is excluded.*

Neither alcohol alone, nor hops, nor indeed both together, preserve beer from becoming acid. The better kinds of ale and porter in England are protected from acidity, but at the loss of the interest of an immense capital. They are placed in large closed wooden vessels, the surfaces of which are covered with sand. In these they are allowed to lie for several years, so that they are treated in a manner exactly similar to wine during its *ripening*.

* The great influence which a rational management of fermentation has upon the quality of beer, is well known in several of the German states. In the grand-duchy of Hesse, for example, a considerable premium is offered for the preparation of beer, according to the Bavarian method; and the premium is to be adjudged to any one who can prove that the beer brewed by him has lain for six months in the *store-vats* without becoming acid. Hundreds of casks of beer became changed to vinegar before an empirical knowledge of those conditions was obtained, the influence of which is rendered intelligible by theory.

A gentle diffusion of air takes place through the pores of the wood, but the quantity of azotised substances being very great in proportion to the oxygen which enters, they consume it, and prevent its union with the alcohol. But the beer treated in this way does not keep for two months without acidifying, if it be placed in smaller vessels, to which free access of the air is permitted.

CHAPTER X.

FERMENTATION ASCRIBED TO THE GROWTH OF FUNGI AND OF INFUSORIA.

The microscopical examination of vegetable and animal matter, in the act of fermentation or putrefaction, has lately given rise to the opinion, that these actions themselves, and the changes suffered by the bodies subjected to them, are produced in consequence of the development of fungi, or of microscopical animals, the germs or eggs of which are supposed to be diffused everywhere, in a manner inappreciable to our senses; they are supposed to be developed when they meet with a medium fitted to afford them nourishment.

Several philosophers have ascribed to this circumstance the fermentation of wort, and of the juice of the grape. They assert, that the decomposition of sugar into alcohol and carbonic acid is effected by the contact of particles of the sugar with the growing plants, which they view as the yeast, or ferment, without studying more closely the final causes of the decomposition of the sugar. It has been supposed that this view is opposed to the theory detailed in the preceding pages, which described contact as the cause of a peculiar activity or power.

In all chemical processes, and in all changes effected by chemical affinity, we observe that contact is essential for the exercise of the acting power. Hence, chemists describe affinity as a force distinct from other powers, because it acts only in immediate contact, or at inappreciable distances. Thus contact plays an important part in every case of combination or decomposition, for without contact these changes would not take place. In this sense, all substances effecting combination or decomposition, are bodies acting by contact.

In the theory of fermentation alluded to, it was not asserted that the yeast or ferment could effect the decomposition of sugar at appreciable distances. In this respect, therefore, the two theories are not opposed to each other. They deviate, however, in this, that the one theory considers yeast as a body, the smallest particles of which are in a state of motion and transposition, and that, by virtue of this state, the particles of sugar in contact with it are thrown into the same state of change; while the other theory asserts, that the particles of yeast are little fungi, which are developed from germs or seeds falling into the fermenting liquid from the air; and that in this they grow at the expense of the substances containing nitrogen, which are thus converted into, and separated as, fungi. The particles of sugar in contact with the fungi are supposed to be converted into carbonic acid and alcohol, which, in other words, signifies, that the act of vegetation effects a disturbance in the chemical attractions of the elements of the sugar, in consequence of which they arrange themselves into new compounds.

Gay-Lussac showed by experiments that the juice of grapes expressed apart from air, under a bell-jar full of mercury, did not enter into putrefaction, although it did so in the course of a few hours when air was admitted. The same chemist also showed, that fermentation immediately commences on the introduction of oxygen gas, of which a quantity is absorbed equal only to the $\frac{1}{120}$th part of the volume of carbonic acid evolved during the fermentation. It can scarcely be supposed, that the germs of fungi exist in chlorate of potash or black oxide of manganese, out of which the oxygen was obtained; and hence, it is difficult to ascribe to a growing vegetation the causes of the decomposition.

Gay-Lussac further showed, that the juice entered into fermentation on being connected with the wires of a galvanic battery, under circumstances, therefore, which quite excluded the introduction of every foreign body. Hence the view, that the fermentation of sugar is effected by contact with growing plants, must presuppose that living beings, plants for example, may be formed and developed without germs or seeds—a circumstance in direct contradiction to all observation regarding the growth of plants.

It is certain that sponges and fungi, growing in places from which light is quite excluded, follow laws of nutrition different from those governing green plants; and it cannot be doubted that their nourishment is derived from putrefying bodies, or from the products of their putrefaction, which pass directly into this kind of plants, and obtain an organised form by the vital powers residing within them. During their growth they constantly emit carbonic acid, increasing in weight at the same time, while all other plants, under similar circumstances, would decrease in weight. Hence it is possible, and indeed probable, that fungi may have the power of growing in fermenting and putrefying substances, in as far as the products arising from the putrefaction are adapted for their nourishment. When a quantity of fungi are exposed to the temperature of boiling water, their vitality and power of germinating become completely destroyed. If they be now kept at a proper temperature, an evolution of gas proceeds in the mass thus treated; they pass over into putrefaction, and, if air be admitted, into decay; and at last nothing remains except their inorganic elements. The putrefaction in this case cannot be viewed as the act of the formation of organic beings, but as the act of the passage of their elements into inorganic compounds.

Observations of another kind,—for example, that flesh and other animal bodies may be kept for several weeks without putrefying, if placed in a vessel containing air previously heated to redness,—have gone far to support the opinion that the process of putrefaction is effected by the growth of organic beings: but all such experiments are of very subordinate value in support of these conclusions. In some experiments instituted by the author, for the purpose of detecting quinine in the urine of a patient in the habit of taking this medicine, he obtained the remarkable result, that this urine kept for several weeks without passing into complete putrefaction, although the urea of urine, under ordinary circumstances, is often completely converted into carbonate of ammonia, in the space of six or eight hours. In the present case, the urine effervesced only slightly with acids after fourteen days. This seemed to give sufficient foundation for the opinion that the quinine must be the cause of this delay in the putrefaction. But further experiments proved that common urine introduced when freshly drawn into perfectly pure vessels behaved in an exactly similar manner. When a little putrefying urine was added to the fresh urine, the putrefaction of the latter was accelerated in a high degree. Wood, in which urine had been retained, exerted this action in a very decided manner, and the white, or yellowish-white deposite from putrefying urine (which does not possess an organised form) effects the conversion of urea into carbonate of ammonia in the course of a few hours.

Fresh flesh remains for several weeks without experiencing appreciable change in a perfectly pure glass vessel, whether the latter contains common air, or air previously heated to redness; but, at the same time, it absorbs oxygen and emits carbonic acid, and passes into putrefaction, if the necessary quantity of water be present, without being prevented or retarded by the air thus treated.

It cannot be supposed, that dung-flies, living upon animal excrements, are the cause of this putrefaction; neither can a similar conclusion be drawn in the case of mites and maggots found so abundantly in old cheese.

When we consider, that the intermediate products formed in the passage of animal and vegetable matters into inorganic compounds possess the power of supporting the life of certain animals and vegetables low in the scale of creation, then, the only mystery is, in what manner the germs of the fungi, or the eggs of the infusoria, reach the place fitted for their development; for this being known, there is no difficulty, since the discoveries of Ehrenberg, in conceiving this extraordinary increase. Now, as it is observed that the infusoria increase in size only to a certain point, it must hence be concluded that their nourishment, even if only from the point at which they are to grow, passes out of their bodies in the form of excrements, precisely as in the higher orders of animals. As is the case with all other excrements, these must possess, in an eminent degree, the property of passing into decay or putrefaction; and this condition must at all events be induced by contact with the original putrefying body. Hence the increase in numbers of the infusoria must induce and accelerate the process of putrefaction in the putrefying body itself. The ultimate products of decay and putrefaction are carbonic acid, ammonia, and water. In order to comprehend the chemical process by which this conversion is effected, it is of much interest to become acquainted with the intermediate compounds formed by the elements. But in regard to the process itself, it is, chemically speaking, quite indifferent whether the first, second, or third product, before they assume the final state, be in the form of fungi, or of living animals (infusoria). These plants and animals are not the causes of the conversion, for they suffer after death the same changes which finally occasion their complete disappearance.

The enormous layers of microscopic animals in the chalk, (the siliceous infusoria) do not contain any organic matter. The lime of their shells, and the silica of their bony coverings, were obtained from the water in which they were developed. If this water had been deficient in lime, or in silica, these animals could not have been produced; and if they had not found nourishment in the products of the putrefaction of former species (the remains of which are found in the *muschelkalk*), they would not have been developed; and without the co-operation of both these causes, they could not have formed such extensive masses and layers as they actually do.

But these animals are not the causes of the formation of the chalk, or of the layers of flint, and as little are they the cause of the decay and putrefaction of those substances, which yielded to them their organic constituents. Without these animals there might not have been chalk, but there would have been marble, or another limestone; and the silica would have been deposited as siliceous schist, or as quarz, after the evaporation of the water. Hence it is only the form which is given to the layers by organic life; but the substance of these strata (chalk) is chemically in no respect different from crystallised calcareous spar: in fact, the same explanation of their origin might be made as that adopted in the case of the older limestone formations.

The conversion of the constituents of an elephant into aerial compounds, is the same process, and is effected by the same causes as those occasioning the destruction of the carcase of the microscopical animals, which themselves obtained their elements from extinct species of other animals. The final products are identical in both cases.

There have been very wonderful and incomprehensible observations made on the behaviour and functions of certain microscopical animals. From these observations, there seem to follow conclusions regarding the nutrition and growth of these creatures, quite at variance with all that we know of the process of nutrition of the higher classes of animals.

In a treatise on the composition of the salt-springs in Hesse-Cassel, Pfannkuch mentions a singular phenomenon, that the slimy mass which deposits in the tubs set to receive the brine percolating through the wells of the graduating-house, contains a gas which is found to be pure oxygen gas. The fresh brine obtained directly from the draw-well is quite clear, and contains 5 per cent. of salt, with gypsum and sulphuretted hydrogen in such considerable quantity that it might be used as a sulphureous water. During the summer months, a slimy transparent mass forms in this brine, covering the bottom of the vessel containing it to the depth of one to two inches. This matter is everywhere filled with bubbles of gas, of a considerable size, often two or three inches broad; these rise to the surface, when the membrane inclosing them is torn with a stick. The quantity of these gas-bubbles is so great, that it would be easy to fill hundreds of bottles with them in a short time. They are so rich in oxygen gas, that a glowing match of wood introduced into the collected gas, bursts into flame, and continues to burn with brilliancy. On being analysed, this gas is found to consist of 51 per cent. of oxygen, and 49 per cent. of nitrogen; but there can be little doubt that the gas originally consisted of pure oxygen, which became mixed with the nitrogen of air by virtue of diffusion, just as it does when confined in an animal membrane. In fact, it is found, that when the water in the tubs is very low, the bubbles existing in the deposite appear to be pure air, owing to the celerity with which the diffusion has taken place (Wöhler).

Wöhler has subjected to microscopical examination the slimy membranous deposite, and has shown that it consists almost entirely of living and moving infusoria, principally species of Navicula and Gallionella, such as occur in the paper-like formations of Freiberg, and in the siliceous fossil strata of Franzensbad. The whole deposit possesses a slight greenish colour, and is intersected with very fine colourless fibres of confervæ. After washing and drying the deposite, a residue like paper is obtained; and this, on being heated, gives distinct indications of ammonia, showing that it contains nitrogen. It yields also a mass resembling paper, which, on incineration, being treated with muriatic acid, leaves behind siliceous skeletons, which preserve the shape of the animal so completely, that it appears as if the original deposite itself were submitted to examination (Wöhler).

These observations are of remarkable interest, for, as Wöhler asks—Whence comes the oxygen gas—from the confervæ or from the infusoria? The quantity of oxygen being so large, and the infusoria being in great preponderance, would lead to the conclusion that the former must be derived from these; and yet this is opposed to all analogy. The water comes out of a depth of 500 feet; and its sulphuretted hydrogen shows that it comes out of a layer of rocks containing putrefying animal matter, which, acting upon the sulphates, produces sulphuretted hydrogen: and in this water is formed, with the aid of solar light, a source of oxygen gas, to all appearances more abundant than we see in the case of green plants. Sir B. Thompson (better known as Count Rumford) published some experiments 56 years since, which are of such a remarkable nature, that we give them in the author's own words. Thompson found that silk, cotton, sheep's wool, eider-down, and other organic substances, evolve oxygen gas, when they are freed from air by washing, and then exposed to sun-light in a glass globe perfectly filled with water. After two or three days, the water assumed a greenish hue, and from that moment the evolution of gas commenced.

"One hundred and twenty grains of cotton, in a bell jar, along with 296 cubic inches of spring water, gave out, during the first four days, $2\frac{3}{4}$ C. I. of gas, containing almost no oxygen. It was not till the sixth day, when the sun was very powerful, that the water suddenly became green, and gave out during the next six days, $44\frac{1}{2}$ C. I. of oxygen nearly pure. On examining the water under the microscope, it was found to contain a multitude of very minute, nearly spherical animalcules. Wherever the water was green, these animalcules were found, insomuch that the green colour seemed to be caused by them." After describing his numerous experiments, Count Rumford adds—

"The phenomena now described may, perhaps, admit of explanation, if we assume that the air produced in the water in the different experiments was derived from the green matter; and that the leaves, silk, cotton, &c., only facilitated its disengagement by furnishing a surface adapted to the collection and escape of the gas-bubbles.

"These phenomena may also be explained by an assumption favourable to the hypothesis of Priestley, namely, that the green matter consists of plants, which adhering to the surface of the bodies placed in the water, there vegetate, and in consequence give rise to the gas."

"I would willingly adopt this opinion, were it not that a most careful and attentive examination of the green water by means of an excellent microscope, at the period when the oxygen was most copiously disengaged, has convinced me, that at this period nothing to which the name of vegetable can be given is present. The colouring matter of the water is of an animal nature, and is nothing else than the accumulation of an infinite number of little moving animals.—*Philosophical Transactions of the Royal Society*, Vol. lxxvii., 1787.

In a very interesting memoir, by Messrs. August and Morren (*Transactions of the Academy of Brussels*, 1841), it is shown that water with organic substances evolve "a gas" which contains 61 per cent. of oxygen; and they conclude their treatise in the following words:—"It follows from the preceding remarks, that the phenomenon of the evolution of oxygen gas is due to the *Chlamidomonas pulvisculus* (Ehrenberg), and to several other green animals still lower in the scale."

The author took the opportunity of convincing himself of the accuracy of this long-observed fact, by means of some water out of a water-trough in his garden, the water being coloured strongly green by different kinds of infusoria. This water was freed by means of a sieve from all particles of vegetable matter, and, being placed in a jar, inverted in a porcelain vessel containing the same water, was exposed for several weeks to the action of solar light. During this time, a continued accumulation of gas took place in the upper part of this jar; after fourteen days $\frac{1}{3}$ of the water in the jar had been pressed out of it, and the gas, which had taken its place, ignited a glowing match of wood, and in all respects behaved like pure oxygen gas. It must be here expressly stated, that the water, before being exposed to the action of solar light, was examined by one of Ploessl's best microscopes, without the detection of confervæ or of any kind of vegetable matter.*

Without venturing upon any opinion on the mode of nutrition of these animals, it is quite certain that water containing living infusoria becomes a source of oxygen gas when exposed to the action of light. It is also certain, that as soon as these animals can be detected in the water, the latter ceases to act injuriously to plants or animals; for it is

* One hundred cubic inches of water saturated with air, contained, in the form of air, according to the experiments of Humboldt and Gay-Lussac, not above 1·6 cubic inches of oxygen gas.

impossible to assume that pure oxygen gas can be evolved from water containing any decaying or putrefying matters, for these possess the property of combining with oxygen. Now, it is obvious, if we add to such water any animal or vegetable matter, in a state of decay, that this being in contact with oxygen will resolve itself into the ultimate products of oxidation, in a much shorter time than if infusoria were not present.

Thus we recognise in these animals, or perhaps only in certain classes of them, by means of the oxygen which in some way as yet incomprehensible accompanies their appearance, a most wise and wonderful provision for removing from water the substances hurtful to the higher classes of animals; and for substituting, in their stead, the food of plants (carbonic acid), and the oxygen gas essential to the respiration of animals. They cannot be viewed as the causes of putrefaction, or of the generation of the products affecting animal and vegetable life; but they make their appearance in order to accelerate the conversion of putrefying organic matter into its ultimate products.

Many fungi grow without light, and, in their growth and life, are characterised by all the phenomena which characterise animal life; they destroy air by absorbing oxygen and evolving carbonic acid, and, in a chemical point of view, behave like animals without motion. (See Appendix to Part II.)

In opposition to this class of beings, which can scarcely be designated as plants, we have living creatures endowed with motion, and with the organs which characterise animals, and yet which behave in the light like green plants; for while they increase in size and number, they furnish sources of oxygen when its access, in the form of air, is excluded or prevented.

CHAPTER XI.

DECAY OF WOODY FIBRE.

THE conversion of woody fibre into the substances termed humus and mould is, on account of its influence on vegetation, one of the most remarkable processes of decomposition in nature.

Decay is not less important in another point of view; for, by means of its influence on dead vegetable matter, the oxygen retained by plants during life is again restored to the atmosphere.

The decomposition of woody fibre is effected in three forms, the results of which are different, so that it is necessary to consider each separately.

The first takes place when it is in the moist condition, and subject to free uninterrupted access of air: the second occurs when the air is excluded; and the third when the wood is covered with water, and in contact with putrefying organic matter.

It is known that woody fibre may be kept under water, or in dry air, for thousands of years, without suffering any appreciable change; but that when brought into contact with air, in the moist condition, it converts the oxygen surrounding it into the same volume of carbonic acid, and is itself gradually changed into a yellowish brown, or black matter, of a loose texture.*

Carbonic acid and mould or humus are therefore the pro-

* According to the experiments of De Saussure, 240 parts of dry sawdust of oak-wood convert 10 cubic inches of oxygen into the same quantity of carbonic acid, which contains 3 parts, by weight, of carbon; while the weight of the sawdust is diminished by 15 parts. Hence, 12 parts, by weight, of water, are at the same time separated from the elements of the wood.

ducts of the decomposition of wood. We have assumed that the water formed by the combination of the hydrogen of the wood with the oxygen of the atmosphere, and that during the process of oxidation carbon and oxygen escape from the wood in the form of carbonic acid.

It has already been mentioned, that pure woody fibre contains carbon and the elements of water. Humus, however, is not produced by the decay of pure woody fibre, but by that of wood which contains foreign soluble and insoluble organic substances, besides its essential constituents.

The relative proportion of the component elements are, on this account, different in oak wood and in beech, and the composition of both of these differs very much from woody fibre, which is the same in all vegetables. The difference, however, is so trivial, that it may be altogether neglected in the consideration of the questions which will now be brought under discussion; besides, the quantity of the foreign substances is not constant, but varies according to the season of the year.

According to the careful analysis of Gay-Lussac and Thénard, 100 parts of oakwood, dried at 212° (100° C.), from which all soluble substances had been extracted by means of water and alcohol, contained 52·53 parts of carbon, and 47·47 parts of hydrogen and oxygen, in the same proportion as they are contained in water.

Now it has been mentioned that moist wood acts in oxygen gas exactly as if its carbon combined directly with oxygen, and that the products of this action are carbonic acid and humus.

If the action of the oxygen were confined to the carbon of the wood, and if nothing but carbon were removed from it, the remaining elements would necessarily be found in the humus, unchanged except in the particular of being combined with less carbon. The final result of the action would therefore be a complete disappearance of the carbon, whilst nothing but the elements of water would remain.

But when decaying wood is subjected to examination

in different stages of its decay, the remarkable result is obtained, that the proportion of carbon in the different products augments. Consequently, if we did not take into consideration the evolution of carbonic acid under the influence of the air, the conversion of wood into humus might be viewed as a removal of the elements of water from the carbon.

The analysis of mouldered oak wood, taken from the interior of the trunk of an oak, and possessing a chocolate-brown colour and the structure of wood, showed that 100 parts of it contained 53·36 parts of carbon and 46·44 parts of hydrogen and oxygen in the same relative proportions as in water. From an examination of mouldered wood of a light-brown colour, easily reducible to a fine powder, and taken from another oak, it appeared that it contained 56·211 carbon and 47·789 water.

These indisputable facts point out the similarity of the decay of wood, with all other instances of the slow combustion or oxidation of bodies containing a large quantity of hydrogen. Viewed as a kind of combustion, it would indeed be a very extraordinary process, if the carbon combined directly with the oxygen; for it would be a combustion in which the carbon of the burning body augmented constantly, instead of diminishing. Hence it is evident that it is the hydrogen which is oxidised at the expense of the oxygen of the air; while the carbonic acid is formed from the elements of the wood. Carbon never combines at common temperatures with oxygen, so as to form carbonic acid.

In whatever stage of decay wood may be, its elements must always be capable of being represented by their equivalent numbers.

The following formula illustrates this fact with great clearness:—

$C_{36}\ H_{22}\ O_{22}$—oak wood, according to Guy-Lussac and Thénard.*
$C_{35}\ H_{20}\ O_{20}$—humus from oak wood (Meyer).†
$C_{34}\ H_{18}\ O_{18}$— " " (Dr. Will).‡

* The calculation from this formula gives 52·5 carbon, and 47·5 water.
† The calculation gives 54 carbon, and 46 water.
‡ The calculation gives 56 carbon, and 44 water.

It is evident from these numbers, that for every two equivalents of hydrogen oxidised, two atoms of oxygen and one of carbon are set free.

Under ordinary circumstances, woody fibre requires a very long time for its decay; but this process is of course much accelerated by an elevated temperature and free unrestrained access of air. The decay, on the contrary, is much retarded by absence of moisture, and by the wood being surrounded with an atmosphere of carbonic acid, which prevents the access of air to the decaying matters.

Sulphurous acid, and all antiseptic substances, arrest the decay of woody fibre. It is well known that corrosive sublimate is employed for the purpose of protecting the timber of ships from decay; it is a substance which completely deprives vegetable or animal matters, the most prone to decomposition, of their property of entering into fermentation, putrefaction, or decay.

But the decay of woody fibre is very much accelerated by contact with alkalies or alkaline earths; for these enable substances to absorb oxygen, although they do not possess this power themselves; alcohol, gallic acid, tannin, the vegetable colouring matters, and several other substances, are thus affected by them. Acids produce quite an opposite effect; they greatly retard decay.

Heavy soils, consisting of loam, retain longest the most important condition for the decay of the vegetable matter contained in them, viz., water; but their impermeable nature prevents contact with the air.

In moist sandy soils, particularly such as are composed of a mixture of sand and carbonate of lime, decay proceeds very quickly, it being aided by the presence of the slightly alkaline lime.

Now let us consider the decay of woody fibre during a very long period of time, and suppose that its cause is the gradual removal of the hydrogen in the form of water, and the separation of its oxygen in that of carbonic acid. It is evident that if we subtract from the formula $C_{36}\ H_{22}\ O_{22}$ the 22 equivalents of oxygen, with 11 equivalents of carbon, and 22 equivalents of hydrogen, which are supposed to be oxidised by the oxygen of the air, and separated in the form of water; then from 1 atom of oak wood, 25 atoms of pure carbon will remain as the final product of the decay. In other words, 100 parts of oak, containing 52·5 parts of carbon, will leave as a residue 36·5 parts of carbon, which must remain unchanged, since carbon does not combine with oxygen at common temperatures.

But this final result is never attained in the decay of wood under common circumstances; and for this reason, that with the increase of the proportion of carbon in the residual humus, as in all decompositions of this kind, its attraction for the hydrogen, which still remains in combination, also increases, until at length the affinity of oxygen for the hydrogen is equalled by that of the carbon for the same element.

In proportion as the decay of woody fibre advances, its property of burning with flame, or, in other words, of developing carburetted hydrogen on the application of heat, diminishes. Decayed wood burns without flame; whence no other conclusion can be drawn, than that the hydrogen, which analysis shows to be present, is not contained in it in the same form as in wood.

Decayed oak contains more carbon than fresh wood, but its hydrogen and oxygen are in the same proportion.

We would naturally expect that the flame given out by decayed wood should be more brilliant in proportion to the increase of its carbon, but we find, on the contrary, that it burns like tinder, exactly as if no hydrogen were present. For the purposes of fuel, decayed or diseased wood is of little value, for it does not possess the property of burning with flame—a property upon which the advantages of common wood depend. The hydrogen of decayed wood must, consequently, be supposed to be in the state of water; for had it any other form, the characters we have described would not be possessed by the decayed wood.

If we suppose decay to proceed in a liquid containing

carbon and hydrogen, then a compound with still more carbon must be formed, in a manner similar to the production of the crystalline colourless naphthalin from a gaseous compound of carbon and hydrogen. And if the compound thus formed were itself to undergo further decay, the final result must be the separation of carbon in a crystalline form.

Science can point to no process capable of accounting for the origin and formation of diamonds, except the process of decay. Diamonds cannot be produced by the action of fire; for a high temperature, and the presence of oxygen gas, would call into play their combustibility. But there is the greatest reason to believe that they are formed in the humid way—that is, in a liquid, and the process of decay is the only cause to which their formation can with probability be ascribed.

Amber, fossil resin, and the acids in mellite, are the products of vegetable matter which has suffered eremacausis. They are found in wood (or brown) coal, and have evidently proceeded from the decomposition of substances which were contained in quite a different form in the living plants. They are all distinguished by their proportionally small quantity of hydrogen. The acid from mellite (mellitic acid) contains precisely the same proportions of carbon and oxygen as that from amber (succinic acid); they differ only in the proportion of their hydrogen. Succinic acid may be obtained by oxidation from wax and all other solid fats.

CHAPTER XII.

VEGETABLE MOULD.

THE term VEGETABLE MOULD, in its general signification, is applied to a mixture of disintegrated minerals, with the remains of animal and vegetable substances. It may be considered as earth in which humus is contained in a state of decomposition. Its action upon the air has been fully investigated by Ingenhouss and De Saussure.

When moist vegetable mould is placed in a vessel full of air, it extracts the oxygen therefrom with greater rapidity than decayed wood, and replaces it by an equal volume of carbonic acid. When this carbonic acid is removed and fresh air admitted, the same action is repeated.

Cold water dissolves only $\frac{1}{10,000}$th of its own weight of vegetable mould; and the residue left on its evaporation consists of common salt with traces of sulphate of potash and lime, and a minute quantity of organic matter, for it is slightly blackened when heated to redness. Boiling water extracts several substances from vegetable mould, and acquires a yellow or yellowish brown colour, which is dissipated by absorption of oxygen from the air, a black flocculent deposite being formed. When the coloured solution is evaporated, a residue is left which becomes black on being heated to redness, and afterwards yields carbonate of potash when treated with water.

A solution of caustic potash becomes black when placed in contact with vegetable mould, and the addition of acetic acid to the coloured solution causes no precipitate or turbidity. But dilute sulphuric acid throws down a light flocculent precipitate of a brown or black colour, from which the acid can be removed with difficulty by means of water. When this precipitate, after having been washed with water, is brought whilst still moist under a receiver filled with oxygen, the gas is absorbed with great rapidity; and the same thing takes place when the precipitate is dried in the air. In the perfectly dry state it has entirely lost its solubility in water, and even alkalies dissolve only traces of it.

It is evident, therefore, that boiling water extracts a matter from vegetable mould, which owes its solubility to the presence of the alkaline salts contained in the remains of plants. This substance is a product of the incomplete decay of woody fibre, and contains a certain quantity of ammonia chemically combined. Its composition is intermediate between woody fibre and humus, into which it is converted, by being exposed in a moist condition to the action of the air.

CHAPTER XIII.

ON THE MOULDERING OF BODIES.—PAPER, BROWN COAL, AND MINERAL COAL.

THE decomposition of wood, woody fibre, and all vegetable bodies when subjected to the action of water, and excluded from the air, is termed *mouldering*.

Wood, (or brown) coal, and mineral coal, are the remains of vegetables of a former world; their appearance and characters show that they are products of the processes of decomposition termed decay and putrefaction. We can easily ascertain by analysis the manner in which their constituents have been changed, if we suppose the greater part of their bulk to have been formed from woody fibre.

But it is necessary, before we can obtain a distinct idea of the manner in which coal is formed, to consider a peculiar change which woody fibre suffers by means of moisture, when partially or entirely excluded from the air.

It is known, that when pure woody fibre, as linen, for example, is placed in contact with water, considerable heat is evolved, and the substance is converted into a soft friable mass, which has lost all coherence. This substance was employed in the fabrication of paper before the use of chlorine, as an agent for bleaching. The rags employed for this purpose were placed in heaps; and it was observed, that on their becoming warm a gas was disengaged, and their weight diminished from 18 to 25 per cent.

When sawdust moistened with water is placed in a closed vessel, carbonic acid gas is evolved in the same manner as when air is admitted. A true putrefaction takes place, the wood assumes a white colour, loses its peculiar texture, and is converted into a rotten friable matter.

The white decayed wood found in the interior of trunks of dead trees which have been in contact with water, is produced in the way just mentioned.

An analysis of wood of this kind, obtained from the interior of the trunk of an oak, yielded, after having been dried at 212°,

Carbon	47·11	48·14
Hydrogen	6·31	6·06
Oxygen	45·31	44·43
Ashes	1·27	1·37
	100·00	100·00

Now, on comparing the proportions obtained from these numbers with the composition of oak wood, according to the analysis of Gay-Lussac and Thénard, it is immediately perceived that a certain quantity of carbon has been separated from the constituents of wood, whilst the hydrogen is, on the contrary, increased. The numbers obtained by the analysis correspond very nearly to the formula $C_{33} H_{27} O_{24}$*.

The elements of water have, therefore, become united with the wood, whilst carbonic acid is disengaged by the absorption of a certain quantity of oxygen.

If the elements of 5 atoms of water and 3 atoms of oxygen be added to the composition of the woody fibre of the oak, and 3 atoms of carbonic acid deducted, the exact formula for white mouldered wood is obtained.

Wood	$C_{36} H_{22} O_{22}$
To this add 5 atoms of water	$H_5 O_5$
3 atoms of oxygen	O_3
	$C_{36} H_{27} O_{30}$
Subtract from this 3 atoms carbonic acid	$C_3 \quad O_6$
	$C_{33} H_{27} O_{24}$

The process of mouldering is, therefore, one of putrefaction and decay, proceeding simultaneously, in which the oxygen of the air and the component parts of water take part. But the composition of mouldered wood must change according

* The calculation from this formula gives in 100 parts 47·9 carbon, 6·1 hydrogen, and 46 oxygen.

as the access of oxygen is more or less prevented. White mouldered beech-wood yielded on analysis 47·67 carbon, 5·67 hydrogen, and 46·68 oxygen; this corresponds to the formula $C_{33}\ H_{25}\ O_{24}$.

The decomposition of wood assumes, therefore, two different forms, according as the access of the air is free or restrained. In both cases carbonic acid is generated; and in the latter case, a certain quantity of water enters into chemical combination.

It is highly probable that in this putrefactive process, as well as in all others, the oxygen of the water assists in the formation of the carbonic acid.

Wood-coal (brown coal of Werner) must have been produced by a process of decomposition similar to that of mouldering. But it is not easy to obtain wood coal suited for analysis, for it is generally impregnated with resinous or earthy substances, by which the composition of those parts which have been formed from woody fibre is essentially changed.

The wood-coal, which forms extensive layers in the Wetterau (a district in Hesse-Darmstadt), is distinguished from that found in other places, by possessing the structure of wood unchanged, and by not containing bituminous matter. This coal was subjected to analysis, a piece being selected upon which the annual circle could be counted. It was obtained from the vicinity of Laubach; 100 parts contained

Carbon	57·28
Hydrogen	6·03
Oxygen	36·10
Ashes	0·59
	100·00

The large amount of carbon, and small quantity of oxygen, constitute the most obvious difference between this analysis and that of wood. It is evident that the wood which has undergone the change into coal must have parted with a certain portion of its oxygen. The proportion of these numbers are expressed by the formula $C_{33}\ H_{21}\ O_{16}$*.

* The calculation gives 57·5 carbon, and 5·98 hydrogen.

When these numbers are compared with those obtained by the analysis of oak, it would appear that the brown coal was produced from woody fibre by the separation of one equivalent of hydrogen, and the elements of three equivalents of carbonic acid.

1 atom wood	$C_{36}\ H_{22}\ O_{22}$
Minus 1 atom hydrogen and 3 atoms carbonic acid	$C_{3}\ H_{1}\ O_{6}$
Wood Coal . .	$C_{33}\ H_{21}\ O_{16}$

All varieties of wood-coal, from whatever strata they may be taken, contain more hydrogen than wood does, and less oxygen than is necessary to form water with this hydrogen; consequently they must all be produced by the same process of decomposition. The excess of hydrogen is either hydrogen of the wood remaining in it unchanged, or it is derived from some exterior source. The analysis of wood-coal from Ringkuhl, near Cassel, where it is seldom found in pieces with the structure of wood, gave, when dried at 212°,

Carbon	62·60	63·83
Hydrogen	5·02	4·80
Oxygen	26·52	25·51
Ashes	5·86	5·86
	100·00	100·00

The proportions derived from these numbers correspond very closely to the formula $C_{32}\ H_{15}\ O_{9}$, or they represent the constituents of wood, from which the elements of carbonic acid, water, and 2 equivalents hydrogen, have been separated.

$C_{36}\ H_{22}\ O_{22}$=Wood.
Subtract $C_{4}\ H_{7}\ O_{13}$=4 atoms carbonic acid+5 atoms of water +2 atoms of hydrogen.

$C_{32}\ H_{15}\ O_{9}$=Wood Coal from Ringkuhl.

The formation of both these specimens of wood-coal appears from these formulæ to have taken place under circumstances which did not entirely exclude the action of the air, and consequent oxidation and removal of a certain quantity of hydrogen. Now the Laubacher coal is covered with a layer of basalt, and the coal of Ringkuhl was taken from the lowest seam of layers, which possess a thickness of from 90 to 120 feet; so that both may be considered as well protected from the air.

During the formation of brown coal, therefore, the elements of carbonic acid have been separated from the wood either alone, or at the same time with a certain quantity of water. It is quite possible that the difference in the process of decomposition may depend upon the high temperature and pressure under which the decomposition took place. At least, a piece of wood assumed the character and appearance of Laubacher coal, after being kept for several weeks in the boiler of a steam-engine, and had then precisely the same composition. The change in this case was effected in water, at a temperature of from 334° to 352° F. (150° to 160° C.), and under a corresponding pressure. The ashes of the wood amounted to 0·51 per cent.; a little less, therefore, than those of the Laubacher coal; but this must be ascribed to the peculiar circumstances under which it was formed. The ashes of plants examined by Berthier amounted always to much more than this.

The peculiar process by which the decomposition of these extinct vegetables has been effected, namely, a disengagement of carbonic acid from their substance, appears still to go on at great depths in all the layers of wood-coal. At all events, it is remarkable that springs impregnated with carbonic acid occur in many places, in the country between the Meissner, in the electorate of Hesse, and the Eifel, which are known to possess large layers of wood-coal. These springs of mineral water are produced on the spot at which they are found; the springs of common water meeting with carbonic acid during their ascent, and becoming impregnated with it.

In the vicinity of the layers of wood-coal at Salzhausen (Hesse-Darmstadt), an excellent acidulous spring of this kind existed a few years ago, and supplied all the inhabitants of that district; but it was considered advantageous to surround the sides of the spring with sandstone, and the consequence was, that all the outlets to the carbonic acid were closed, for this gas generally gains access to the water from the sides of the spring. From that time to the present this valuable mineral water has disappeared, and in its place is found a spring of common water.

Springs of water impregnated with carbonic acid occur at Schwalheim, at a very short distance from the layers of wood-coal at Dorheim. M. Wilhelmi observed some time since, that they are formed of common spring water which ascends from below, and of carbonic acid which issues from the sides of the spring. The same fact has been shown to be the case in the famed Fachinger spring, by M. Schapper.

The carbonic acid gas from the springs in the Eifel is, according to Bischoff, seldom mixed with nitrogen or oxygen, and is probably produced in a manner similar to that just described. At any rate the air does not appear to take any part in the formation of these acidulous springs. Their carbonic acid has evidently not been formed either by a combustion at high or low temperatures; for if it were so, the gas resulting from the combustion would necessarily be mixed with $\frac{4}{5}$ of nitrogen, but it does not contain a trace of this element. The bubbles of gas which escape from these springs are absorbed by caustic potash, with the exception of a residuum too small to be appreciated.

The wood-coal of Dorheim and Salzhausen must have been formed in the same way as that of the neighbouring village of Laubach; and since the latter contains the exact elements of woody fibre, minus a certain quantity of carbonic acid, its composition indicates very plainly the manner in which it has been produced.

The coal of the upper bed is subjected to an incessant decay by the action of the air, by means of which its hydrogen is removed in the same manner as in the decay of wood. This is recognised by the way in which it burns, and by the formation of carbonic acid in the mines.

The gases which are formed in mines of wood-coal, and cause danger in their working, are not combustible or inflammable as in mines of mineral coal; but they consist generally of carbonic acid gas, and are very seldom intermixed with combustible gases.

Wood-coal from the middle bed of the strata at Ringkuhl gave on analysis 65·40—64·01 carbon and 4·75—4·76* hydrogen; the proportion of carbon here is the same as in specimens procured from greater depths, but that of the hydrogen is much less.

Wood and mineral coal are always accompanied by iron pyrites (sulphuret of iron) or zinc blende (sulphuret of zinc); which minerals are still formed from salts of sulphuric acid, with iron or zinc, during the putrefaction of all vegetable matter. It is possible that the oxygen of the sulphates in the layers of wood-coal is the means by which the removal of the hydrogen is effected, since wood-coal contains less of this element than wood.

According to the analysis of Richardson and Regnault, the composition of the combustible materials in splint coal from Newcastle, and cannel coal from Lancashire, is expressed by the formula $C_{24}\ H_{13}\ O$. When this is compared with the composition of woody fibre, it appears that these coals are formed from its elements, by the removal of a certain quantity of carburetted hydrogen and carbonic acid, in the form of combustible oils. The composition of both of these coals is obtained by the subtraction of 3 atoms of carburetted hydrogen, 3 atoms of water, and 9 atoms of carbonic acid from the formula of wood.

		$C_{36}\ H_{22}\ O_{22}$=wood
3 atoms of carburetted hydrogen	$C_3\ H_6$	
3 atoms of water . .	$H_3\ O_3$	
9 atoms of carbonic acid . .	$C_9\ O_{18}$	$C_{12}\ H_9\ O_{21}$
	Mineral coal	$C_{24}\ H_{13}\ O$

Carburetted hydrogen generally accompanies all mineral coal; other varieties of coal contain volatile oils which may

* The analysis of brown coal from Ringkuhl, as well as all those of the same substance given in this work, have been executed in this laboratory by M. Khünert, of Cassel.

be separated by distillation with water. (Reichenbach.) The origin of naphtha is owing to a similar process of decomposition. Caking coal from Caresfield, near Newcastle, contains the elements of cannel coal, *minus* the constituents of olefiant gas $C_4\ H_4$.

The inflammable gases which stream out of clefts in the strata of mineral coal, or in rocks of the coal formations, always contain carbonic acid, according to a recent examination by Bischoff, and also carburetted hydrogen, nitrogen, and olefiant gas; the last of which had not been observed, until its existence in these gases was pointed out by Bischoff. The analysis of *fire-damp*, after it had been treated with caustic potash, showed its constituents to be—

	Gas from an abandoned mine near Wallesweiler. *Vol.*	Gerbard's passage near Luisenthal. *Vol.*	Gas from a mine near Liekwege. *Vol.*
Light carburetted hydrogen	91·36	83·08	79·10
Olefiant gas	6·32	1·98	16·11
Nitrogen gas	2·32	14·94	4·79
	100·00	100·00	100·00

The evolution of these gases proves that changes are constantly proceeding in the coal.

It is obvious from this, that a continual removal of oxygen in the form of carbonic acid is effected from layers of wood-coal, in consequence of which the wood must approach gradually to the composition of mineral coal. Hydrogen, on the contrary, is disengaged from the constituents of mineral coal in the form of a compound of hydro-carbon; a complete removal of all the hydrogen would convert coal into anthracite.

The formula $C_{36}\ H_{22}\ O_{22}$, which is given for wood, has been chosen as the empirical expression of the analysis, for the purpose of bringing all the transformations which woody fibre is capable of undergoing under one common point of view.

Now, although the correctness of this formula must be doubted, until we know with certainty the true constitution of woody fibre, this cannot have the smallest influence on the account given of the changes to which woody fibre must necessarily be subjected in order to be converted into wood or mineral coal. The theoretical expression refers to the absolute quantity, the empirical merely to the relative *proportion* in which the elements of a body are united. Whatever form the first may assume, the empirical expression must always remain unchanged.

CHAPTER XIV.

ON POISONS, CONTAGIONS, AND MIASMS.

A GREAT many chemical compounds, some derived from inorganic nature, and others formed in animals and plants, produce peculiar changes or diseases in the living animal organism. They destroy the vital functions of individual organs; and when their action attains a certain degree of intensity, death is the consequence.

The action of inorganic compounds, such as acids, alkalies, metallic oxides, and salts, can in most cases be easily explained. They either destroy the continuity of particular organs, or they enter into combination with their substance. The action of sulphuric, muriatic, and oxalic acids, hydrate of potash, and all those substances which produce the direct destruction of the organs with which they come into contact, may be compared to a piece of iron, which can cause death by inflicting an injury on particular organs, either when heated to redness, or when in the form of a sharp knife. Such substances are not poisons in the limited sense of the word, for their injurious action depends merely upon their condition.

The action of the proper inorganic poisons is owing, in most cases, to the formation of a chemical compound by the union of the poison with the constituents of the organ upon which it acts; it is owing to an exercise of a chemical affinity more powerful than the vitality of the organ.

It is well to consider the action of inorganic substances in general, in order to obtain a clear conception of the mode of action of those which are poisonous. We find that certain soluble compounds, when presented to different parts of the body, are absorbed by the blood, whence they are again eliminated by the organs of secretion, either in a changed or in an unchanged state.

Iodide of potassium, sulpho-cyanuret of potassium, ferro-cyanuret of potassium, chlorate of potash, silicate of potash, and all salts with alkaline bases, when administered internally to man and animals in dilute solutions, or applied externally, may be again detected in the blood, sweat, chyle, gall, and splenic veins; but all of them are finally excreted from the body through the urinary passages.

Each of these substances, in its transit, produces a peculiar disturbance in the organism—in other words, they exercise a medicinal action upon it, but they themselves suffer no decomposition. If any of these substances enter into combination with any part of the body, the union cannot be of a permanent kind; for their re-appearance in the urine shows that any compounds thus formed must have been again decomposed by the vital processes.

Neutral citrates, acetates, and tartrates of the alkalies, suffer change in their passage through the organism. Their bases can indeed be detected in the urine, but the acids have entirely disappeared, and are replaced by carbonic acid which has united with the bases. (Gilbert Blane and Wöhler.)

The conversion of these salts of organic acids into carbonates, indicates that a considerable quantity of oxygen must have united with their elements. In order to convert one equivalent of acetate of potash into the carbonate of the same base, 8 equivalents of oxygen must combine with it, of which either 2 or 4 equivalents (according as an acid or neutral salt is produced) remain in combination with the alkali; whilst the remaining 6 or 4 equivalents are disengaged as free carbonic acid. There is no evidence presented by the organism itself, to which these salts have been administered, that any of its proper constituents have yielded so great a quantity of oxygen as is necessary for their conversion into carbonates. Their oxidation can, therefore, only be ascribed to the oxygen of the air.

During the passage of these salts through the lungs, their acids take part in the peculiar process of eremacausis proceeding in that organ; a certain quantity of the oxygen gas inspired unites with their constituents, and converts their hydrogen into water, and their carbon into carbonic acid. Part of this latter product (1 or 2 equivalents) remains in combination with the alkaline base, forming a salt which suffers no further change by the process of oxidation; and it is this salt which is separated by the kidneys or liver.

It is manifest that the presence of these organic salts in the blood must produce a change in the process of respiration. A part of the oxygen inspired, which usually combines with the constituents of the blood, must, when they are present, combine with their acids, and thus be prevented from performing its usual office. The immediate consequence of this must be the formation of arterial blood in less quantity, or in other words, the process of respiration must be retarded.

Neutral acetates, tartrates, and citrates placed in contact with the air, and at the same time with animal or vegetable bodies in a state of eremacausis, produce exactly the same effects as we have described them to produce in the lungs. They participate in the process of decay, and are converted into carbonates just as in the living body. If impure solutions of these salts in water are left exposed to the air for any length of time, their acids are gradually decomposed, and at length entirely disappear.

Free mineral acids, or organic acids without volatility, and salts of mineral acids with alkaline bases, completely arrest decay when added to decaying matter in sufficient quantity; and when their quantity is small, the process of decay is protracted and retarded. They produce in living bodies the same phenomena as the neutral organic salts, but their action depends upon a different cause.

The absorption by the blood of a quantity of an inorganic salt sufficient to arrest the process of eremacausis in the lungs, is prevented by a very remarkable property of all animal membranes, skin, cellular tissue, muscular fibre, &c.; namely, by their incapability of being permeated by concentrated saline solutions. It is only when these solutions are diluted to a certain degree with water that they are absorbed by animal tissues.

A dry bladder remains more or less dry in saturated solutions of common salt, nitre, ferro-cyanuret of potassium, sulpho-cyanuret of potassium, sulphate of magnesia, chloride of potassium, and sulphate of soda. These solutions run off its surface in the same manner as water runs from a plate of glass besmeared with tallow.

Fresh flesh, over which salt has been strewed, is found, after 24 hours, swimming in brine, although not a drop of water has been added. The water has been yielded by the muscular fibre itself, and having dissolved the salt in immediate contact with it, and thereby lost the power of penetrating animal substances, it has on this account separated from the flesh. The water still retained by the flesh contains a proportionally small quantity of salt, having that degree of dilution at which a saline fluid is capable of penetrating animal substances.

This property of animal tissues is taken advantage of in domestic economy for the purpose of removing so much water from meat that a sufficient quantity is not left to enable it to enter into putrefaction.

In respect of this physical property of animal tissues, alcohol resembles the inorganic salts. It is incapable of moistening, that is, of penetrating, animal tissues, and possesses such an affinity for water as to extract it from moist substances.

When a solution of a salt, in a certain degree of dilution, is introduced into the stomach, it is absorbed; but a concentrated saline solution, in place of being itself absorbed, extracts water from the organ, and a violent thirst ensues. Some interchange of water and salt takes place in the stomach; the coats of this viscus yield water to the solution, a part of which, having previously become sufficiently diluted, is, on the other hand, absorbed. But the greater part of the concentrated solution of salt remains unabsorbed, and is not removed by the urinary passages; it consequently enters the intestines and intestinal canal, where it causes a dilution of the solid substances deposited there, and thus acts as a *purgative.*

Each of the salts just mentioned possesses this purgative action, which depends on a physical property shared by all of them; but, besides this, they exercise a medicinal action, because every part of the organism with which they come in contact absorbs a certain quantity of them.

The composition of the salts has nothing to do with their purgative action; it is quite a matter of indifference as far as the mere production of this action is concerned (not as to its intensity), whether the base be potash or soda, or in many cases lime and magnesia; and whether the acid be phosphoric, sulphuric, nitric, or hydrochloric.

If we drink, fasting, a glass of common spring water every ten minutes, a strong diuretic action becomes apparent, the quantity of salts in the water being much less than that in the blood.

When the second glass is taken, a quantity of urine is eliminated, the weight and volume of which corresponds nearly to that of the first glass; and by drinking twenty successive glasses of water, nineteen evacuations of urine take place, the last of which is colourless, and scarcely differs in its amount of saline ingredients from the spring water itself.

When the same experiment is made with a water containing exactly the amount of salts as in blood ($\frac{3}{4}$ to 1 per cent. of common salt for example), a separation of urine is not effected, and it becomes almost impossible to drink more than three glasses of such water. A sensation of fulness in the stomach, of pressure and weight, seem to show that water containing an equal amount of saline ingredients as blood, requires a much longer time to be taken up by the blood-vessels.

When the water taken contains a larger amount of salts than that existing in blood, a more or less active purgative action ensues. Hence, we see that three kinds of action take place, according to the quantities of salt existing in the water.

Besides these salts, the action of which does not depend upon their power of entering into combination with the component parts of the organism, there is a large class of others which, when introduced into the living body, effect changes of a very different kind, and produce diseases or death, according to the nature of these changes, without effecting a visible lesion of any organs.

These are the true inorganic poisons, the action of which depends upon their power of forming permanent compounds with the substance of the membranes and muscular fibre.

Salts of lead, iron, bismuth, copper, and mercury, belong to this class.

When solutions of these salts are treated with a sufficient quantity of albumen, milk, muscular fibre, and animal membranes, they enter into combination with those substances, and lose their own solubility; while the water in which they were dissolved loses all the salt which it contained.

The salts of alkaline bases extract water from animal substances; whilst the salts of the heavy metallic oxides are, on the contrary, extracted from the water, for they enter into combination with the animal matters.

Now, when these substances are administered to an animal, they lose their solubility by entering into combination with the membranes, cellular tissue, and muscular fibre; but in very few cases can they reach the blood. According to all the experiments yet made on the subject, it appears, that after the lapse of the same time as is required for the appearance of alkaline salts in the urine, the metallic salts above mentioned cannot be detected in that fluid. In fact, during their passage through the organism, they come into contact with many substances by which they are retained. By degrees, however, the constituents of the tissues with which they have combined are altered by the change of matter; their nitrogen appears in the urine, and along with it the mineral elements previously combined with the organic matter, such as mercury, copper, &c. When such substances enter into combination with organised parts, the functions of those parts must be disturbed, and must take an abnormal direction, producing morbid phenomena.

The action of corrosive sublimate and arsenious acid is very remarkable in this respect. It is known that these substances possess, in an eminent degree, the property of entering into combination with all parts of animal and vegetable bodies, rendering them at the same time insusceptible of decay or putrefaction. Wood and cerebral substance are both bodies which undergo change with great rapidity and facility when subject to the influence of air and water; but if they are digested for some time with arsenious acid or corrosive sublimate, they may subsequently be exposed to all the influences of the atmosphere without altering in colour or appearance.

It is further known that the parts of a body which come in contact with these substances during poisoning, and which therefore enter into combination with them, do not afterwards putrefy; so that there can be no doubt regarding the cause of their poisonous qualities.

It is obvious that if arsenious acid and corrosive sublimate are not prevented by the vital principle from entering into combination with the component parts of the body, and consequently from rendering them incapable of decay and putrefaction, they must deprive the organs of the principal property which appertains to their vital condition, viz., that of suffering and effecting transformations; or, in other words, organic life must be destroyed. If the poisoning is merely superficial, and the quantity of the poison so small that only individual parts of the body capable of being regenerated have entered into combination with it, then eschars are produced—a phenomenon of a secondary kind—the compounds of the dead tissues with the poison being thrown off by the healthy parts. From these considerations it may readily be inferred that all internal signs of poisoning are variable and uncertain; for cases may happen, in which no apparent indication of change can be detected by simple observations of the parts, because, as has been already remarked, death may occur without the destruction of any organs.

When arsenious acid is administered in solution, it may enter into the blood. If a vein is exposed and surrounded with a solution of this acid, every blood-globule will combine with it, that is, will become poisoned.

The compounds of arsenic, which have not the property of entering into combination with the tissues of the organism, are without influence on life, even in large doses. Many insoluble basic salts of arsenious acid are known not to be poisonous. The substance called alkargen, discovered by Bunsen, has not the slightest injurious action upon the organism; yet it contains a very large quantity of arsenic, and approaches very closely in composition to the organic arsenious compounds found in the body.

These considerations enable us to fix with tolerable certainty the limit at which the above substances cease to act as poisons. For since their combination with organic matters must be regulated by chemical laws, death will inevitably result, when the organ in contact with the poison finds sufficient of it to unite with atom for atom; whilst if the poison is present in smaller quantity, a part of the organ will retain its vital functions.

According to the experiments of Mulder,* the equivalent in which fibrin combines with muriatic acid, and with the oxides of lead and copper, is expressed by the number 6361. It may be assumed therefore approximatively, that a quantity of fibrin corresponding to the number 6361 combines with 1 equivalent of arsenious acid, or 1 equivalent of corrosive sublimate.

When 6361 parts of anhydrous fibrin are combined with 30,000 parts of water, it is in the state in which it is contained in muscular fibre or blood in the human body. 100 grains of fibrin in this condition would form a neutral compound of equal equivalents with $3\frac{1}{10}$ grains of arsenious acid, and 5 grains of corrosive sublimate.

* Poggendorff's Annalen, Band xl. S. 259.

The atomic weight of the albumen of eggs and of the blood deduced from the analysis of the compound which it forms with oxide of silver is 7447, and that of animal gelatin 5652.

One hundred grains of albumen containing all the water with which it is combined in the living body, should consequently combine with $1\frac{1}{4}$ grain of arsenious acid.

These proportions, which may be considered as the highest that can be adopted, indicate the remarkably high atomic weights of animal substances, and at the same time teach us what very small quantities of arsenious acid or corrosive sublimate are requisite to produce fatal effects.

All substances administered as antidotes in cases of poisoning, act by destroying the power which arsenious acid and corrosive sublimate possess, of entering into combination with animal matters, and of thus acting as poisons. Unfortunately no other body surpasses them in that power, and the compounds which they form can only be broken up by affinities so energetic, that their action is as injurious as that of the above-named poisons themselves. The duty of the physician consists, therefore, in his causing those parts of the poison which may be free and still uncombined, to enter into combination with some other body, so as to produce a compound incapable of being decomposed or digested in the same conditions. Hydrated peroxide of iron is an invaluable substance for this purpose.

When the action of arsenious acid or corrosive sublimate is confined to the surface of an organ, those parts only are destroyed which enter into combination with it; an eschar is formed, and is gradually thrown off.

Soluble salts of silver would be quite as deadly a poison as corrosive sublimate, did not a cause exist in the human body by which their action is prevented, unless their quantity is very great. This cause is the presence of common salt in all animal liquids. Nitrate of silver, it is well known, combines with animal substances, in the same manner as corrosive sublimate, and the compounds formed by both are exactly similar in the character of being incapable of decay or putrefaction.

When nitrate of silver in a state of solution is applied to skin or muscular fibre, it combines with them instantaneously; animal substances dissolved in any liquid are precipitated by it, and rendered insoluble, or, as it is usually termed, they are coagulated. The compounds thus formed are colourless, and so stable, that they cannot be decomposed by other powerful chemical agents. They are blackened by exposure to light, like all other compounds of silver, in consequence of a part of their oxide of silver being reduced to the metallic state. Parts of the body united to salts of silver no longer belong to the living organism, for their vital functions have been arrested by combination with oxide of silver; and if they are capable of being reproduced, the neighbouring living structures throw them off in the form of an eschar.

When nitrate of silver is introduced into the stomach, it meets with common salt and free muriatic acid; and if its quantity is not too great, it is immediately converted into chloride of silver—a substance absolutely insoluble in pure water. In a solution of salt or muriatic acid, however, chloride of silver does dissolve in extremely minute quantity; and it is this small part which exercises a medicinal influence when nitrate of silver is administered; the remaining chloride of silver is eliminated from the body in the ordinary way.

Without solubility, or the power of being carried to every part of the circulation, no substance possesses activity in reference to the animal organism.

The soluble salts of lead possess many properties in common with the salts of silver and mercury; but all compounds of lead with organic matters are capable of decomposition by dilute sulphuric acid. The disease called *painter's colic* is unknown in all manufactories of white-lead in which the workmen are accustomed to take as a preservative *sulphuric acid-lemonade* (a solution of sugar rendered acid by sulphuric acid).

The organic substances which have combined in the living body with metallic oxides or metallic salts, lose their property of imbibing water and retaining it, without at the same time being rendered incapable of permitting liquids to penetrate through their pores. A strong contraction and shrinking of the surface is the general effect of contact with these metallic bodies. But corrosive sublimate, and several of the salts of lead, possess a peculiar property, in addition to those already mentioned. When they are present in excess, they dissolve the first formed insoluble compounds, and thus produce an effect quite the reverse of contraction, namely, a softening of the part of the body on which they have acted.

Salts of oxide of copper, even when in combination with the most powerful acids, are reduced by many vegetable substances, particularly such as sugar and honey, either into metallic copper, or into the red suboxide, neither of which enters into combination with animal matter. It is well known that sugar has been long employed as the most convenient antidote for poisoning by copper.

With respect to some other poisons, namely, hydrocyanic acid, and the organic bases *strychnia, brucia, &c.*, we are not acquainted with facts calculated to elucidate the nature of their action. It may, however, be presumed with much certainty, that experiments upon their mode of action on different animal substances would very quickly lead to the most satisfactory conclusions regarding the cause of their poisonous effects.

There is a peculiar class of substances, which are generated during certain processes of decomposition, and which act upon the animal economy as deadly poisons, not on account of their power of entering into combination with it, or by reason of their containing a poisonous material, but solely by virtue of their peculiar condition.

In order to attain a clear conception of the mode of action of these bodies, it is necessary to call to mind the cause on which we have shown the phenomena of fermentation, decay, and putrefaction to depend.

This cause may be expressed by the following law, long since proposed by La Place and Berthollet, although its truth with respect to chemical phenomena has only lately been proved. "A MOLECULE SET IN MOTION BY ANY POWER CAN IMPART ITS OWN MOTION TO ANOTHER MOLECULE WITH WHICH IT MAY BE IN CONTACT."

This is a law of dynamics, the operation of which is manifest in all cases, in which the resistance (*force, affinity,* or *cohesion*) opposed to the motion is not sufficient to overcome it.

We have seen that ferment or yeast is a body in the state of decomposition, the atoms of which, consequently, are in a state of motion or transposition. Yeast placed in contact with sugar communicates to the elements of that compound the same state, in consequence of which, the constituents of the sugar arrange themselves into new and simpler forms, namely, into alcohol and carbonic acid. In these new compounds the elements are united together by stronger affinities than they were in the sugar, and therefore under the conditions in which they were produced further decomposition is arrested.

We know, also, that the elements of sugar assume totally different arrangements, when the substances which excite their transposition are in a different state of decomposition from the yeast just mentioned. Thus, when sugar is acted on by rennet or putrefying vegetable juices, it is not converted into alcohol and carbonic acid, but into lactic acid, mannite, and gum, or into butyric acid.

Again, it has been shown, that yeast added to a solution of pure sugar gradually disappears, but that, when added to vegetable juices which contain gluten as well as sugar, it is reproduced by the decomposition of the former substance.

The yeast with which these liquids are made to ferment, has itself been originally produced from gluten.

The conversion of gluten into yeast in these vegetable juices is dependent on the decomposition (fermentation) of sugar; for, when the sugar has completely disappeared, any gluten still remaining in the liquid does not suffer change from contact with the newly-deposited yeast, but retains all the characters of gluten.

Yeast is a product of the decomposition of gluten; but it

passes into a second stage of decomposition when in contact with water. On account of its being in this state of further change, yeast excites fermentation in a fresh solution of sugar; and if this second saccharine fluid should contain gluten (should it be *wort*, for example), yeast is again generated, in consequence of the transposition of the elements of the sugar exciting a similar change in this gluten.

After this explanation, the idea that yeast reproduces itself, as seeds reproduce seeds, cannot for a moment be entertained.

From the foregoing facts it follows, that a body in the act of decomposition (it may be named the *exciter*), added to a mixed fluid in which its constituents are contained, can reproduce itself in that fluid, exactly in the same manner as new yeast is produced when yeast is added to saccharine liquids containing gluten. This must be more certainly effected when the liquid acted upon contains the body by the metamorphosis of which the *exciter* has been originally formed.

It is also obvious that if the *exciter* be able to impart its own state of transformation to one only of the component parts of the mixed liquid acted upon, its own reproduction may be the consequence of the decomposition of this one body.

This law may be applied to organic substances forming part of the animal organism. We know that all the constituents of these substances are formed from the blood, and that the blood by its nature and constitution is the most complex of all existing matters.

Nature has adapted the blood for the reproduction of every individual part of the organism; its principal character consists in its component parts being subordinate to every attraction. These are in a perpetual state of change or transformation, which is effected in the most various ways through the influence of the different organs.

The blood does not possess the power of causing transformations; on the contrary, its principal character consists in its readily suffering transformations; and no other matter can be compared with it in this respect.

Now it is a well-known fact, that when blood, cerebral substance, gall, pus, and other substances in a state of putrefaction, are laid upon fresh wounds, vomiting, debility, and at length death, are occasioned. It is also well known that bodies in anatomical rooms frequently pass into a state of decomposition capable of imparting itself to the living body, the smallest cut with a knife which has been used in their dissection producing in these cases dangerous consequences.

The poison of bad sausages belongs to this class of noxious substances. Several hundred cases are known in which death has occurred from the use of this kind of food. In Würtemberg especially these cases are very frequent, for there the sausages are prepared from very various materials. Blood, liver, bacon, brains, milk, meal, and bread, are mixed together with salt and spices; the mixture is then put into bladders or intestines, and after being boiled is smoked.

When these sausages are well prepared, they may be preserved for months, and furnish a nourishing savoury food; but when the spices and salt are deficient, and particularly when they are smoked too late or not sufficiently, they undergo a peculiar kind of putrefaction which begins at the centre of the sausage. Without any appreciable escape of gas taking place they become paler in colour, and more soft and greasy in those parts which have undergone putrefaction, and they are found to contain free lactic acid, or lactate of ammonia; products seldom absent from putrefying bodies, especially vegetable matter.

The cause of the poisonous nature of these sausages was ascribed at first to hydrocyanic acid, and afterwards to sebacic acid, although neither of these substances had been detected in them. But sebacic acid is no more poisonous than benzoic acid, with which it has so many properties in common; and the symptoms produced are sufficient to show that hydrocyanic acid is not the poison.

The death which is the consequence of poisoning by putrefied sausages succeeds very lingering and remarkable symptoms. There is a gradual wasting of muscular fibre, and of all the constituents of the body similarly composed; the patient becomes much emaciated, dries to a complete mummy, and finally dies. The carcase is stiff as if frozen, and is not subject to putrefaction. During the progress of the disease the saliva becomes viscous and acquires an offensive smell.

Experiments have been made for the purpose of ascertaining the presence of some matter in the sausages to which their poisonous action could be ascribed; but no such matter has been detected. Boiling water and alcohol completely destroy the poisonous properties of the sausages, without themselves acquiring similar properties.

Now this is the peculiar character of all substances which exert an action by virtue of their existing condition—of those bodies the elements of which are in the state of decomposition or transposition; a state which is destroyed by boiling water and alcohol without the cause of the influence being imparted to those liquids; for a state of action or power cannot be preserved in a liquid.

Sausages, in the state here described, exercise an action upon the organism, in consequence of the stomach and other parts with which they come in contact not having the power to arrest their decomposition; and entering the blood in some way or other, while still possessing their whole power, they impart their peculiar action to the constituents of that fluid.

The poisonous properties of decayed sausages are not destroyed by the stomach as those of the small-pox virus are. All the substances in the body capable of putrefaction are gradually decomposed during the course of the disease, and after death nothing remains except fat, tendons, bones, and a few other substances incapable of putrefying in the conditions afforded by the body.

It is impossible to mistake the *modus operandi* of this poison, for Colin has clearly proved that muscle, urine, cheese, cerebral substance, and other matters, in a state of putrefaction, communicate their own state of decomposition to substances much less prone to change of composition than the blood. When placed in contact with a solution of sugar, they cause its putrefaction, or the transposition of its elements into carbonic acid and alcohol.

When putrefying muscle or pus is placed upon a fresh wound, it occasions disease and death. It is obvious that these substances communicate their own state of putrefaction to the sound blood FROM WHICH THEY WERE PRODUCED, exactly in the same manner as gluten in a state of decay or putrefaction causes a similar transformation in a solution of sugar.

Poisons of this kind are even generated by the body itself in particular diseases. In small-pox, plague, and syphilis, substances of a peculiar nature are formed from the constituents of the blood. These matters are capable of inducing in the blood of a healthy individual a decomposition similar to that of which they themselves are the subjects; in other words, they produce the same disease. The morbid virus appears to reproduce itself just as seeds appear to reproduce seeds.

The mode of action of a morbid virus exhibits such a strong similarity to the action of yeast upon liquids containing sugar and gluten, that the two processes have been long since compared to one another, although merely for the purpose of illustration. But when the phenomena attending the action of each respectively are considered more closely, it will in reality be seen that their influence depends upon the same cause.

In dry air, and in the absence of moisture, all these poisons remain for a long time unchanged; but when exposed to the air in the moist condition, they lose very rapidly their peculiar properties. In the former case, those conditions are afforded which arrest their decomposition without destroying it; in the latter, all the circumstances necessary for the completion of their decomposition are presented.

The temperature at which water boils, and contact with alcohol, render such poisons inert. Acids, salts of mercury, sulphurous acid, chlorine, iodine, bromine, aromatic substances, volatile oils, and particularly empyreumatic oils, smoke, and a decoction of coffee, completely destroy their contagious properties, in some cases combining with them or otherwise effecting their decomposition. Now all these agents, without exception, retard fermentation, putrefaction, and decay, and when present in sufficient quantity, completely arrest these processes of decomposition.

A peculiar matter to which the poisonous action is due, cannot, we have seen, be extracted from decayed sausages; and it is equally impossible to obtain such a principle from the virus of small-pox or plague, and for this reason, that their peculiar power is due to an active condition, only recognisable by our senses through the phenomena which it produces.

In order to explain the effects of contagious matters, a peculiar principle of life has been ascribed to them—a life similar to that possessed by the germ of a seed, which enables it under favourable conditions to develop and multiply itself. It would be impossible to find a more correct figurative representation of these phenomena; it is one which is applicable to contagions, as well as to ferment, to animal and vegetable substances in a state of fermentation, putrefaction, or decay, and even to a piece of decaying wood, which, by mere contact with fresh wood, causes the latter to undergo gradually the same changes, and become decayed and mouldered.

If the property possessed by a body of producing such a change in any other substance as causes the reproduction of itself, with all its properties, be regarded as life, then, indeed, all the above phenomena must be ascribed to life. But in that case they must not be considered as the only processes due to vitality, for the above interpretration of the expression embraces the majority of the phenomena which occur in organic chemistry. Life would, according to that view, be admitted to exist in every body in which chemical forces act.

If a body A, for example OXAMIDE (a substance scarcely soluble in water, and without the slightest taste), be brought into contact with another compound B, which is to be reproduced; and if this second body be OXALIC ACID DISSOLVED in water, then the following changes are observed to take place:—The oxamide is decomposed by the oxalic acid, provided the conditions necessary for their exercising an action upon one another are present. The elements of water unite with the constituents of oxamide and AMMONIA is one product formed, and OXALIC ACID the other, both in exactly the proper proportions to combine and form a neutral salt.

Here the contact of oxamide and oxalic acid induces a transformation of the oxamide, which is decomposed into oxalic acid and ammonia. The oxalic acid thus formed, as well as that originally added, are shared by the ammonia—or in other words, as much free oxalic acid exists after the decomposition as before it, and is of course still possessed of its original power. It matters not whether the free oxalic acid is that originally added, or that newly produced; it is certain that it has been reproduced in an equal quantity by the decomposition.

If we now add to the same mixture a fresh portion of oxamide, exactly equal in quantity to that first used, and treat it in the same manner, the same decomposition is repeated; the free oxalic acid enters into combination whilst another portion is liberated. In this manner a very minute quantity of oxalic acid may be made to effect the decomposition of several hundred pounds of oxamide; and one grain of the acid to reproduce itself in unlimited quantity.

We know that the contact of the virus of small-pox causes such a change in the blood, as gives rise to the reproduction of the poison from the constituents of the fluid. This transformation is not arrested until all the particles of the blood susceptible of the decomposition have undergone the metamorphosis. We have just seen that the contact of oxalic acid with oxamide caused the production of fresh oxalic acid, which in its turn exercised the same action on a new portion of oxamide. The transformation was only arrested in consequence of the quantity of oxamide present being limited. In their form both these transformations belong to the same class; but although what here takes place exactly corresponds to the definition of life above assumed, no unprejudiced mind would admit vitality in either process; since they are obviously chemical processes dependent upon the common chemical forces.

The best definition of life involves something more than mere reproduction, namely, the idea of an active power exercised BY VIRTUE OF A DEFINITE FORM, and production and generation IN A DEFINITE form. By chemical agency we can produce the constituents of muscular fibre, skin, and hair; but we cannot form by their means an organised tissue, or an organic cell.

The production of organs, the co-operation of a system of organs, and their power not only to produce their component parts from the food presented to them, but to generate *themselves* in their original form and with all their properties, are characters belonging exclusively to organic life, and constitute a form of reproduction independent of chemical powers.

The chemical forces are subject to the invisible cause by which this form is produced. Of the existence of this cause itself we are made aware only by the phenomena which it produces. Its laws must be investigated just as we investigate those of the other powers which effect motion and changes in matter.

The chemical forces are subordinate to this cause of life, just as they are to electricity, heat, mechanical motion, and friction. By the influence of the latter forces, they suffer changes in their direction, an increase or diminution of their intensity, or a complete cessation or reversal of their action.

Such an influence and no other is exercised by the vital principle over the chemical forces; but in every case where combination or decomposition takes place, chemical affinity and cohesion are in action.

The vital principle is only known to us through the peculiar form of its instruments, that is, through the organs in which it resides. Hence, whatever kind of energy a substance may possess, if it is amorphous and destitute of organs from which the impulse of motion or change proceeds, it does not live. Its energy depends in this case on a chemical action. Light, heat, electricity, or other influences may increase, diminish, or arrest this action, but they are not its efficient cause.

In the same way the vital principle governs the chemical powers in the living body, and this is particularly apparent with regard to vegetable life. All those substances to which we apply the general name of food, and all the bodies formed from them in the organism, are chemical compounds. The vital principle has, therefore, no other resistance to overcome, in order to convert these substances into component parts of the organism, than the chemical powers by which their constituents are held together. If the food possessed life, not merely the chemical forces, but this vitality, would offer resistance to the vital force of the organism it nourished.

The equilibrium in the chemical attractions of the constituents of the food is disturbed by the vital principle of the plant, as we know it may be by many other causes. But the union of its elements, so as to produce new combinations and forms, indicates the presence of a peculiar mode of attraction, and the existence of a power distinct from all other powers of nature, namely, the vital principle.

The vital principle opposes to the continual action of the atmosphere, moisture and temperature upon the organism, a resistance which is, in a certain degree, invincible. It is by the constant neutralisation and renewal of these external influences that life and motion are maintained.

The greatest wonder in the living organism is the fact that an unfathomable Wisdom has made the cause of a continual decomposition or destruction, namely, the support of the process of respiration, to be the means of renewing the organism, and of resisting all the other atmospheric influences, such as those of moisture and changes of temperature.

When a chemical compound of simple constitution is introduced into the stomach, or any other part of the organism, it must exercise a chemical action upon all substances with which it comes in contact; for we know the peculiar character of such a body to be an aptitude and power to enter into combinations and effect decompositions.

The chemical action of such a compound is, of course, opposed by the vital principle. The results produced depend upon the strength of their respective actions: either an equilibrium of both powers is attained, a change being effected without the destruction of the vital principle, in which case A MEDICINAL EFFECT is occasioned; or the acting body yields to the superior force of vitality, that is, IT IS DIGESTED; or, lastly, the chemical action obtains the ascendancy, AND IT ACTS AS A POISON.

Every substance may be considered as *nutriment* which loses its former properties when acted on by the vital principle, and does not exercise a chemical action upon the living organ.

Another class of bodies change the direction, the strength, and intensity of the resisting force (the vital principle), and thus exert a modifying influence upon the functions of its organs. They produce a disturbance in the system, either by their presence, or by themselves undergoing a change; these are MEDICAMENTS.

A third class of compounds are called POISONS, when they possess the property of uniting with organs or with their component parts, and when their power of effecting this is stronger than the resistance offered by the vital principle.

The quantity of a substance and its condition must obviously completely change the mode of its chemical action.

Increase of quantity is known to be equivalent to superior affinity. Hence a *medicament* administered in excessive quantity may act as a *poison*, and a poison in small doses as a *medicament*.

Food will act as a poison, that is, it will produce disease, when it is able to exercise a chemical action by virtue of its quantity; or when either its condition or its presence retards, prevents, or arrests the motion of any organ.

A compound acts as a poison when all the parts of an organ with which it is brought into contact enter into chemical combination with it, while it may operate as a medicine when it produces only a partial change.

No other component part of the organism can be compared to the blood, in respect of the feeble resistance which it offers to exterior influences. The blood is not an organ which is formed, but an organ in the act of formation; indeed, it is the sum of all the organs which are being formed. The chemical force and the vital principle hold each other in such perfect equilibrium, that every disturbance, however trifling, or from whatever cause it may proceed, effects a change in the blood. This liquid possesses so little of permanence that it cannot be removed from the body without immediately suffering a change, and cannot come in contact with any organ in the body, without yielding to its attraction.

The slightest action of a chemical agent upon the blood exercises an injurious influence; even the momentary contact with the air in the lungs, although effected through the medium of cells and membranes, alters the colour and other qualities of the blood. Every chemical action propagates itself through the mass of the blood; for example, the active chemical condition of the constituents of a body undergoing decomposition, fermentation, putrefaction, or decay, disturbs the equilibrium between the chemical force and the vital principle in the circulating fluid, and overcomes the latter. Numerous modifications in the composition and condition of the compounds produced from the elements of the blood, result from the conflict of the vital force with chemical affinity, in their incessant endeavour to overcome one another.

All the characters of the phenomena of contagion tend to disprove the existence of vitality in contagious matters. They without doubt exercise an influence very similar to some processes in the living organism; but the cause of this influence is chemical action, which is capable of being subdued by other chemical actions, by opposed agencies.

Several of the poisons generated in the body by disease lose all their power when introduced into the stomach, but others are not thus destroyed.

It is a fact very decisive of their chemical nature and mode of action, that those poisons which are neutral or alkaline, such as the poisonous matter of the contagious fever in cattle (*typhus contagiosus ruminantium*), or that of the small-pox, lose their whole power of contagion in the stomach; whilst that of sausages, which has an acid reaction, retains all its frightful properties under the same circumstances.

In the former of these cases, the free acid present in the stomach destroys the action of the poison, the chemical properties of which are opposed to it; whilst in the latter it strengthens, or at all events does not offer any impediment to poisonous action.

Microscopical examination has detected peculiar bodies resembling the globules of the blood in malignant putrefying pus, in the matter of vaccine, &c. The presence of these bodies has given weight to the opinion, that contagion proceeds from the development of a diseased organic life; and these formations have been regarded as the living seeds of disease.

This view, which does not admit of discussion, has led those philosophers who are accustomed to search for explanations of phenomena in forms, to consider the yeast produced by the fermentation of beer as possessed of life. They have imagined it to be composed of animals or plants, which nourish themselves from the sugar in which they are placed, and at the same time yield alcohol and carbonic acid as excrementitious matters.*

It would perhaps appear wonderful if bodies, possessing a crystalline structure and geometrical figure, were formed during the processes of fermentation and putrefaction from the organic substances and tissues of organs. We know, on the contrary, that the complete dissolution into organic compounds is preceded by a series of transformations, in which the organic structures gradually resign their forms.

* Annalen der Pharmacie, Band xxix. S. 93 und 100.

Blood, in a state of decomposition, may appear to the eye unchanged; and when we recognise the globules of blood in a liquid contagious matter, the utmost that we can thence infer is, that those globules have taken no part in the process of decomposition. All the phosphate of lime may be removed from bones, leaving them transparent and flexible like leather, without the form of the bones being in the smallest degree lost. Again: bones may be burned until they be quite white, and consist merely of a skeleton of phosphate of lime, but they will still possess their original form. In the same way processes of decomposition in the blood may affect individual constituents only of that fluid, which will become destroyed and disappear, whilst its other parts will maintain the original form.

Several kinds of contagion are propagated through the air: so that, according to the view already mentioned, we must ascribe life to a gas, that is, to an aeriform body.

All the supposed proofs of the vitality of contagions are merely ideas and figurative representations, fitted to render the phenomena more easy of apprehension by our senses, without explaining them. These figurative expressions, with which we are so willingly and easily satisfied in all sciences, are the foes of all inquiries into the mysteries of nature; they are like the *fata morgana*, which show us deceitful views of seas, fertile fields, and luscious fruits, but leave us languishing when we have most need of what they promise.

It is certain that the action of contagions is the result of a peculiar influence dependent on chemical forces, and in no way connected with the vital principle. This influence is destroyed by chemical actions, and manifests itself wherever it is not subdued by some antagonist power. Its existence is recognised in a connected series of changes and transformations, in which it causes all substances capable of undergoing similar changes to participate.

An animal substance in the act of decomposition, or a substance generated from the component parts of a living body by disease, communicates its own condition to all parts of the system capable of entering into the same state, if no cause exist in these parts by which the change is counteracted or destroyed.

DISEASE is thus excited by contagion.

The transformations produced by the disease assume a series of forms.

In order to obtain a clear conception of these transformations, we may consider the changes which substances, more simply composed than the living body, suffer from the influence of similar causes. When putrefying blood or yeast in the act of transformation is placed in contact with a solution of sugar, the elements of the latter substance are transposed, so as to form alcohol and carbonic acid.

A piece of the rennet-stomach of a calf in a state of decomposition occasions the elements of sugar to assume a different arrangement. The sugar is converted into lactic acid without the addition or loss of any element. One atom of sugar of grapes $C_{12}\ H_{12}\ O_{12}$ yields two atoms of lactic acid $= {}_2\,(C\ \ H_6\ O_6)$.

When the juice of onions or of beet-root is made to ferment at high temperatures, lactic acid, mannite, and gum are formed. Thus, according to the different states of the transposition of the elements of the exciting body, the elements of the sugar arrange themselves in different manners, that is, different products are formed.

The immediate contact of the decomposing substance with the sugar is the cause by which its particles are made to assume new forms and natures. The removal of that substance occasions the cessation of the decomposition of the sugar, so that should its transformation be completed before the sugar, the latter can suffer no further change.

In none of these processes of decomposition is the exciting body reproduced; for the conditions necessary to its reproduction do not exist in the elements of the sugar.

Just as yeast, putrefying flesh, and the stomach of a calf in a state of decomposition, when introduced into solutions of sugar, effect the transformation of this substance, without being themselves regenerated; in the same manner, miasms and certain contagious matters produce diseases in the human organism, by communicating the state of decomposition, of which they themselves are the subject, to certain parts of the organism, without themselves being reproduced in their peculiar form and nature during the progress of the decomposition.

The disease in this case is not *contagious.*

Now, when yeast is introduced into a mixed liquid containing both sugar and gluten, such as wort, the act of decomposition of the sugar effects a change in the form and nature of the gluten, which is, in consequence, also subjected to transformation. As long as some of the fermenting sugar remains, gluten continues to be separated as yeast, and this new matter in its turn excites fermentation in a fresh solution of sugar or wort. If the sugar, however, should be first decomposed, the gluten remaining in solution is not converted into yeast. We see, therefore, that the reproduction of the exciting body here depends—

1. Upon the presence of that substance from which it was originally formed;

2. Upon the presence of a compound capable of being decomposed by contact with the exciting body.

If we express in the same terms the reproduction of contagious matter in contagious diseases, since it is quite certain that they must have their origin in the blood, we must admit that the blood of a healthy individual contains substances, by the decomposition of which the exciting body or contagion can be produced. It must further be admitted, when contagion results, that the blood contains a second constituent capable of being decomposed by the exciting body. It is only in consequence of the conversion of the second constituent, that the original exciting body can be reproduced.

A susceptibility of contagion indicates the presence of a certain quantity of this second body in the blood of a healthy individual. The susceptibility for the disease and its intensity must augment according to the quantity of that body present in the blood; and in proportion to its diminution or disappearance, the course of the disease will change.

When a quantity, however small, of contagious matter, that is of the exciting body, is introduced into the blood of a healthy individual, it will be again generated in the blood, just as yeast is reproduced from wort. Its condition of transformation will be communicated to a constituent of the blood; and in consequence of the transformation suffered by this substance, a body identical with or similar to the exciting or contagious matter will be produced from another constituent substance of the blood. The quantity of the exciting body newly produced must constantly augment, if its further transformation or decomposition proceeds more slowly than that of the compound in the blood, the decomposition of which it effects.

If the transformation of the yeast generated in the fermentation of wort proceeded with the same rapidity as that of the particles of the sugar contained in it, both would simultaneously disappear when the fermentation was completed. But yeast requires a much longer time for decomposition than sugar, so that after the latter has completely disappeared, there remains a much larger quantity of yeast than existed in the fluid at the commencement of the fermentation, —yeast which is still in a state of incessant progressive transformation, and therefore possessed of all its peculiar properties.

The state of change or decomposition which affects one particle of blood, is imparted to a second, a third, and at last to all the particles of blood in the whole body. It is communicated in like manner to the blood of another individual, to that of a third person, and so on—or in other words, the disease is excited in them also.

It is quite certain that a number of peculiar substances exist in the blood of some men and animals, which are absent from the blood of others.

The blood of the same individual contains, in childhood

and youth, variable quantities of substances, which are absent from it in other stages of growth. The susceptibility of contagion by peculiar exciting bodies in childhood, indicates a propagation and regeneration of the exciting bodies, in consequence of the transformation of certain substances present in the blood, and in the absence of which no contagion could ensue. The form of a disease is termed BENIGNANT, when the transformations OF TWO constituents of the body not essential to life, are simultaneously completed without the other parts taking a share in the decomposition; it is termed MALIGNANT when they affect essential organs.

It cannot be supposed that the different changes in the substance of the existing organs, by which their constituents are converted into fat, muscular fibre, substance of the brain and nerves, bones, hair, &c., and the transformation of food into blood, can take place without the simultaneous formation of new compounds which require to be removed from the body by the organs of excretion.

In an adult these excretions do not vary much either in their nature or quantity. The food taken is not employed in increasing the size of the body, but merely for the purpose of replacing any substances which may be consumed by the various actions in the organism; every motion, every manifestation of organic properties, and every organic action being attended by a change in the material of the body, and by the assumption of a new form by its constituents.*

But in a child this normal condition of sustenance is accompanied by an abnormal condition of growth and increase in the size of the body, and of each individual part of it. Hence, there must be a much larger quantity of foreign substances, not belonging to the organism, diffused through every part of the blood in the body of a young individual.

* The experiments of Barruel upon the different odours emitted from blood on the addition of sulphuric acid, prove that peculiar substances are contained in the blood of different individuals; the blood of a man of a fair complexion and that of a man of dark complexion were found to yield different odours; the blood of animals also differed in this respect very perceptibly from that of man.

When the organs of secretion are in proper action, these substances will be removed from the system; but when the functions of those organs are impeded, they will remain in the blood, or become accumulated in particular parts of the body. The skin, lungs, and other organs, assume the functions of the diseased secreting organs, and the accumulated substances are eliminated by them. If, when thus exhaled, these substances happen to be in the state of progressive transformation, they are contagious; that is, they are able to produce the same state of disease in another healthy organism, provided the latter organism is susceptible of their action—or in other words, contains a matter capable of suffering the same process of decomposition.

The production of matters of this kind, which render the body susceptible of contagion, may be occasioned by the manner of living, or by the nutriment taken by an individual. A superabundance of strong and otherwise wholesome food may produce them, as well as a deficiency of nutriment, uncleanliness, or even the use of decayed substances as food.

All these conditions for contagion must be considered as accidental. Their formation and accumulation in the body may be prevented, and they may even be removed from it without disturbing its most important functions or health. Their presence is not necessary to life.

The action, as well as the generation of the matter of contagion is, according to this view, a chemical process participated in by all substances in the living body, and by all the constituents of those organs in which the vital principle does not overcome the chemical action. The contagion, accordingly, either spreads itself over every part of the body, or is confined particularly to certain organs, that is, the disease attacks all the organs, or only a few of them, according to the feebleness or intensity of their resistance.

In the abstract chemical sense, reproduction of a contagion depends upon the presence of two substances, one of which becomes completely decomposed, but communicates its own state of transformation to the second. The second substance thus thrown into a state of decomposition is the newly-formed contagion.

The second substance must have been originally a constituent of the blood: the first may be a body accidentally present; but it may also be a matter necessary to life. If both be constituents indispensable for the support of the vital functions of certain principal organs, death is the consequence of their transformation. But if the absence of the one substance which was a constituent of the blood do not cause an immediate cessation of the functions of the most important organs, if they continue in their action, although in an abnormal condition, convalescence ensues. In this case the products of the transformations still existing in the blood are used for assimilation, and at this period secretions of a peculiar nature are produced.

When the constituent removed from the blood is a product of an unnatural manner of living, or when its formation takes place only at a certain age, the susceptibility of contagion ceases upon its disappearance.

The effects of vaccine matter indicate that an accidental constituent of the blood is destroyed by a peculiar process of decomposition, which does not affect the other constituents of the circulating fluid.

If the manner in which the precipitated *yeast* of Bavarian beer acts be called to mind, the *modus operandi* of vaccine lymph can scarcely be matter of doubt.

Both the kind of yeast here referred to and the ordinary ferment are formed from gluten, just as the vaccine virus and the matter of small-pox are produced from the blood. Ordinary yeast and the virus of human small-pox, however, effect a violent tumultuous transformation, the former in vegetable juices, the latter in blood, in both of which fluids respectively their constituents are contained, and they are reproduced from these fluids with all their characteristic properties. The precipitated yeast of Bavarian beer, on the other hand, acts entirely upon the sugar of the fermenting liquid, and occasions a very protracted decomposition of it, in which the gluten which is also present takes no part. But the air exercises an influence upon the latter substance, and causes it to assume a new form and nature, in consequence of which this kind of yeast also is reproduced.

The action of the virus of cow-pox is analogous to that of the low yeast; it communicates its own state of decomposition to a matter in the blood, and from a second matter is itself regenerated, but by a totally different mode of decomposition; the product possesses the mild form, and all the properties of the lymph of cow-pox.

The susceptibility of infection by the virus of human small-pox must cease after vaccination, for the substance to the presence of which this susceptibility is owing has been removed from the body by a peculiar process of decomposition artificially excited. But this substance may be again generated in the same individual, so that he may again become liable to contagion; and a second or a third vaccination will again remove the peculiar substance from the system.

Chemical actions are propagated in no organs so easily as in the lungs; and it is well known that diseases of the lungs are, above all others, frequent and dangerous.

If it is assumed that chemical action and the vital principle mutually balance each other in the blood, it must further be supposed that the chemical powers will have a certain degree of preponderance in the lungs, where the air and blood are in immediate contact; for these organs are fitted by nature to favour chemical action; they do not offer resistance to the changes experienced by the venous blood.

The contact of air with venous blood is limited to a very short period of time by the motion of the heart, and any change beyond a determinate point is, in a certain degree, prevented by the rapid removal of the blood which has become arterialised. Any disturbance in the functions of the heart, and any chemical action from without, even though weak, occasions a change in the process of respiration. Solid substances also, such as dust from vegetable, animal, or inorganic bodies, act in the same way as they do in a saturated solution of a salt in the act of crystallisation, that is, they occasion a deposition of solid matters from the blood, by which the action of the air upon the latter is altered or prevented.

When gaseous and decomposing substances, or those which exercise a chemical action, such as sulphuretted hydrogen and carbonic acid, obtain access to the lungs, they meet with less resistance in this organ than in any other. The chemical process of slow combustion in the lungs is accelerated by all substances in a state of decay or putrefaction, by ammonia and alkalies; but it is retarded by empyreumatic substances, volatile oils, and acids. Sulphuretted hydrogen produces immediate decomposition of the blood, and sulphurous acid combines with the substance of the tissues, the cells, and membranes.

When the process of respiration is modified by contact with a matter in the progress of decay, when this matter communicates the state of decomposition, of which it is the subject, to the blood, disease is produced.

If the matter undergoing decomposition is the product of a disease, it is called contagion; but if it is a product of the decay or putrefaction of animal and vegetable substances, or if it acts by its chemical properties, (not by the state in which it is,) and therefore enters into combination with parts of the body, or causes their decomposition, it is termed MIASM.

Gaseous contagious matter is a miasm emitted from blood, and capable of generating itself again in blood.

But miasm properly so called, causes disease without being itself reproduced.

All the observations hitherto made upon gaseous contagious matters prove, that they also are substances in a state of decomposition. When vessels filled with ice are placed in air impregnated with gaseous contagious matter, their outer surfaces become covered with water containing a certain quantity of this matter in solution. This water soon becomes turbid, and in common language putrefies, or, to describe the change more correctly, the process of decomposition of the dissolved contagious matter is completed in the water.

All gases emitted from putrefying animal and vegetable substances in processes of disease, generally possess a peculiar nauseous offensive smell, a circumstance which, in most cases, proves the presence of a body in a state of decomposition. Smell itself may in many cases be considered as a reaction of the nerves of smell, or as a resistance offered by the vital powers to chemical action.

Many metals emit a peculiar odour when rubbed, but this is the case with none of the precious metals,—those which suffer no change when exposed to air and moisture. Arsenic, phosphorus, musk, the oil of linseed, lemons, turpentine, rue, and peppermint, possess an odour only when they are in the act of eremacausis (oxidation at common temperatures).

The odour of gaseous contagious matters is owing to the same cause; but it is also generally accompanied by ammonia, which may be considered in many cases as the means through which the contagious matter receives a gaseous form, just as it is the means of causing the smell of innumerable substances of little volatility, and of many which have no odour. (Robiquet.)*

Ammonia is very generally produced in cases of disease; it is always emitted in those in which contagion is generated, and is an invariable product of the decomposition of animal matter. The presence of ammonia in the air of chambers in which diseased patients lie, particularly of those afflicted with a contagious disease, may be readily detected; for the moisture condensed by ice in the manner just described, produces a white precipitate in a solution of corrosive sublimate, just as a solution of ammonia does. The ammoniacal salts also, obtained by the evaporation of rain-water after an acid has been added, when treated with lime so as to set free their ammonia, emit an odour most closely resembling that of corpses, or the peculiar smell of dunghills.

* Ann. de Chim. et de Phys. XV. 27.

By evaporating acids in air containing gaseous contagions, the ammonia is neutralised, and we thus prevent further decomposition, and destroy the power of the contagion, that is, its state of chemical change. Muriatic and acetic acids, and in several cases nitric acid, are to be preferred for this purpose before all others. Chlorine also is a substance which destroys ammonia and organic bodies with much facility; but it exerts such an injurious influence upon the lungs, that it may be classed amongst the most poisonous bodies known, and should never be employed in places in which men breathe.

Carbonic acid and sulphuretted hydrogen, which are frequently evolved from the earth in cellars, mines, wells, sewers, and other places, are amongst the most pernicious miasms. The former may be removed from the air by alkalies; the latter, by burning sulphur (sulphurous acid), or by the evaporation of nitric acid.

The characters of many organic compounds are well worthy of the attention and study both of physiologists and pathologists, more especially in relation to the mode of action of medicines and poisons.

Several of such compounds are known, which to all appearance are quite indifferent substances, and yet cannot be brought into contact with one another in water without suffering a complete transformation. All substances which thus suffer a mutual decomposition, possess complex atoms; they belong to the highest order of chemical compounds. For example, amygdalin, a constituent of bitter almonds, is a perfectly neutral body, of a slightly bitter taste, and very easily soluble in water. But when it is introduced into a watery solution of synaptas, (a constituent of sweet almonds,) it disappears completely without the disengagement of any gas, and the water is found to contain free hydrocyanic acid, hydruret of benzule (oil of bitter almonds), a peculiar acid and sugar, all substances of which merely the elements existed in the amygdalin. The same decomposition is effected when bitter almonds, which contain the same white matter as the sweet, are rubbed into a powder and moistened with water. Hence it happens that bitter almonds pounded and digested in alcohol, do not yield oil of bitter almonds containing hydrocyanic acid, by distillation with water; for the substance which occasions the formation of those volatile substances, is dissolved by alcohol without change, and is therefore extracted from the pounded almonds. Pounded bitter almonds do not contain amygdalin, after having been moistened with water, for that substance is completely decomposed when they are thus treated.

Volatile compounds cannot be detected by their smell in the seeds of the *Sinapis alba* and *S. nigra.* A fixed oil of a mild taste is obtained from them by pressure, but no trace of a volatile substance. If, however, the seeds are rubbed to a fine powder, and subjected to distillation with water, a volatile oil of a very pungent taste and smell passes over along with the steam. But if, on the contrary, the seeds are treated with alcohol previously to their distillation with water, the residue does not yield a volatile oil. The alcohol contains a crystalline body called sinapin, and several other bodies. These do not possess the characteristic pungency of the oil, but it is by the contact of them with water, and with the albuminous constituents of the seeds, that the volatile oil is formed.

Thus bodies which would be regarded as absolutely indifferent in inorganic chemistry, on account of their possessing no prominent chemical characters, when placed in contact with one another, are mutually decomposed. Their constituents arrange themselves in a peculiar manner, so as to form new combinations; a complex atom dividing into two or more atoms of less complex constitution, in consequence of a mere disturbance in the attraction of their elements.

The white constituents of the almonds and mustard, which resemble coagulated albumen, must be in a peculiar state in order to exert their action upon amygdalin, and upon those constituents of mustard from which the volatile pungent oil is produced. If almonds, after being blanched and pounded, are thrown into boiling water, or treated with hot alcohol, with mineral acids, or with salts of mercury, their power to effect a decomposition in amygdalin is completely destroyed. Synaptas is an azotised body which cannot be preserved when dissolved in water. Its solution becomes rapidly turbid, deposits a white precipitate, and acquires the offensive smell of putrefying bodies.

It is exceedingly probable that the peculiar state of transposition into which the elements of synaptas are thrown when dissolved in water, may be the cause of the decomposition of amygdalin, and formation of the new products arising from it. The action of synaptis in this respect is very similar to that of rennet upon sugar.

Malt, and the germinating seeds of corn in general, contain a substance called DIASTASE, which is formed from the gluten contained in them, and cannot be brought in contact with starch and water, without effecting a change in the starch.

When bruised malt is strewed upon warm paste of starch, the paste after a few minutes becomes quite liquid, and the water is found to contain, in place of starch, a substance in many respects similar to gum. But when more malt is added, and the heat longer continued, the liquid acquires a sweet taste, and all the starch is found to be converted into sugar of grapes.

The elements of diastase have at the same time arranged themselves into new combinations.

The conversion of the starch contained in food into sugar of grapes, in diabetes mellitus, indicates that amongst the constituents of some one organ of the body a substance or substances exist in a state of chemical action, to which the vital principle of the diseased organ does not oppose resistance. The component parts of the organ must suffer changes simultaneously with the starch, so that the more starch is furnished to it, the more energetic and intense the disease must become; while if only food incapable of suffering such transformations from the same cause is supplied, and the vital energy is strengthened by stimulant remedies and strong nourishment, the chemical action may finally be subdued, or in other words, the disease cured.

The conversion of starch into sugar may also be effected by pure gluten, and by dilute mineral acids.

From all the preceding facts, we see that very various transpositions, and changes of composition and properties, may be produced in complex organic molecules, by every cause which occasions a disturbance in the attraction of their elements.

When moist copper is exposed to air containing carbonic acid, the contact of this acid increases the affinity of the metal for the oxygen of the air in so great a degree that they combine, and the surface of the copper becomes covered with green carbonate of copper. Two bodies which possess the power of combining together, assume, however, opposite electric conditions at the moment at which they come in contact.

When copper is placed in contact with iron, a peculiar electric condition is excited, in consequence of which the property of the copper to unite with oxygen is destroyed, and the metal remains quite bright.

When formate of ammonia is exposed to a temperature of 388° F. (180° C.) the intensity and direction of the chemical force undergo a change, and the conditions under which the elements of this compound are enabled to remain in the same form cease to be present. The elements, therefore, arrange themselves in a new form; hydrocyanic acid and water being the results of the change.

Mechanical motion, friction, or agitation, is sufficient to cause a new disposition of the constituents of fulminating silver and mercury, that is, to effect another arrangement of their elements, or to cause the production of new compounds in a liquid.

We know that electricity and heat possess a decided influence upon the exercise of chemical affinity; and that the attractions of substances for one another are subordinate to numerous causes which change the condition of these substances by altering the direction of their attractions. In the same manner, therefore, the exercise of chemical powers in the living organism is dependent upon the vital principle.

The power of elements to unite together, and to form peculiar compounds, which are generated in animals and vegetables, is chemical affinity; but the cause by which they are prevented from arranging themselves according to the degrees of their natural attractions—the cause, therefore, by which they are made to assume their peculiar order and form in the body, is the vital principle.

After the removal of the cause which forced their union—that is, after the extinction of life—most organic atoms retain their condition, form, and nature, only by a *vis inertiæ;* for a great law of nature proves that matter does not possess the power of spontaneous action. A body in motion loses its motion only when a resistance is opposed to it; and a body at rest cannot be put in motion, or into any action whatever, without the operation of some exterior cause.

The same numerous causes which are opposed to the formation of complex organic molecules, under ordinary circumstances, occasion their decomposition and transformations when the only antagonist power, the vital principle, no longer counteracts the influence of those causes. Contact with air and the most feeble chemical action now effect changes in the complex molecules; even the presence of any body the particles of which are undergoing motion or transposition, is often sufficient to destroy their state of rest, and to disturb the statical equilibrium in the attractions of their constituent elements. An immediate consequence of this is, that they arrange themselves according to the different degrees of their mutual attractions, and that new compounds are formed in which chemical affinity has the ascendancy, and opposes any further change, while the conditions under which these compounds were formed remain unaltered.

APPENDIX TO PART II.

Some potatoes, which had been wrapped in several folds of paper, placed in a box, and kept in a dark but moderately warm place in the laboratory, were found in March to be enveloped in a kind of net, formed of sprouts of two lines in thickness. On these sprouts there were several hundred small tubers, of $\frac{1}{8}$ to $\frac{1}{4}$ of an inch in thickness. The sprouts and the tubers possessed a white colour, and did not exhibit any signs of leaves. On examining the parent potato with a microscope, it was found that its exterior cells were still partly filled with granules of starch; but the interior part was quite empty, and its substance soft and elastic. The sprouts and the cells of the young potatoes abounded in starch.

The growth of these sprouts and the formation of the tubers at the expense of the constituents of the potatoes is analogous to the formation and nutrition of fungi. The organic substance present in the potato obtains a new form by means of the active power resident in the germ; for in this case, it cannot be supposed that the food was extracted from the air. Now, just as the constituents of the old potato entered into, and were again found unchanged in the sprouts of the young ones, in like manner animal and vegetable substances in a state of decay enter into the fungi arising from them. Thus the ingredients of these bodies as the products of their putrefaction, pass over into the fungi, exactly as the interior substance of the parent potato enters into the sprouts and young tubers. For this conversive organic power alone is sufficient, and light and other conditions of vegetable life may be entirely excluded.

TABLE

SHEWING THE PROPORTION BETWEEN THE ENGLISH AND HESSIAN STANDARD OF WEIGHTS AND MEASURES.

1 lb. English is equal to 0·90719 lbs. Hessian.

1 Hessian acre is equal to 26,910 English square feet.

1 English square foot is equal to 1·4864 Hessian square feet.

1 English cubic foot contains 1·81218 of a Hessian cubic foot.

INDEX.

A.

Absorption, by roots, 66
 Of salts, 74
Acid, acetic, transformation of, 275
 —— formation of, 299—303
 Boracic, 80
 Carbonic, 3
 ——— contained in the atmosphere, 15
 ——— decomposed by plants, 16
 ——— disintegrates rocks, 117
 ——— is furnished by humus,
 ——— is expired by animals, 17
 ——— is a product of decay, 295
 ——— why necessary to plants 66
 Cyanic, 279
 Formic, 42
 Hippuric, 50
 Humic, 5
 —— properties of, 5
 —— contains ammonia, 6
 Hydrocyanic, 260
 Kinic, 73
 Meconic, 73
 Melanic, 295
 Nitric, 225, 303
 Phosphoric, in ashes of plants, 121
 Rocellic, in plants, 67
 Succinic, 341
 Sulphuric, a source of sulphur, 63
 —— retains ammonia, 189
 —— action on soils, 194
Acids, action of upon sugar, 273
 Arrest decay,
 Capacity for saturation, 68
 Organic, in plants, 3, 67
 —— how formed, 145
 —— essential to formation of sugar, 144
Agave Americana, absorbs oxygen, 23
Agriculture, object of, 113
 —— how attained, 113
 Its importance, 112
 A principle in, 187
 Science necessary to, 129
Air, access of, favoured, 30
 Ammonia in, 44
 Carbonic acid in, 15
 Effect of upon juices, 299
 —— on soils, 88
 Improved by plants, 17
 Necessary to respiration, 174
 —— to plants, 99
 Nitric acid contained in, 229
Albumen, 60, 141
Alcohol, effect of heat on, 276
 Products of its oxidation, 295
 From sugar, 282
Alkalies, contained in soils, 86
 Essential to the formation of sugar, starch, and gum, 144
 Necessity for restoring to soils, 118
 Promotes decay of wood, 339
 Quantity of in aluminous minerals, 115
 Use of in plants, 126
Alkaline Bases, in plants, on what their existence depends, 72
 Salts in plants, sources of, 118
 —— contained in fertile soils, 119
 —— liberated from soils by the action of air and of lime, 134
Alloxan, 321
Alloxantin, 321
Alumina, in fertile soils, 114
 Its influence on vegetation, 115
Ammonia, carbonate of, in air, 44
 ——— how fixed, 189
 Cause of nitrification, 305
 Changes colours, 42
 Condensed by charcoal, 58
 Conversion of, into nitric acid, 304
 Early existence of, 216
 Evolved from manure, 53

AMMONIA—*continued.*
Fixed by gypsum, 189
Contained in beet-root, 47
———— maple juice, 47
———— privies, 189
———— stables, 190
Furnishes nitrogen, 59
Deductions of Boussingault concerning, 208
Boussingault's deductions are erroneous, 209
Is not an essential constituent of manure, 212
Is always of use in manure, 213
Loss from evaporation, 189
Inorganic origin of, 80
Product of decay, 43
Properties of, 42
Quantity absorbed by charcoal, 58
————————by decayed wood, 58
In rain-water, 45
In humus, 6
In snow-water, 45
Separated from soils by rain, 58
Solubility of, 44
Transformation of, 42
Product of disease, 383
ANALYSIS of ashes of great importance, 213
ANIMAL food, preservation of, 300
Life, dependent on plants, 59
—— processes of, 174
Matter, products of decay, 59
——— essential to nitrification, 225
ANIMALS, excrements of, how formed, 51, 52
Fæces of, contain little nitrogen, 51
Liquid excrements of, rich in nitrogen, 49
Obtain their substances from plants, 62
ANNUAL plants, how nourished, 103
ANTHOXANTHUM Odoratum, acid in, 50
ANTHRACITE, 351
ANTIDOTES to poisons, 361
APATITE, 122
ARABLE land, formation of, 84
AROMATICS, their influence on fermentation, 312
ARGILLACEOUS earth, 115
ARSENIOUS acid, action of, 359
ASHES, analysis of various plants, *See pages* 70, 149, 156, 213, *and Appendix to Part I.*

ASHES—*continued.*
As a manure, 181
Of bones, 193
Of peat, 194
Of coals, 186
Of wood, 184
ASSIMILATION of carbon, 3, 28
Of carbonic acid and ammonia, 101
Of hydrogen, 36, 40
Of nitrogen, 41, 59
ATMOSPHERE, ammonia in, 44
Carbonic acid in, 15
Composition is invariable, 13
Motion of, 19
ATOMS, motions of, 266
Permanence in position of, 266
AZOTISED matter in juices of plants, 106
Substances, combustion of, 304

B.

BARK of trees viewed as excrementitious, 171
Of fir, analysis of its ashes, 171
BARLEY, analysis of its ashes, 156
Experiment on, 239
BASALT, analysis of, 91
BASES, alkaline in plants, on what their existence depends, 72
Organic, 4, 73
Oxygen contained in, 69
In plants, 75,
Substitution of, 70, 72
BEANS, analysis of ashes of, 149
Contain casein, 141
Nutritive power of, 140
BEECH, analysis of its ashes, 247
BEER, 308
Bavarian, 316
BEET, analysis of, 12
Ammonia contained in, 47
From sandy soils, 108
BENIGNANT disease, 380
BENZOIC acid, formed, 50
BIRCH tree, ammonia in, 47
BLOOD, analysis of its ashes, 149
Action of chemical agents upon, 374
Its feeble resistance to exterior influences, 374
Organic salts in, 355
Its character, 365
BLOSSOMS, when produced, 33
Increased, 101
BONES, dust of, 185
Durability of, 193
Gelatine in, 193

Bouquet of wines, 314
Boracic acid, 80
Brandy from corn, 312
Oil of, 313
Brazil, wheat in, 119
Brown Coal, 346
Buckwheat, 240

C.

Cactus, 35
Calcium, fluoride of, 124
Chloride of, 189
Caoutchouc, in plants, 35
Carbon, assimilation of, 3, 28
Of decaying substances, seldom affected by oxygen, 295
Derived from air, 15
In sea-water, 82
Produce of, in land, 12
——— in beet, 12
——— in straw, 12
Restored to soil, 33
Received by leaves, 16
Carbonate of ammonia, contained in rain-water, 45
Decomposed by gypsum, 189
Of lime in caverns and vaults, 97
Carbonic acid in the atmosphere, 15
Changes in the leaves, 109
Decomposed by plants, 16
Decomposes soils, 35
Emission of, at night, 26
Evaporation of, 27
Evolution from decaying bodies, 297
From humus, 30
—— respiration, 174
—— woody fibre, 337
Increase of, prevented, 17
Influence of light on its decomposition, 109
Carburetted hydrogen, with coal, 350
Caverns, stalactites in, 97
Charcoal, condenses ammonia, 58
Promotes growth of plants, 194
Chemical effects of light, 109
Processes in the nutrition of vegetables, 2
Transformations, 266
Chemistry, organic, what it is, 1
Chloride of calcium, 189
Of potassium, 74
Of sodium, its volatility, 81
Clay slate, 91, 92, 123
Clays, formation of, 93
From porphyry, 93
From felspar, 94
Potash in, 116
Clay, burned, how it acts as manure, 57
Coal, formation of, 344—362
Inflammable gases from, 351
Of humus, 5
Wood or brown, 348
Colours of flowers, 42
Combustion at low temperatures, 296
Of decayed wood, 336—341
Induction of, 301
Respiration, viewed as, 176
Spontaneous, 293
Concretions from horses, 123
Constituents of the blood exist in plants, 62
The formation of the main object of agriculture, 54
Contagion, reproduction of, on what dependent, 369
Susceptibility to, how occasioned, 378
Contagions, how produced, 367
Propagation of, 376
Contagious matters, action of, 376, 372, 390
Their effects explained, 369
Life in, disproved, 369
Reproduction of, 369
Copper, oxide of, in clay slate, 123
Corn, how cultivated in Italy, 118
Corn brandy, 312
Corrosive sublimate, action of, 361
Cow-pox, action of virus of, 383
Cow, urine of, analysis, 176
Crops, rotation of, 139, 172
Cultivation, its benefits, 19
Different methods of, 112
Object of, 113
Culture, art of, 96, 127
Cyanic acid, transformation of, 279
Cyanogen, combustion of, 305
transformation of, 280

D.

Darwin on the formation of soils, 84
——— descriptions of the gold ores in Chili, 131
Death, the source of life, 59
Decay, 291
A source of ammonia, 43
Of wood, 336
And putrefaction, 263
Decomposition, 263
Diamond, its origin, 341
Diastase, 123
Contains nitrogen, 124
Disease, how excited, 365

DISINTEGRATION of rocks, 92
Of ores, 131
DUNG of the nightingale, 181

E.

EBONY WOOD, oxygen and hydrogen in, 25
ELEMENTS of plants, 3
EREMACAUSIS, 291, 307
Analogous to putrefaction, 298
Arrested, 292
Definition of, 291
Necessary to nitrification, 304
Of bodies containing nitrogen, 304
Of bodies destitute of nitrogen, 299
ETHER, œnanthic, 313
EXCREMENTS, manure in which they are found, 176
EXCREMENTS of animals, contain the same amount of nitrogen as that present in the food, 176
EXCREMENTITIOUS MATTER, production of illustrated, 175
EXCREMENT, animal, its chemical nature, 176
Of the cow, horse, &c., 176, 177
EXCREMENTS of plants, 33
Conversion of, into humus, 33
Of man, amount of, 179, 180
EXCRETION, organs of, 174
Of plants, theory of, 33

F.

FALLOW, 128—138
FELSPAR, decomposition of, 93
Analysis of, 89
Various kinds of, 89
Decomposed analysis of, 93
FERMENT, 284
FERMENTATION, 282, 308—335
Ascribed to fungi and infusoria, 326
Of Bavarian beer, 317, 325
Of beer, 309
Gay-Lussac's experiments in, 327
Of sugar, 282
Of vegetable juices, 283
Vinous, 308
Of wort, 309
FERTILITY of fields, how preserved, 181
FIBRIN, 60, 141
FIRES, plants on localities of, 120
FIR-BARK, analysis of its ashes, 171
—— WOOD, analysis of its ashes, 171
FISHES in salt-pans, 79
FLESH, effect of salt on, 356
FLESH preserved under certain circumstances, 328, 329
FLUORINE in ancient bones, 124
FOOD, effect on products of plants, 108
Of young plants, 100
Transformation and assimilation of, 33
FOOD, knowledge of its composition essential, 139
FOOD, undergoes combustion in the body, 175
FORMATION OF WOOD, 106
FRANCONIA, caverns in, 97
FRUIT, increased, 101
Ripening of, 39
————, changes attending, 102
FUCUS GIGANTEUS, 236
FUNGI, supposed to cause fermentation, 326

G.

GASEOUS substances in the lungs, effect of, 384
GASTEROSTEUS ACULEATUS, in salt-pans, 79
GAY-LUSSAC, his experiments, 327
GERMINATION of potatoes, 102
Of grain, 105
GLUE, manure from, 186
GLUTEN, conversion of into yeast, 320
Decomposition of, 289
Gas from, 308
GRAIN, germination of, 105
GRAPES, fermentation of, 308
Juice of, differences in, 315
Potash in, 72
GRAUWACKE, soil from, 117
GUANO, 49, 165, 181
GYPSUM, experiment with, 55
Decomposed by carbonate of ammonia, 189
Decomposed by salt, 65
Its influence, 55
Use of, 189

H.

HAVANNAH TOBACCO, analysis of its ashes, 247
HANOVER TOBACCO, 247
HAY, analysis of ashes, 156
Carbon in, 11
HESSIAN and English weights and measures, 391
HYBERNATING animals, 103
HORSE, urine of the, 176
Concretions in the, 123
HORSE-DUNG, analysis of, 176

Humate of lime, quantity received by plants, 9
Humic acid, 5
Sometimes contains ammonia, 6
Action of, 97
Properties of, 5
Is not contained in soils, 7
Quantity received by plants, 9
Insolubility of, 97
Humus, 5
Action of, 97
Analysis of, 6
Erroneous opinions concerning, 7
Action upon oxygen, 96, 97
Coal of, 5
Conversion of woody fibre into, 335
How produced, 5
Its insolubility, 97
Properties of, 5
Sources of carbonic acid, 97
Theory of its action, 97
Hydrogen, assimilation of, 36—40
Excess of in wood accounted for, 37
Of decayed wood, 838
Of plants, source of, 37
Peroxide of, 264
Hyett, Mr., on nitrate of soda, 233

I.

Ice, bubbles of gas in, 27
Ingenhouss, his experiments, 22
Ingredients of soil removed by crops, 154
Inorganic constituents of plants, 66—83

L.

Lava, soil from, 114
Lead, salts of, compounds with organic matter, 362
Leaves, absorb carbonic acid, 16
Ashes of, contain alkalies, 83
Leaves of pine and fir—
Cessation of their functions, 33
Change colour from absorption of oxygen, 39
Decompose carbonic acid, 16
Their office, 16
Power of absorbing nutriment, how increased, 31
Life, notion of, 369
Light, absence of, its effect, 22
Chemical effects of, 109
Influences decomposition of carbonic acid, 109
Lime, phosphate of, 185
Lime-plants, 156
Lime, action of, 133—136
Linden-tree yields sugar, 107
Limestones, hydraulic, 94, 95, 136

M.

Magnesia, phosphate of, in seeds, 196
Manure, 173—194
In the ashes of food burned in the body, 153, 176
Of bones, 193
The form of, important, 137
Waste of, in England, 165
Animal, yields ammonia, 176
Maple juice, ammonia from, 47
Trees, sugar of, 47
Mesotype, properties of, 90
Miasm, defined, 384
Morbid poisons, 368
Mosses grow luxuriantly with green light, 110
Motion, its influence on chemical forces, 266
Mould, vegetable, 342
Mouldering of bodies, 344—362

N.

Nitrate of soda as a manure, 233
Nitric acid from ammonia, 228
How formed, 226
Nitrification, 304—307
Nitrogen, assimilation of, 41—59
In excrements, 176
In plants, 4
Production of, the object of agriculture, 54
Transformation of bodies containing, 304
Nutrition, inorganic substances requisite in, 66
Superfluous, how employed, 101
Of young plants, 100

O.

Oaks, ashes of, 249
Dwarf, 31
Oak-wood, composition of, 249
Odour of gaseous contagious matter, 385
Œnanthic ether, 313
Organic acids, 146
Decomposition of, 146
Chemistry, 1
Oxamide, decomposition of,
Oxygen, absorption of, at night, 22
——— by leaves, 39
——— by respiration, 174

OXYGEN—*continued.*
Absorption of by wood, 337
Action upon woody fibre, 337
Emitted by leaves, 16
In air, 13
Consumption of, 14
In water, 37
Separated during the formation of acids, 146
Is furnished by the decomposition of water, 37

P.

PEAS, 149
PEROXIDE OF HYDROGEN, 264
PHONOLITE, analysis of, 91
PHOSPHATES are constituents of plants, 66, 69
PHOSPHORIC ACID in ashes of plants, 66, 69
PINE-TREE ashes, 70
PLANTS absorb oxygen, 22
Analysis of ashes, 149
Characterised by their principal ingredients, 155
Decompose carbonic acid, 27, 28
Effect of, on rocks, 92
Elements of, 4
Exhalation of carbonic acid from, 22
Functions of, 17
Improve the air, 17
Influence of gases on, 23
Ingredients of, 155
Inorganic constituents of, 66, 83
Life of, connected with that of animals, 59
Marine, food of, 163
Milky-juiced, in barren soils, 35
Size of, proportioned to organs of nourishment, 31
Rotation of, its advantage, 139
PLOUGHING, its use, 132
POISONS, generated by disease, 353
Inorganic, 358
Peculiar class of, 363
Rendered inert by heat, 368
POISONING, superficial, 359
By sausages, 365
POMPEII, bones from, 124
PORPHYRY, by disintegrating, forms clay, 93
POTASH, in grapes, 72
Plants, 156
Replaced by soda, 72
Quantity in soils, 115
POTATOES, germination of, *Appendix to Part II.*
PURGATIVE effect of salts explained, 357
PUS, globules in, 375
PUTREFACTION, 26
Communicated, 367
Source of ammonia, 59
—— of carbonic acid, 59
PUTREFYING sausages, death from, 365
Their mode of action, 366
Substances, their effect on wounds, 367
——— alkaline, 375
——— acid, 375

R.

RAIN-WATER, contains ammonia, 45
RAIN, necessity for, to furnish alkalies to plants, 124
Want of, or excess in, producing diseases in plants, 125
REMOVAL OF BRANCHES, effects of, 101
RHODODENDRON FERRUGINEUM, 101
RIPENING OF FRUIT, 39
ROOTS, excrements of, 75
ROTATION OF CROPS, 139, 172
RYE, 149

S.

SALINE PLANTS, 73
SALT, volatilisation of, 81
SALTS, absorption of, 75
Effects of, on the organism, 354
——— on flesh, 356
——— on the stomach, 356
Organic, in the blood, 355
Passage of, through the lungs, 354
SALT-WORKS, loss in, 81
SAND, disintegrates when exposed to the action of carbonic acid, 90
SATURATION, capacity of, 68
SAUSAGES, poisonous, 365
SAUSSURE, his experiments on air, 15
On the mineral ingredients of plants, 66
SCIENCE not opposed to practice, 129
SEA-WATER, analysis of, 81
Contains carbon, 82
Contains ammonia, 82
SILICA, properties of, 86, 88
SILICATES, disintegration of, 161
SILVER, salts, poisonous effects of, 361
SINAPIS ALBA, 387
SIZE of plants proportional to organs of nourishment, 31
SNOW-WATER, ammonia in, 45
SODA, may replace potash, 72
SOILS, advantages of loosening, 132

Soils—*continued.*
Analyses of, 253
Exhaustion of, 117
Ferruginous, improved, 99
Fertile, of Vesuvius, 137
From lava, 137
Imbibe ammonia, 53
Physical properties of,
Important, 157
Exhaustion of, 151
Stagnant water, effect of, 99
Stalactites in caverns, 97
Starch, accumulation of, in plants, 102
Composition of, 38
Development of plants influenced by, 103
Product of, the life of plants, 22
In willows, 102
Straw, analysis of, 12
Of rye, 248
Struve, experiments of, 117
Substitution of bases, 68
Succinic acid, 341
Sugar, formed from acids, 143
Composition of, 282
Carbon in sugar, 12
Contained in the maple-tree, 47
In Clerodendron fragrans, 107
Development of plants, influence on, 103
Fermentation of, 282
In beet-roots, 47
Metamorphosis of, 283
Organic compounds, all form sugar, 273
Product of, the life of plants, 22
Transformation of, 275
When produced, 32
Sulphur, crystallised, dimorphous,
Proportion of, to nitrogen in plants, 64
Source of, in plants, 60
Sulphate of ammonia, well adapted to furnish plants with sulphur, 63
Sulphates in water of springs, 63
Yield sulphur, 64
Sulphuric acid, action of, on soils, 194
Sulphurous acid arrests decay, 339
Synaptas, 388

T.

Tables of Hessian and English weights and measures, 391
Tannic acid, 38
Tartaric acid, 38
Tartaric acid—*continued.*
Converted into sugar, 39
In wine, 311
Teltow Parsnip, 31
Thenard, his experiments on yeast, 286
Tin, action on nitric acid, 263
Tobacco juice, contains ammonia, 48
Nitric acid, 50
In Virginia, 117
Transformation, by heat, 275
Chemical, 259
Of acetic acid, 275
Of carbonic acid, 110
Of meconic acid, 275
Of bodies containing nitrogen, 277
Of bodies destitute of nitrogen, 275
Results of, 33
Of wood, 276
Of cyanic acid, 279
Of cyanogen, 280
Of gluten, 309
Transplantation, effect of, 101
Trees, diseases of, 106
Require alkalies, 119

U.

Ulmin, 5
Urea, converted into carbonate of ammonia, 50
Urine, contains nitrogen, 50
Its use as manure, 49
Of men, 180
Of horses, 176
Human, analysis of, 180
Of cows, 176

V.

Vaccination, its effect, 382
Vegetable albumen, 50
Mould, always contains carbonate of ammonia, 100
Vegetation, tropical, 167
Vesuvius, fertile soil of, 116
Vines, Juice of, yields ammonia, 48
Vinous fermentation, 308
Virginia, early products of its soils, 118
Virus, of small pox, 382
Vaccine, 382
Vital principle, how balanced in the blood, 371

W.

Water, carbonic acid of, absorbed, 17
——— decomposes rocks, 117

WATER—*continued.*
Composition of, 37
Dissolves mould, 342
Plants, their action upon, 27
Rain, contains ammonia, 45
—— required by gypsum, 57
Salt, analysis of, 81
WAVELLITE, 121
WHEAT, exhausts, 118
Gluten of, 48
Why it does not thrive on certain soils, 119
In Virginia, 118
Red, 149
White, 189
WILLOWS, growth of, 102
WINE, effect of gluten upon, 315
Fermentation of, 315
Properties of, 314
Substances in, 310
Taste and smell, 311
Varieties of, 311
WOAD, decomposition of, 289
WOOD, Decayed combustion of, 340
Absorbs ammonia, 58
Analysis of, 25
Conversion of, into humus, 337
Decay of, 336
Requires air, 336
Decomposition of, 259, 289
Effect of moisture and air on, 336
WOOD—*continued.*
Elements of, 337
Formation of, 106
Source of its carbon, 12
Transformation of, 276
WOOD-COAL, how produced, 345
Analysis of, 346
WOODY-FIBRE, changes in, 336
Composition of, 337
Decomposition of, 336
Difference between it and wood, 25
Formation of, 21
Moist, evolves carbonic acid, 336
Mould from, 343
WORT, fermentation of, 318
WOUNDS, effect of putrefying substances on, 365

Y.

YEAST, 284
Destroyed, 309
Experiments on, 284
Formed, 309
Its mode of action, 287
Its production, 363
Two kinds of, 318

Z.

ZEOLITE, analysis of, 91

THE END.

LONDON:
BRADBURY AND EVANS, PRINTERS, WHITEFRIARS.

Printed in Dunstable, United Kingdom

91219910R00238